Climate-Responsive Plant Biology

Hem Chandra Joshi
Dhanushka Udayanga
Saurabh Gangola • Mohd Rafatullah
Editors

Climate-Responsive Plant Biology

Environmental Crosstalk, Sustainability, and Future Technologies

Springer

Editors
Hem Chandra Joshi
Department of Botany Dev Singh
Bisht Campus
Kumaun University
Nainital, Uttarakhand, India

Dhanushka Udayanga
Applied Sciences
Sri Lanka Institute of Information
Technology
Malabe, Sri Lanka

Saurabh Gangola
Department of Microbiology
Graphic Era (Deemed to be University)
Dehradun, Uttarakhand, India

Mohd Rafatullah
Department of Mathematics and Natural
Sciences
Prince Mohammad Bin Fahd University
Al-Khobar, Saudi Arabia

ISBN 978-981-95-7353-0 ISBN 978-981-95-7354-7 (eBook)
https://doi.org/10.1007/978-981-95-7354-7

This Springer imprint is published by the registered company Springer Nature Singapore Pte Ltd.
The registered company address is: 152 Beach Road, #21-01/04 Gateway East, Singapore 189721, Singapore

This book is dedicated to the global community of plant scientists, agronomists, ecologists, environmental researchers, and students whose tireless efforts continue to deepen our understanding of the plant world. Their relentless pursuit of knowledge, from exploring the intricate mechanisms of plant physiology to developing innovative strategies for climate resilience, forms the foundation of progress in the face of unprecedented environmental challenges. We also extend this dedication to the farmers and field workers who, through their daily experiences and traditional wisdom, remind us of the irreplaceable bond between humans, plants, and the environment. Their commitment to nurturing the earth, even in times of climate uncertainty, continues to inspire scientific inquiry and sustainable innovation. To the educators and mentors who ignite curiosity in young minds, and to the emerging researchers who carry the torch of discovery forward your passion and perseverance illuminate the path toward a greener, more sustainable, and climate-secure future.

Editors
Dr. Hem Chandra Joshi
Dr. Dhanushka Udayanga
Dr. Saurabh Gangola
Dr. Mohd Rafatullah

Foreword

Climate change represents one of the greatest challenges humanity has ever faced, and its impacts on the world's vegetation are both profound and far-reaching. Plants—the foundation of terrestrial ecosystems and agriculture—stand at the frontline of these environmental shifts. Understanding how plants perceive, respond to, and adapt to a changing climate is not only an academic pursuit but a global necessity. *Climate-Responsive Plant Biology* offers a comprehensive and timely exploration of the biological, ecological, technological, and socioeconomic dimensions of plant-environment interactions under climate stress. Bringing together diverse chapters authored by experts from around the world, this volume captures the breadth of current research-from plant physiology and epigenetic regulation to phenomics, gene editing, sustainable agriculture, and green nanotechnology. By integrating classical knowledge with cutting-edge innovations, the editors have crafted a resource that will support researchers, educators, policymakers, and students alike. This book not only enhances our scientific understanding but also strengthens our collective effort to build climate-resilient ecosystems and food systems.

I congratulate the editors—Dr. Hem Ch. Joshi, Dr. Saurabh Gangola, Dr. Dhanushka Udayanga, and Dr. Mohd Rafatullah—for curating this important scientific contribution. Their dedication will undoubtedly inspire future work in plant sciences and climate resilience.

Associate Professor, University of Arkansas
Fayetteville, AR, USA

Neelendra K. Joshi

Preface

The rapidly accelerating impacts of climate change have brought unprecedented challenges to global plant systems, affecting agriculture, biodiversity, ecosystem stability, and food security. In this context, there is a critical need for consolidated scientific knowledge that explains how plants sense, respond to, and adapt to environmental stressors. This edited volume, *Climate-Responsive Plant Biology*, was envisioned to address this need. The book brings together fourteen carefully curated chapters that span fundamental plant-environment interactions, physiological and molecular responses to climate stress, state-of-the-art technologies, and future-oriented strategies for sustainable agriculture. Each chapter offers unique insights while collectively presenting a holistic understanding of plant resilience in the face of climate change. We are grateful to all contributing authors for sharing their expertise and for adhering to the timeline with enthusiasm and scholarly dedication. Their contributions have enriched this volume far beyond our initial expectations. We also appreciate the support from Springer Nature in guiding the publication process with professionalism and clarity.

It is our hope that this book will serve as a valuable reference for researchers, academicians, students, agronomists, environmentalists, and policymakers. May it inspire new ideas, collaborative efforts, and innovative solutions for building climate-resilient plant systems worldwide.

Nainital, Uttarakhand, India — Hem Chandra Joshi
Malabe, Sri Lanka — Dhanushka Udayanga
Dehradun, Uttarakhand, India — Saurabh Gangola
Al-Khobar, Saudi Arabia — Mohd Rafatullah

Acknowledgments

We express our heartfelt gratitude to all the chapter contributors whose scientific rigor, dedication, and timely efforts have made this book possible. Their expertise forms the backbone of this volume and reflects the collaborative spirit of the global plant research community. We extend our sincere thanks to Springer Nature for providing us the platform and guidance to bring this book to fruition. The meticulous editorial support and assistance throughout the process have been invaluable. We are deeply thankful to our institutions, colleagues, and research groups for their constant encouragement and support. Special appreciation goes to our families for their patience, understanding, and unwavering motivation throughout the preparation of this book. Lastly, we acknowledge the broader scientific community whose ongoing research and commitment to climate resilience continue to inspire us. This book is a collective contribution to that shared mission to enhance plant resilience and safeguard the future of our ecosystems and agriculture.

Introduction

Climate-Responsive Plant Biology emerges at a time when the urgency to understand plant responses to climate change has never been greater. Plants form the foundation of terrestrial ecosystems, food security, and global biodiversity. However, climate-induced stressors such as rising temperatures, altered precipitation patterns, increased frequency of droughts, salinity intrusion, and atmospheric irregularities pose unprecedented threats to plant systems worldwide.

Plants possess remarkable physiological, biochemical, and molecular mechanisms that enable them to perceive and adapt to environmental fluctuations. Advances in omics technologies, biotechnology, nanotechnology, and computational tools have revolutionized our ability to unravel these complex regulatory networks. At the same time, socioeconomic and ecological considerations highlight the need for holistic, sustainable strategies for climate-resilient agriculture. This book integrates diverse yet interconnected themes to provide a comprehensive understanding of plant-environment interactions under climate change. The chapters in this volume explore:

- Fundamental principles of plant physiology and adaptation
- Evolutionary and ecological perspectives on plant distribution
- Molecular and epigenetic mechanisms underlying stress tolerance
- Technological innovations such as CRISPR, phenomics, and nanotechnology
- Plant-microbe interactions and their role in resilience
- Socioeconomic dimensions of climate-driven agricultural challenges

By bringing together experts from multiple disciplines, this volume presents a multilayered perspective on how plants respond to environmental stressors and how scientific advancements can inform sustainable solutions.

Contents

Editors and Contributors

About the Editors

Hem Chandra Joshi, Assistant Professor of Botany at D.S.B. Campus, Kumaun University, Nainital, specializes in plant physiology and nutrient-environment interactions. He earned his Ph.D. in Plant Physiology from G.B. Pant University of Agriculture and Technology, Pantnagar, and completed his M.Sc. in Botany as a University Gold Medalist. With more than 9 years of research and teaching experience, his work integrates mineral nutrition, environmental biotechnology, bio-nanotechnology, and biosensor applications to support sustainable agriculture and precise plant health monitoring. Dr. Joshi has published over 35 peer-reviewed research papers, presented at national and international scientific forums, and supervised postgraduate dissertations. He actively engages in academic development through workshops, FDPs, scientific webinars, and membership in professional scientific societies. In recognition of his research contributions, he received the Young Scientist Award from the State Science and Technology Department. He also serves as a reviewer for several reputed international journals, including review contributions for *3 Biotech*, *Agricultural Water Management*, *Environmental Research Communications*, *Materials Research Express*, *Nano Express*, and *Plant Nano Biology*. His scientific and academic leadership aims to promote interdisciplinary research and technological innovation for environmentally resilient and resource-efficient agricultural systems.

Dhanushka Udayanga is affiliated with the Department of Applied Sciences, Faculty of Humanities and Sciences, Sri Lanka Institute of Information Technology (SLIIT), where he teaches and conducts research in biotechnology and mycology. He has extensive research experience, with a focus on molecular systematics of fungi and their applications in agriculture and environmental sciences. He obtained his Ph.D. from Mae Fah Luang University, Thailand, and has served as a visiting scientist at the United States Department of Agriculture, Agricultural Research Services (USDA-ARS) in Beltsville, Maryland, USA. Prof. Udayanga has authored over 50 research articles in reputed international journals and is actively involved in peer review and editorial duties for several academic journals.

Saurabh Gangola is an assistant professor in the Department of Microbiology at Graphic Era University, Dehradun, India. He brings more than six years of teaching and research experience, particularly in environmental microbiology. He holds a doctorate in Microbiology and his core research interests include the biodegradation and bioremediation of xenobiotic compounds such as pesticides, dyes, heavy metals, and polyaromatic hydrocarbons. Dr. Gangola has been recognized with the Governor Research Award in 2019 by the state of Uttarakhand. He has published 60 research and review papers, contributed 37 book chapters, edited 3 academic books, and holds 5 patents in related fields.

Mohd Rafatullah is currently an academic staff member in Department of Mathematics and Natural Sciences, Prince Mohammad Bin Fahd University, Saudi Arabia. He began his journey at USM as a Postdoctoral Fellow in 2008. He earned his Ph.D. in Environmental Chemistry from Aligarh Muslim University (AMU), India, in 2007. His areas of specialization include environmental pollutants, nanomaterials for environmental applications, water and wastewater treatment, adsorption and ion exchange mechanisms, and instrumental analytical techniques. Dr. Rafatullah has published numerous research and review articles in international journals and has presented his findings at both national and international conferences. He actively participates in professional development through various workshops and seminars worldwide.

Contributors

Rishika Ajay Department of Botany, IPGGPGCC Haldwani, Kumaun University, Nainital, Uttarakhand, India

Shahzad Anwar Plant Biotechnology Section, Department of Botany, Aligarh Muslim University, Aligarh, India

Kausar Arisha Plant Biotechnology Section, Department of Botany, Aligarh Muslim University, Aligarh, India

H. M. C. V. Bandara Department of Botany, Faculty of Applied Sciences, University of Sri Jayewardenepura, Gangodawila, Nugegoda, Sri Lanka

Santan Barthwal ICFRE-Forest Research Institute, Dehradun, India

Sonakshi Beri Department of Botany, IPGGPGCC Haldwani, Kumaun University, Nainital, Uttarakhand, India

Narendra Singh Bhandari Shree Guru Gobind Singh Tricentenary University, Gurugram, Haryana, India

Palak Chauhan Department of Biotechnology, School of Applied and Life Sciences, Uttaranchal University, Dehradun, Uttarakhand, India

Kanika Dulta Center of Advanced Innovation Technologies, VŠB-Technical University of Ostrava, Ostrava-Poruba, Czech Republic

Widad Fadhullah Environmental Technology Division, School of Industrial Technology, Universiti Sains Malaysia, Penang, Malaysia

Renewable Biomass Transformation Cluster, School of Industrial Technology, Universiti Sains Malaysia, George Town, Pulau Pinang, Malaysia

Saurabh Gangola Department of Microbiology, Graphic Era (Deemed to be University), Dehradun, Uttarakhand, India

Pengqi Gao Department of Civil and Environmental Engineering, The Faculty of Engineering, The University of Auckland, Auckland, New Zealand

Sanjay Gupta School of Biosciences, Swami Rama Himalayan University, Jollygrant, Dehradun, Uttarakhand, India

S. K. Guru Department of Plant Physiology, College of Basic Sciences and Humanities, G.B. Pant University of Agriculture and Technology, Pantnagar, Uttarakhand, India

Demus Matheus Huang Environmental Technology Division, School of Industrial Technology, Universiti Sains Malaysia, Penang, Malaysia

Weihong Hu Environmental Technology Division, School of Industrial Technology, Universiti Sains Malaysia, Penang, Malaysia

Hem C. Joshi Department of Botany, D.S.B. Campus, Kumaun University, Nainital, Uttarakhand, India

Neeraj Joshi L'institut Agro Montpellier, Montpellier, France

Samiksha Joshi School of Biosciences, Swami Rama Himalayan University, Jollygrant, Dehradun, Uttarakhand, India

Anroop Kaur School of Agricultural Biotechnology, Punjab Agricultural University, Ludhiana, Punjab, India

Kapil Khulbe Department of Botany, DSB CAMPUS Kumaun University, Nainital, Uttarakhand, India

Munmun Kothari Department of Plant Physiology, College of Basic Sciences and Humanities, G.B. Pant University of Agriculture and Technology, Pantnagar, Uttarakhand, India

Abhishek Kumar Department of Biotechnology, School of Applied and Life Sciences, Uttaranchal University, Dehradun, Uttarakhand, India

Kishan Kumar ICFRE-Forest Research Institute, Dehradun, India

Pardeep Kumar Department of Agriculture, Motherhood University, Roorkee, Uttarakhand, India

Vivek Kumar Department of Botany, IPGGPGCC Haldwani, Kumaun University, Nainital, Uttarakhand, India

Maneesh Lingwan Donald Danforth Plant Science Center, Saint Louis, MO, USA

Apurva Malik ICFRE-Forest Research Institute, Dehradun, India

Dimuthu S. Manamgoda Department of Botany, Faculty of Applied Sciences, University of Sri Jayewardenepura, Gangodawila, Nugegoda, Sri Lanka

Manushree Metalytics, Inc., Cary, NC, USA

Pal Neha Plant Biotechnology Section, Department of Botany, Aligarh Muslim University, Aligarh, India

Anamika Pal Department of Biotechnology, School of Applied and Life Sciences, Uttaranchal University, Dehradun, Uttarakhand, India

Deep Pandey Department of Botany, M.B. Government P.G. College, Haldwani, Uttarakhand, India

Ganesh Pandey Faculty of Agricultural Sciences, Shree Guru Gobind Singh Tricentenary University, Chandu-Budhera, Gurugram, Haryana, India

Niharika Pandey Department of Molecular Biology and Genetic Engineering, College of Basic Sciences and Humanities, G.B. Pant University of Agriculture and Technology, Pantnagar, Uttarakhand, India

Yogesh Pant Indian Institute of Technology, Mandi, HP, India

Prem Prakash Department of Botany, M.B. Government P.G. College, Haldwani, Uttarakhand, India

Mohd Rafatullah Department of Mathematics and Natural Sciences, Prince Mohammad Bin Fahd University, Al-Khobar, Al-Khobar, Saudi Arabia

Aashu Rajput School of Agriculture, Uttaranchal University, Dehradun, Uttarakhand, India

Sanju Department of Botany, IPGGPGCC Haldwani, Kumaun University, Nainital, Uttarakhand, India

E. D. Supipi Sewwandi Department of Botany, Faculty of Applied Sciences, University of Sri Jayewardenepura, Gangodawila, Nugegoda, Sri Lanka

A. K. Sharma Department of Biological Sciences, College of Basic Sciences and Humanities, G. B. Pant University of Agriculture and Technology, Pantnagar, Uttarakhand, India

Diksha Sharma Department of Plant Physiology, College of Basic Sciences and Humanities, G.B. Pant University of Agriculture and Technology, Pantnagar, Uttarakhand, India

Megha Sharma School of Biosciences, Swami Rama Himalayan University, Jollygrant, Dehradun, Uttarakhand, India

Akanksha Singh Department of Food Technology, School of Applied and Life Sciences, Uttaranchal University, Dehradun, Uttarakhand, India

Gaurav Ayodhya Singh EF Polymer Pvt. Ltd, Udaipur, Rajasthan, India

Hukum Singh ICFRE-Forest Research Institute, Dehradun, India

Jyoti Singh Department of Biotechnology, School of Applied and Life Sciences, Uttaranchal University, Dehradun, Uttarakhand, India

Tribhuwan Singh Centre for Oilseeds Research, S.D. Agricultural University, Gujarat, India

Rashmi Srivastava Department of Biological Sciences, College of Basic Sciences and Humanities, G. B. Pant University of Agriculture and Technology, Pantnagar, Uttarakhand, India

Sushma Tamta Department of Botany, D.S.B Campus, Kumaun University, Nainital, Uttarakhand, India

Shikha Thakur Department of Biotechnology, School of Applied and Life Sciences, Uttaranchal University, Dehradun, Uttarakhand, India

Shambhavi Tiwari Department of Plant Pathology, SVPUAT, Meerut, Uttar Pradesh, India

Department of Agriculture, Tulas Institute, Dehradun, India

Dhanushka Udayanga Department of Applied Sciences, Faculty of Humanities and Sciences, Sri Lanka Institute of Information Technology, Malabe, Sri Lanka

Yinqiu Wang Environmental Technology Division, School of Industrial Technology, Universiti Sains Malaysia, Penang, Malaysia

Lv Wan School of Life Sciences and Chemical Engineering, Jiangsu Second Normal University, Jiangsu, Nanjing, China

1 Current Trends in Plant–Environment Interactions: Bridging Abiotic and Biotic Frontiers

Dimuthu S. Manamgoda, H. M. C. V. Bandara, E. D. Supipi Sewwandi, and Dhanushka Udayanga

Abstract

Plant–environment interactions are significant in shaping ecosystem dynamics, agricultural productivity, and plant resilience under changing climatic conditions. In this chapter, we explore emerging trends in plant–environment interactions, emphasizing novel mechanisms by which plants perceive and respond to environmental changes. Recent advances highlight the role of root exudates in modulating rhizosphere microbiomes, the integration of epigenetic regulation in stress memory, and the bidirectional influence of climate change on plant–microbe symbioses. We also discuss cutting-edge research on how plants employ signaling networks to adapt to abiotic stressors, such as drought, salinity, and extreme temperatures, while leveraging microbial partnerships for enhanced nutrient acquisition and defense. This chapter provides a current perspective on harnessing plant–environment interactions for sustainable agriculture and ecosystem restoration in the face of global environmental challenges.

Keywords

Abiotic stress · Biotic stress · Epigenetic stress memory · Climate resilience · Sustainable agriculture

---

D. S. Manamgoda (✉) · H. M. C. V. Bandara · E. D. S. Sewwandi
Department of Botany, Faculty of Applied Sciences, University of Sri Jayewardenepura, Gangodawila, Nugegoda, Sri Lanka
e-mail: dsmanamgoda@sjp.ac.lk

D. Udayanga
Department of Applied Sciences, Faculty of Humanities and Sciences, Sri Lanka Institute of Information Technology, SLIIT Malabe Campus, New Kandy Road, Malabe, Sri Lanka

H. C. Joshi et al. (eds.), *Climate-Responsive Plant Biology*,
https://doi.org/10.1007/978-981-95-7354-7_1

1.1 Introduction

Plant–environment interactions are fundamental to ecosystem stability, and agriculture under rapidly changing climatic conditions. Plants employ diverse adaptations ranging from structural and physiological adjustments to biochemical, molecular, and epigenetic responses to survive under abiotic and biotic stresses (Traubenik et al. 2024; Zhang et al. 2023). For instance, drought triggers abscisic acid (ABA)-mediated stomatal regulation, deeper root growth, and reduced leaf area to maintain water balance (Chen et al. 2020; Kou et al. 2022; Devi et al. 2023). Salt tolerance depends on ion transporters that exclude or sequester toxic ions (Zhou et al. 2024a, b, c), while heat stress is alleviated by heat shock proteins and DREB2A transcription factors that refold damaged proteins (Pattnaik et al. 2021; Liu et al. 2025).

Plants also respond to heatwaves, flooding, and cold snaps through phenological shifts, production of protective compounds, and structural adaptations like aerenchyma formation in roots to maintain oxygen supply during flooding (Lahlali et al. 2024). Cold stress activates specific gene networks that remodel membrane composition and enhance reactive oxygen species (ROS) scavenging to minimize damage (Qian et al. 2024). Elevated CO_2 generally enhances photosynthesis and water-use efficiency but can alter nutrient uptake and carbon–nitrogen balance, influencing growth and defense traits (Li et al. 2023), whereas increased ozone reduces photosynthetic capacity, induces oxidative damage, and disrupts plant–microbe and plant–insect interactions by interfering with volatile organic compound (VOC) signaling (Nowroz et al. 2024; Otieno et al. 2023).

Heavy-metal contaminations will be mitigated by phytoremediation mechanisms (Yan et al. 2020; Ejaz et al. 2023) while epigenetic modifications and stress memory enable plants to “remember” prior stress events, enhancing future resilience and, in some cases, transmitting adaptive responses to progeny (Ramakrishnan et al. 2022).

Plants also rely on interactions with their microbiomes to improve nutrient acquisition and stress tolerance. Climate change increases abiotic and biotic pressures, altering plant–pathogen and plant–insect dynamics, reducing plant immunity, and affecting microbial communities (Pedroncelli et al. 2019; Singh et al. 2023). Beneficial microbes, including 1-aminocyclopropane-1-carboxylate (ACC) deaminase-producing plant growth-promoting rhizobacteria (PGPR) and arbuscular mycorrhizal fungi, mitigate these stresses by modulating ethylene, phytohormones, nutrient uptake, and antioxidant systems, and this would offer a sustainable solution for climate-resilient agriculture (Banerjee et al. 2023; Chieb and Gachomo 2023). Plants coordinate pathogen responses to simultaneous stresses via signaling networks involving reactive oxygen species (ROS) and phytohormones. ABA regulates abiotic stress responses, while salicylic acid (SA) and jasmonic acid (JA) govern defense against pathogens. Root exudates shape rhizosphere communities, attracting beneficial microbes such as nitrogen-fixing bacteria and phosphate-solubilizing microbes, while mycorrhizal fungi extend root networks and enhance nutrient uptake (Lin et al. 2022). Microbial partners contribute via antioxidant regulation, phytohormone modulation, and osmolyte production, supporting plant resilience under combined abiotic and biotic stresses (Tan et al. 2022).

Sustainable strategies increasingly focus on integrating plant–microbe interactions with breeding and biotechnology. Microbial inoculants, multifactor stress-tolerant crops, and ecological restoration approaches enhance resilience under drought, heat, salinity, and pathogen pressures. Advanced breeding, genome editing, and microbiome-informed selection enable crops to maintain growth and productivity under multifactorial stresses (Liu et al. 2025). Emerging monitoring technologies, such as hyperspectral imaging, thermal sensors, chlorophyll fluorescence, and IoT-based analytics, allow real-time assessment of plant responses and precision management under complex environmental conditions (Walsh et al. 2024; Zandi et al. 2025; Mansoor et al. 2025). Future directions require affordable, robust, and user-friendly systems, multi-scale data integration, and interdisciplinary collaboration to translate novel insights into sustainable, climate-resilient agricultural and ecological systems.

1.2 Plant Responses to Abiotic Stressors

1.2.1 Drought and Temperature Extremes

Drought and high-temperature episodes frequently occur together and are among the most damaging environmental stresses on crops. They limit photosynthetic efficiency, disrupt water and nutrient relations, and accelerate oxidative injury, ultimately reducing plant growth and yield. Climate change is predicted to increase both the frequency and severity of these events, making a clear understanding of plant adaptation mechanisms essential for developing stress-resilient cultivars (Teshome et al. 2020; Muhammad et al. 2024). Plant responses to drought and temperature extremes can be further clarified as physiological and molecular adaptations.

1.2.1.1 Physiological Adaptions

One of the well-studied physiological responses is limitation of transpirational water loss by stomatal closure, which is primarily regulated by the phytohormone abscisic acid (ABA) (Chen et al. 2020; Kou et al. 2022; Devi et al. 2023). Drought and dehydration conditions induce ABA biosynthesis that initiates signal transduction pathways, activating multiple stress responses. Through the modulation of ion channels in guard cells, ABA promotes ion efflux and consequent turgor loss, leading to guard-cell closure (Chen et al. 2020). The mechanism balances the fundamental trade-off between sustaining CO_2 uptake for photosynthesis and limiting water loss, thereby enhancing plant fitness under water scarcity (Peng et al. 2022; Mukherjee et al. 2023).

Root system change is an adaptive trait that facilitates drought tolerance in plants. When soil moisture lacks, plants modify root architecture through elongation of primary roots, proliferation of lateral roots, and deeper rooting to access water reserves unavailable to shallow root systems (Kou et al. 2022). Such structural modifications not only optimize water acquisition but also improve nutrient uptake, which will strengthen plant resilience under drought and thermal extremes. In addition to root and stomatal responses, leaf morphological adaptations further contribute to limiting water loss and thermal damage. For instance, reduced leaf size, deposition of thicker cuticular layers, increased trichome density, and mechanical responses such as leaf rolling or wilting reduce the effective leaf surface exposed to solar radiation (Liu et al. 2023; Yavas et al. 2024).

Drought tolerance is largely controlled by transcriptional regulation of stress-responsive genes. The transcription factor DREB2A binds dehydration-responsive elements (DREs) in promoters such as *RD29A*, activating cascades that maintain osmotic balance, enhance antioxidant activity, and protect cells under water deficit (Liu et al. 2023). Heat shock proteins (HSPs), major molecular chaperones, preserve proteostasis under high temperatures by preventing aggregation and refolding damaged proteins (Pattnaik et al. 2021; Liu et al. 2025).

1.2.1.2 Molecular-Level Adaptations

Drought tolerance is primarily regulated through transcriptional control of stress-responsive genes. The transcription factor DREB2A binds dehydration-responsive elements (DREs) in promoters such as *RD29A*, triggering cascades that maintain osmotic balance, enhance antioxidant capacity, and protect cells from dehydration (Liu et al. 2023). Heat shock proteins (HSPs) act as molecular chaperones that prevent protein aggregation and refold damaged polypeptides, maintaining proteostasis under heat stress (Liu et al. 2023; Devi et al. 2023). Drought and temperature extremes also elevate reactive oxygen species (ROS), causing oxidative damage to lipids, proteins, and nucleic acids. Plants counteract this via enzymatic antioxidants superoxide dismutase (SOD), catalase (CAT), and peroxidases (POD), and non-enzymatic antioxidants like ascorbate and glutathione, preserving cellular homeostasis (Dvořák et al. 2021).

Plants respond to drought and thermal stress by synthesizing osmo protectants or compatible solutes such as proline, glycine betaine, and trehalose to mitigate osmotic imbalance. These molecules preserve protein structure, maintain membrane integrity, protect enzymatic activities, and act as reactive oxygen species, contributing to cellular resilience under adverse environmental conditions (Singh et al. 2022). The biosynthesis and intracellular accumulation of osmo protectants are under precise genetic control, mediated by stress-responsive genes encoding key enzymes like pyrroline-5-carboxylate synthetase (P5CS) for proline and betaine aldehyde dehydrogenase (BADH) for glycine betaine. Transcription factors from the DREB and NAC families modulate these genes, coordinating stress responses via complex signaling networks involving abscisic acid and other phytohormones (Pattnaik et al. 2021). Plants employ an intricate network that coordinates the interactions among osmoprotectants, transcriptional regulators, and hormonal signaling that allows plants to maintain osmotic balance while protecting cellular structures under simultaneous drought and heat stress (Hou et al. 2023).

1.2.2 Soil Salinity and Heavy Metal Contamination

Soil salinity and heavy metal accumulation are critical constraints for sustainable agriculture. The contamination could happen through natural processes, irrigation practices, industrial effluents, and intensive fertilizer usage. High salinity imposes osmotic stress and ionic imbalance that impair nutrient uptake, while heavy metals disrupt cellular redox homeostasis and interfere with essential metabolic processes. Oil salinity and heavy metal contamination frequently co-occur as major environmental stresses, particularly in arid and semi-arid regions. Salinity alters the chemical speciation and mobility of metals, often enhancing their bioavailability and uptake by plants, while heavy metals such as cadmium (Cd), lead (Pb), and copper (Cu) impose oxidative, osmotic, and ionic stresses that disrupt plant metabolism. Together these stresses reduce germination, growth, and yield, and may also affect soil microbial communities and plant–microbe interactions (Fu and Yang 2023; Ejaz et al. 2023).

Plants cope with salinity and heavy metal stress through ion exclusion, vacuolar sequestration, chelation by peptides such as phytochelatins, and activation of antioxidant defenses. Halophytes, naturally adapted to high salinity, often exhibit dual tolerance to salts and toxic metals. For instance, *Suaeda salsa* can accumulate Cd, Zn, and Pb while sustaining growth, making it a promising candidate for phytoremediation of saline, metal-contaminated soils. The interaction between salinity and metal toxicity varies with species and conditions—salt stress may either reduce metal uptake or enhance its translocation—indicating synergistic or antagonistic effects. Understanding these mechanisms is essential for ecological risk assessment and sustainable land management. A few of such mechanisms are discussed below Sects. 1.2.2.1 and 1.2.2.2.

1.2.2.1 Ion Homeostasis and Exclusion Mechanisms

Under salinity stress, plants regulate sodium (Na^+) and chloride (Cl^-) fluxes, including ion uptake, vascular compartmentalization, and active exclusion to maintain ion homeostasis (Fu et al. 2023; Liu et al. 2023). Selective ion transporters in root cells, including high-affinity potassium transporters (HKTs), sodium-proton antiporters (SOS1), and vacuolar NHX antipotors, facilitate sequestration of sodium ions into vacuoles or remove them from the cytoplasm, minimizing cellular toxicity (Liu et al. 2023).

The coordinated activity of these transport systems ensures proper K^+/Na^+ ratios necessary for metabolic and osmotic stability. Heavy metal stress interferes with ion homeostasis by affecting essential nutrient uptake and inducing ROS generation (Fu et al. 2022; Ejaz et al. 2023). Plants synthesize metal-chelating peptides like phytochelatins and metallothioneins, which compartmentalize metals within vacuoles, minimizing cellular disruption due to metal toxicity (Ejaz et al. 2023; Hasan et al. 2025), while enzymatic and non-enzymatic antioxidant systems mitigate ROS accumulation (Goncharuk and Zagoskina 2023). Integration of signaling pathways triggers transcriptional regulation of ion transporters, chelators, and antioxidative systems, allowing plants to sense and adapt to combined salinity and heavy metal stresses, improving stress resilience (Chen et al. 2020).

1.2.2.2 Phytoremediation Potential in Plants

Plants employ various phytoremediation strategies to tolerate and mitigate heavy metal toxicity and other environmental pollutants. Hyperaccumulator plant species can concentrate substantial amounts of elements like cadmium, lead, and arsenic in their biomass through phytoextraction, wherein plants uptake and translocate heavy metals from contaminated rhizosphere into aerial plant tissue such as shoots and roots. Subsequent harvesting of this biomass physically removes toxic substances, progressively lowering metal concentrations in the plant body (Sharma et al. 2023). Phytostabilization is an important plant-based remediation approach, in which plants mitigate heavy metal uptake by reducing their mobility within root tissue or the rhizosphere. This is primarily achieved through the secretion of organic acids such as citric, oxalic, and malic acids and amino acids, which are capable of chelating heavy metals and altering their chemical specification, leading to precipitation or absorption onto soil particles (Yan et al. 2020; Ejaz et al. 2023).

In addition to chemical chelation via root exudates, plant cell walls contain high levels of negatively charged pectins and polysaccharides serving as metal-binding sites (KrzesBowska et al. 2013). This structural immobilization within root structures effectively restricts metal translocation beyond the roots to sensitive aerial plant tissues. Enzymatic degradation of contaminants into less toxic compounds, via metabolic and enzymatic pathways, is known as phytodegradation. Plants produce enzymes such as oxygenases, peroxidases, and dehalogenases, which catalyze oxidation, reduction, or hydrolysis reactions, converting pollutants into simpler compounds (Kafle et al. 2022; Sharma et al. 2023).

1.2.3 Epigenetic Regulation of Stress Memory

Plants exhibit the capacity to retain a molecular memory of prior abiotic stress events, such as drought, salinity, and extreme temperatures, enabling them to perform faster and more robust responses during future exposures (Ramakrishnan et al. 2022; Gallusci et al. 2023). This phenomenon is mediated by epigenetic processes, including DNA methylation, histone modifications, and the action of small RNAs, which modulate the expression of stress-responsive genes while leaving the DNA sequence unchanged (Ramakrishnan et al. 2022).

Transcriptional activity of stress-responsive gene expression is influenced by DNA methylation at promoter and gene body regions. These modifications can be stably inherited through cell divisions, establishing somatic stress memory that enhances plant resilience to recurring stress conditions (Ramakrishnan et al. 2022; Gallusci et al. 2023). Histone tail modifications, such as methylation and acetylation of specific lysine residues, contribute to transcriptional memory by altering chromatin structure to activate or repress stress-related gene expressions (Yu et al. 2025). Small RNAs, including microRNAs and siRNAs, further modulate epigenetic marks and gene silencing, refining the plant's adaptive stress responses and memory (Ramakrishnan et al. 2022; Yu et al. 2025). Certain stress-induced epigenetic marks can be inherited via sexual reproduction, allowing their transmission to subsequent generations and providing a progeny with pre-established adaptive responses (Abdulraheem et al. 2024).

1.3 Biotic Interactions in the Plant Environment

1.3.1 Root Exudates as Mediators of Rhizosphere

Plant root exudates consist of a complex yet diverse array of organic and inorganic compounds. These include low-molecular-weight compounds, such as organic acids, sugars, and amino acids, as well as high-molecular-weight substances, including polysaccharides (mucilage) and proteins (Ma et al. 2022). Root exudates play a significant role in the rhizosphere by creating a biochemically dynamic and nutrient-rich environment, serving both as a food source and a signaling platform for microbes (Wu et al. 2023).

Among root exudates, proteins act as major contributors to nutrient acquisition and defense by activating immune responses in plants. Most importantly, root exudates regulate microbial activities, including colonization in the rhizosphere (Adeniji et al. 2025). This process strengthens mutualistic interactions between microbes and plants. For example, some root exudate compounds, such as arginine and methyl stearate, act as chemoattractants for beneficial bacteria capable of nitrogen fixation or phosphate solubilization (Lin et al. 2022).

Recent studies reveal that the root exudates of maize contain malic acid, which attracts phosphate-solubilizing bacteria like *Bacillus* spp., improving phosphorus uptake (Lin et al. 2022). Similarly, microbial recruitment improves plant stress tolerance under drought conditions. Chickpea roots increase secretion of proline and sugars, recruiting drought-tolerant microbes that enhance water-use efficiency (Chilakala et al. 2022). The plant microbiota, acting as a "second genome," enables the holobiont to function as an integrated unit with greater fitness and ecological success (Li et al. 2024).

Understanding the chemical ecology of root exudates reveals novel pathways and mechanisms. For example, flavonoid signaling in legume roots specifically attracts symbiotic rhizobia; thus, the pathway could be engineered to enhance nodulation (Concha and Doerner 2020). These pathways demonstrate how plants influence their microbial partners. Also, evidence suggests that such root-derived signals contribute to broader community-level interactions. By directing microbial composition, root exudates indirectly influence competition, cooperation, and niche differentiation among microbes, which in turn governs nutrient cycling, organic matter decomposition, and pathogen suppression in the rhizosphere. These processes exemplify how plants, through targeted chemical signaling, extend their influence beyond individual symbionts to modulate entire microbial networks. This reinforces the concept of the plant holobiont, where roots, microbes, and exudates form an integrated ecological unit that enhances overall plant fitness, ecosystem stability, and adaptability to stress (Voolstra and Ziegler 2020; Lyu et al. 2021). Thus, root exudates are not passive byproducts of metabolism but act as active mediators of biotic interactions, shaping plant, microbe, and soil dynamics essential for sustainable agriculture.

1.3.2 Plant–Microbe Symbioses Under Environmental Stress

Mycorrhizal associations are ecologically important, widespread symbioses between plants and fungi, essential for nutrient uptake in many ecosystems (Wahab et al. 2023). Principal types include arbuscular mycorrhizae (AM), ectomycorrhizae (ECM), and specialized forms such as ericoid and orchid mycorrhizae (Zhang et al. 2025). AM fungi form highly branched intracellular structures called arbuscules within root cortical cells, enabling efficient nutrient exchange (Cargill et al. 2025). This structure allows access to complex organic nutrients through enzymatic decomposition (Dalling and Willey 2025).

Mycorrhizal fungi are able to increase phosphorus availability to plants by extending root absorptive surface area through extensive extraradical hyphal networks, accessing insoluble phosphorus pools beyond root depletion zones. They secrete phosphatases and organic acids that solubilize phosphorus from minerals and organic compounds, making it accessible to plants (Qi et al. 2022). Additionally, they enhance uptake of micronutrients like zinc, copper, and iron, critical for nitrogen fixation, enzymatic cofactors, and other physiological functions (Ahmed et al. 2025). Iron obtained through mycorrhizal pathways is indispensable as a cofactor for nitrogenase enzymes in rhizobia, directly linking micronutrient uptake to nitrogen fixation efficiency (Brear et al. 2013). In return, plants supply photosynthetically fixed carbon to fungi as sugars and lipids, vital for fungal survival and nutrient mobilization (Leng et al. 2023).

Microbial partners in the hyphosphere, such as *Streptomyces* species, contribute by producing enzymes and antibiotics that foster nutrient solubilization and protect the symbiotic system from competing microbes (Jin et al. 2024). The main nitrogen-fixing symbiotic genera such as *Rhizobium* (legume nodulators), *Frankia* (actinorhizal symbionts), and free-living or associative nitrogen fixers (Mus et al. 2016). Rhizobia initiate symbiosis via elaborate molecular recognition; legumes secrete flavonoid compounds that induce nod factor production, triggering nitrogen and phosphorus limitations, which exert a strong selective pressure to enhance symbiotic efficiency. Under nutrient limitations, mycorrhizal colonization and nodulation are supported by upregulated plant genes that reallocate carbon to sustain these mutualisms (Mo et al. 2025). Under the above-mentioned conditions, plants also display plastic responses such as altered root architecture and increased root surface area to improve symbiont access. Conversely, high inorganic nitrogen levels repress nodulation and nitrogenase gene expression to reduce costly symbiotic maintenance, mediated by intricate regulatory feedback loops. Co-limitation scenarios affect microbial community structures and plant responses, requiring precise synchronization of nutrient absorption, symbiosis, and growth (Xue et al. 2022).

Plant–microbe interactions are also strongly influenced by environmental stressors other than nutrient deficiency, broadly categorized as abiotic and biotic factors. Abiotic stresses like drought, salinity, and temperature extremes cause osmotic imbalances, oxidative stress, and structural or functional damage to plants and microbes (Muhammad et al. 2024). For example, high salinity causes ionic toxicity and osmotic pressure that hinder microbial metabolic processes, while drought decreases root exudate diversity and quantity, reshaping rhizosphere community composition (Zhou et al. 2024a, b, c). Also, temperature variability affects microbial resilience by influencing enzymatic activity and growth rates (Cruz-Paredes et al. 2023).

Biotic stresses, such as pathogen attacks and microbial competition, often cause shifts toward opportunistic or pathogenic taxa, destabilizing beneficial microbial consortia through antagonistic pressures (Soth et al. 2025). Environmental stressors typically reduce microbial diversity and functional redundancy, destabilizing key ecosystem processes. Drought- or salinity-affected soils often exhibit reduced microbial abundance, which facilitates opportunistic taxa, including some pathogenic or disease-promoting species. For example, drought-driven microbial community reductions correlate with increased root rot pathogens in wheat (Gao et al. 2025).

Microorganisms deploy various adaptations and resilience mechanisms, including osmolyte synthesis, biofilm formation, and stress-responsive gene expression to tolerate stress (Tan et al. 2022). Horizontal gene transfer spreads beneficial traits such as antibiotic resistance, osmotolerance, and heavy metal detoxification, enabling rapid adaptation after environmental challenges (Woods et al. 2020). Microbial communities engage in feedback loops with host plants, enhancing tolerance by reducing oxidative damage via antioxidative enzymes and modulating hormones like ABA and cytokinins. For example, drought-tolerant maize microbiomes elevate abscisic acid synthesis, improving water-use efficiency and stomatal conductance (Ait-El-Mokhtar et al. 2023).

1.4 Interplay of Abiotic and Biotic Factors

1.4.1 Climate Change as a Driver of Interaction Shifts

Climate change intensifies fluctuations in temperature, precipitation, and atmospheric composition, creating novel combinations of abiotic and biotic pressures. These altered conditions disrupt the timing, intensity, and geographic range of plant–pathogen and plant–insect interactions, often weakening plant defense networks while enhancing pest or disease pressure. Elevated CO_2, heatwaves, and prolonged droughts can shift host–microbe balances, favor opportunistic pathogens, and promote new pest outbreaks or disease vectors in previously unaffected regions (Szczepaniec and Finke 2019; Singh et al. 2023). As illustrated in Table 1.1,

Table 1.1 Common stress pairings, mechanistic interactions, and plant outcomes

Stress pairing	Interaction type	Mechanism (key pathways)	Outcome/example	References
Drought + Heat	Synergistic	ABA-induced closure reduces cooling; ROS impairs PSII	Early senescence, fewer seeds	Fu et al. (2022), Singh et al. (2023)
Heat + Pathogen	Antagonistic	Heat suppresses SA/JA defense; R-protein instability	More Fusarium wilt	Pedroncelli et al. (2019)
Drought + Pathogen	Mixed	ABA suppresses SA; mild drought primes JA/SA	Variable resistance	Hönig et al. (2023)
Salinity + Nutrient Limitation	Synergistic	Na^+ competes with K^+, lowers nitrate assimilation	Chlorosis, stunted growth	Ahmad et al. (2022)
Heat + Insects	Synergistic	High T accelerates life cycle and feeding	More aphids, virus spread	Szczepaniec and Finke (2019)
Drought + PGPR/AMF	Beneficial	ACC deaminase lowers ethylene; AM improves uptake	Higher root growth, WUE	Murali et al. (2021)
Salinity + Heavy Metals	Synergistic	Salts raise metal solubility; ROS bursts	Root/shoot oxidative damage	Fu et al. (2022)
O_3 + Pathogen	Mixed	ROS primes SA defense or damages tissue	Species-specific effects	Xu et al. (2024)
Flooding + Pathogen	Synergistic	Hypoxia weakens roots; microbiome shifts	↑ Phytophthora rot	Petrillo (2023)
UV-B + Drought	Synergistic	UV-B induces flavonoids; ↑ ABA, ROS	↓ photosynthesis, ↑ metabolites	Nawaz et al. (2023)

common stress combinations ultimately determine plant resilience or susceptibility under multifactorial stress conditions.

1.4.1.1 Altered Pathogen and Insect Dynamics Under Drought or Heat

Abiotic stressors are becoming more frequent and intense due to climate change, reshaping biotic interactions within ecosystems. In natural environments, plants typically experience complex combinations of abiotic and biotic challenges that require integrative adaptive strategies.

Drought and high-temperature stress associated with climate change can significantly influence plant-pathogen and plant–insect interactions. Drought stress reduces plant immunity by disrupting hormone signaling pathways and limiting the biosynthesis of defensive metabolites, increasing vulnerability to pathogens and herbivorous insects (Szczepaniec and Finke 2019; Singh et al. 2023). On the other hand, elevated temperatures accelerate the developmental rate of certain insect pests, increasing their population density and feeding activities, ultimately amplifying plant damage (Pedroncelli et al. 2019). Furthermore, drought and elevated temperatures affect pathogen virulence and distribution. Warmer, drier climates have been linked to increased proliferation of fungal pathogens such as Fusarium spp. Colletotrichum spp., intensifying disease incidence (Singh et al. 2023), while shifting temperature regimes facilitate insect vectors to colonize new regions, increasing the risk of vector-borne disease transmission (Szczepaniec and Finke 2019).

1.4.1.2 Microbe-Assisted Abiotic Stress Mitigation

Beneficial plant-associated microorganisms play a crucial role in improving plant survival against abiotic stresses, particularly drought and salinity (Table 1.1). A key mechanism involves the enzyme ACC deaminase, produced by bacteria. The enzyme degrades ACC, the immediate precursor molecule of ethylene, thereby lowering ethylene production and mitigating its growth-inhibitory effects during drought and salinity (Murali et al. 2021).

Through this activity, these microbes promote plant growth, enhance nutrient uptake, and water-use efficiency in adverse environments (Chieb and Gachomo 2023). Plant growth-promoting rhizobacteria (PGPR) such as *Pseudomonas* spp., *Bacillus* spp., and *Azospirillum* spp., influence not only ethylene but also phytohormones such as auxins, cytokinins, and gibberellins, leading to improved root development and enhanced stress tolerance (Ahmad et al. 2022; Chieb and Gachomo 2023). As summarized in Table 1.2, distinguishing plant responses to simultaneous and sequential stress exposures provides a valuable framework for optimizing microbial applications and developing resilient cropping systems under variable environmental conditions.

1.4.2 Signaling Networks at the Crossroads

Plants facing concurrent abiotic and biotic stresses rely on interconnected signaling networks rather than isolated pathways. These networks integrate environmental cues, hormonal signals, and secondary messengers to fine-tune defense and adaptation processes at cellular and whole-plant levels. Crosstalk among signaling hubs such as reactive oxygen species (ROS), calcium fluxes, mitogen-activated protein kinase (MAPK) cascades, and phytohormones allows plants to sense stress combinations, prioritize responses, and coordinate systemic signaling between roots and shoots (Chen et al. 2020; Fu et al. 2022).

1.4.2.1 Shared Pathways in Abiotic and Biotic Responses

Plant responses to multiple stresses operate sophisticated signaling networks where reactive oxygen species (ROS) and phytohormones play central roles in adapting both abiotic and biotic stress responses. Once ROS was viewed as a toxic byproduct, it is now recognized as a key second messenger that initiates stress signaling cascades (Chen et al. 2020). Under stress, controlled ROS bursts activate

Table 1.2 Distinguishing features of simultaneous versus sequential stress exposures in plants

Feature	Simultaneous stress	Sequential stress	Mechanism/Example	References
Signal Integration	Multiple pathways active; MAPK, WRKY crosstalk	First stress primes chromatin for rapid response	MAPK→ROS/Ca^{2+} → epigenetic priming	Fu et al. (2022)
Hormonal Crosstalk	ABA dominates; suppresses SA/JA	Mild ABA preconditions SA/JA	ABA–SA–JA triangle; ethylene modulator	Hönig et al. (2023)
Physiol. Impact	Synergistic membrane damage, ROS surge	Recovery repairs PSII, resets antioxidants	ROS vs. SOD/CAT balance	Nawaz et al. (2023)
Metabolic Trade-off	Growth→defense shift; yield loss	Priming improves growth-defense balance	SnRK1/TOR sensors	Petrillo (2023)
Recovery Window	Minimal overlap; low resilience	Rest resets redox/hormone status	Key breeding trait	Nawaz et al. (2023)
Example Outcome	Drought+heat→senescence	Mild drought→stronger pathogen defense	Stress memory persists	Hönig et al. (2023)

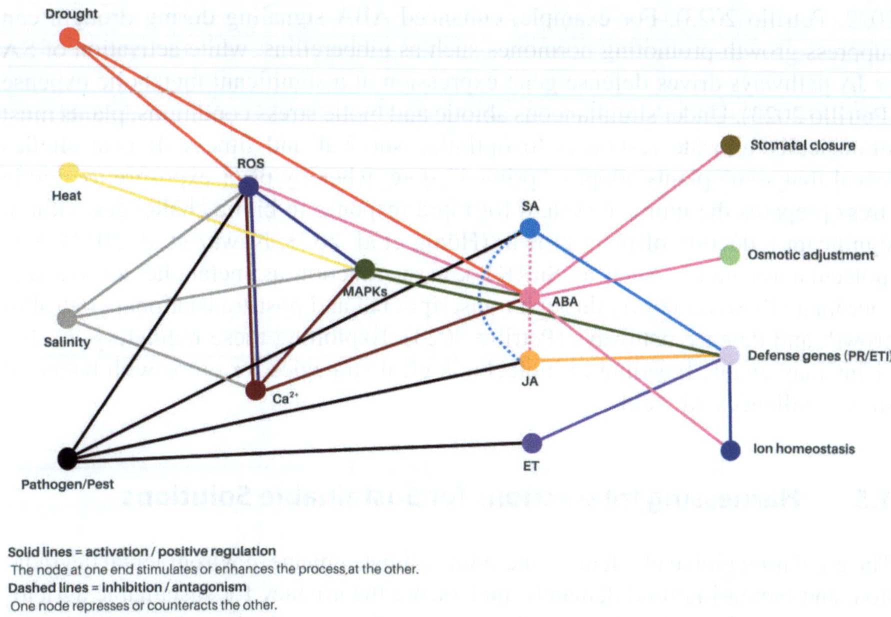

Fig. 1.1 Node-edge diagram showing how abiotic (drought, heat, salinity) and biotic (pathogen/pest) factors interconnected with plant stress responses through Ca^{2+}, MAPKs, ABA, SA, JA, ET and downstream processes like stomatal closure, osmotic adjustment, defense genes and ion homeostasis

mitogen-activated protein kinases (MAPKs) and transcription factors such as WRKY and NAC families, modulating stress-responsive gene expression across multiple stress types (Fu et al. 2022). The integration of abiotic and biotic stress responses is mainly governed by phytohormonal crosstalk. Abscisic acid (ABA) primarily mediates adaptation to abiotic stresses through regulation of stomatal closure and osmotic homeostasis, whereas salicylic acid (SA) and jasmonic acid (JA) govern defense against biotrophic and necrotrophic pathogens, respectively (Xu et al. 2024). However, interactions among these pathways are way more complex and highly depend on the context. ABA can suppress or enhance SA and JA signaling, thereby modulating plant resilience under combined stresses (Fu et al. 2022). This hormonal balance optimizes resource allocation and stress specificity, enabling plants to integrate multiple responses effectively, as illustrated in Fig. 1.1, which depicts the interconnected signaling networks linking abiotic and biotic stress responses through Ca^{2+}, MAPKs, and phytohormonal pathways.

1.4.2.2 Trade-Offs Between Defense and Growth Under Combined Stresses

Plants often maintain a critical balance between growth and defense under stress. During this trade-off, stress-induced hormonal signaling reallocates energy and resources from growth processes toward defense mechanisms, resulting in reduced biomass accumulation and reproductive success under stress conditions (Fu et al.

2022; Petrillo 2023). For example, enhanced ABA signaling during drought can suppress growth-promoting hormones such as gibberellins, while activation of SA or JA pathways drives defense gene expression at a significant metabolic expense (Petrillo 2023). Under simultaneous abiotic and biotic stress conditions, plants must strategically allocate resources to optimize survival and fitness. Recent studies reveal that some plants adopt a "primed" state, whereby prior exposure to abiotic stress prepares the immune system for rapid response to biotic challenges without significant reduction of plant growth (Hönig et al. 2023; Nawaz et al. 2023). Key molecular regulators, such as SnRK1 kinase, function as metabolic sensors that coordinate these trade-offs through transcriptional and post-translational control of growth and defense pathways (Petrillo 2023). Exploiting these regulatory mechanisms may enable breeding or biotechnological strategies for crops with balanced stress resilience and yield.

1.5 Harnessing Interactions for Sustainable Solutions

The escalating global challenges, including climate change, environmental degradation, and increasing food demands, underscore the urgency for sustainable agricultural and ecological solutions. Leveraging the complexity of plant–microbe interactions offers innovative pathways to restore degraded environments, enhance crop resilience, and sustainably optimize plant productivity. Recently advanced microbial biotechnology, plant breeding, and ecological restoration are converging to bridge abiotic and biotic frontiers, fostering resilient systems capable of withstanding diverse environmental pressures.

As illustrated in Fig. 1.2, crop–microbe interactions can result in both beneficial and detrimental effects, emphasizing the importance of careful evaluation in different stress contexts.

1.5.1 Microbial Inoculants for Climate-Resilient Agriculture

Plant growth-promoting rhizobacteria (PGPR), mycorrhizal fungi, and beneficial endophytes, collectively referred to as microbial inoculants, have emerged as critical allies in building climate-resilient crop systems (Saad et al. 2020). These microbial consortia help crops adapt to a range of abiotic stresses, including drought, salinity, high temperatures, and nutrient limitations, by reshaping host plant physiology and biochemical processes. Key mechanisms conferring stress tolerance involve improved nutrient uptake, modulation of phytohormone production (e.g., indole-3-acetic acid), ACC deaminase activity that lowers ethylene-induced stress, and upregulation of antioxidant defenses to prevent oxidative damage (Muhammad et al. 2024). Moreover, some microbes facilitate osmolyte build-up and enhance water use efficiency under drought conditions (Bouremani et al. 2024). Arbuscular mycorrhizal fungi (AMF), particularly *Glomus* spp., improve plant tolerance to heat

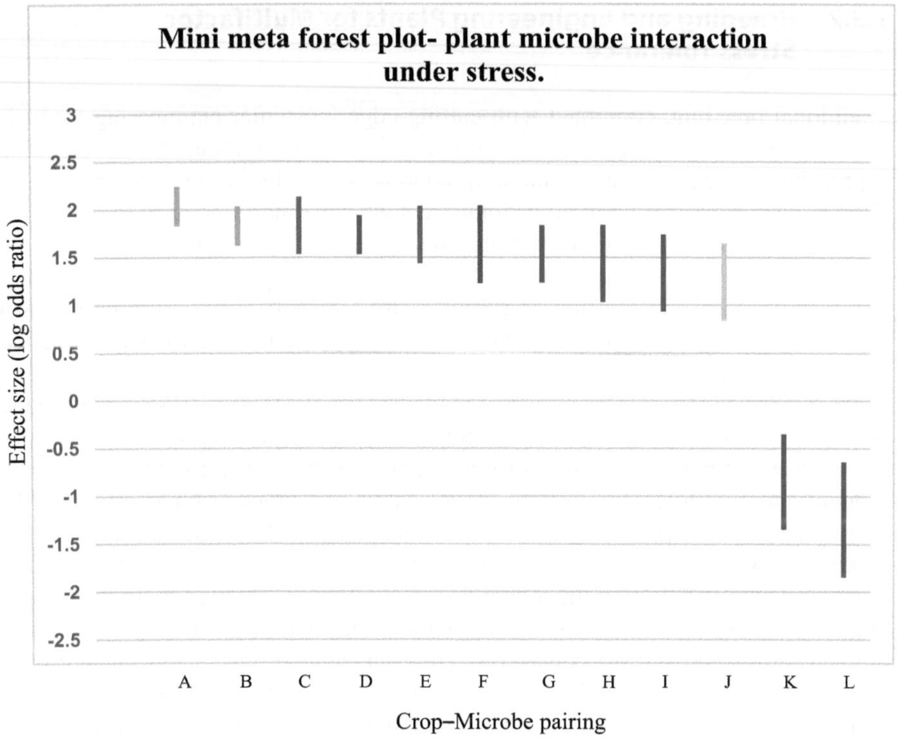

Fig. 1.2 Forest plot of pooled effect sizes (log odds ratios, ln OR) for selected crop–microbe interactions under abiotic and biotic stress. Rows (**a–l**) show crop–microbe pairs: (**a**) Legumes–Rhizobia; (**b**) Rice–N-fixing + P-solubilizing microbes; (**c**) Wheat–AM fungi; (**d**) Maize–drought-tolerant microbiota; (**e**) Maize–*Bacillus* spp.; (**f**) Tomato–*Pseudomonas* spp.; (**g**) Wheat–*Bacillus subtilis*; (**h**) Chickpea–drought-tolerant microbiota; (**i**) Sorghum–*Pseudomonas* spp.; (**j**) Barley–salt-tolerant microbiota; (**k**) Wheat–root-rot pathogens; (**l**) Mixed crops–opportunistic microbes. The x-axis (ln OR) denotes effect direction: positive = beneficial (e.g., nutrient gain, stress tolerance); negative = detrimental (e.g., disease, yield loss). Points show pooled ln OR with 95% CIs; the vertical dashed line at 0 indicates no effect. Marker colors: green = nutrient limitation, blue = P deficiency, red = drought, purple = pathogen, yellow = salt. Strong positive effects appear for legumes–Rhizobia (A) and wheat–AM fungi (C); negative for wheat–root-rot (K) and mixed systems (L). Interpret effect magnitude with CI width and stress context

and drought by increasing root surface area and improving nutrient acquisition under high temperatures.

Endophytes and mycorrhizae act as vital stress alleviators by functioning as bioinoculants and plant growth-promoting rhizobacteria (PGPR) (Kamran et al. 2022). Endophytes enhance plant tolerance to abiotic stresses such as drought and salinity by producing phytohormones, antioxidants, and osmolytes while modulating host stress-responsive genes (Kamran et al. 2022). These symbionts form a sustainable network that promotes plant health and productivity under challenging environmental conditions (Branco et al. 2022).

1.5.2 Breeding and Engineering Plants for Multifactor Stress Tolerance

Traditional breeding, combined with cutting-edge molecular engineering tools, is becoming essential for generating stress-resilient crop varieties. Some breeding approaches, including genetic mapping approaches, including QTL mapping and GWAS, have pinpointed critical genomic regions governing tolerance to combined stresses, such as drought and fungal diseases in staple crops like wheat and rice (Xu et al. n.d.). Genomic selection, particularly when combined with multi-trait prediction models, allows integration of complex stress responses into elite germplasm faster than conventional breeding. Marker-assisted selection (MAS) has expedited the incorporation of multiple stress-resistance traits, e.g., simultaneous tolerance to heat and blast disease in rice cultivars (Mohanavel et al. 2024). CRISPR–Cas9 gene editing has enabled targeted modifications in genes regulating stomatal conductance, ROS detoxification, and hormone signaling, thereby improving combined drought and heat tolerance in both model plants and crops (Kumar et al. 2023). Engineering transcription factors from the DREB and NAC families enhances the expression of downstream stress-responsive genes, conferring multifactor resilience (Shepherd et al. 2020). Genetic manipulation of genes involved in root architecture improves microbial colonization and symbiotic efficiency, promoting sustainable nutrient acquisition and stress buffering (Justamante et al. 2025). Emerging efforts focus on breeding plants with enhanced capacity to recruit beneficial microbiomes more effectively, closing the loop between plant genetics and microbial inoculant technologies (Hanif et al. 2024). CRISPR-based editing of a sodium transporter gene in tomato conferred enhanced salt tolerance while maintaining resistance to bacterial wilt, a biotic stress (Saikia et al. 2024). Advanced breeding strategies in barley have produced lines exhibiting combined resistance to powdery mildew and heat stress, validated in multi-location trials (Fodor et al. 2024). Integrating molecular breeding with microbiome-informed phenotyping opens promising pathways to develop crops capable of thriving under the multifactorial stress conditions anticipated in future agricultural systems.

Microbiome engineering harnesses synthetic microbial communities and metagenomic insights to enhance sustainable agriculture. Such consortia often include complementary microbes like *Bacillus*, *Pseudomonas*, and *Trichoderma* species that synergistically colonize the rhizosphere and enhance plant performance. By designing beneficial microbial consortia tailored to specific crops and environments, this approach improves nutrient uptake, stress tolerance, and disease resistance (Behr et al. 2023). Metagenomic technologies reveal the composition and functions of native microbial communities, enabling targeted manipulation for optimized plant–microbe interactions (Lucaciu et al. 2019). These strategies reduce reliance on chemical inputs, promote soil health, and boost productivity with minimal environmental impact. Although challenges remain in microbial establishment and regulation, microbiome engineering represents a promising solution to global agricultural sustainability and food security (Lucaciu et al. 2019).

1.5.3 Ecological Restoration: Synergies of Plants and Microbes in Degraded Soils

Globally, degraded soils compromise ecosystem stability and thereby reduce the availability of productive land. Consequently, ecological restoration that leverages the synergies of plants and microbes has emerged as a promising strategy to rehabilitate soil health, particularly in challenging environments such as post-mining sites, saline regions, and severely eroded landscapes (Raupp et al. 2024). Within these systems, interactions among plants, soils, and microbial communities collectively enhance soil structure, nutrient cycling, and organic matter accumulation, thereby increasing biological activity, which is critical for restoration success (Du et al. 2025).

In this context, pioneer plant species and their associated microbiomes act as ecosystem engineers, enabling colonization by other species while simultaneously promoting progressive soil recovery. When considering the key functional mechanisms, nitrogen-fixing bacteria associated with pioneer legumes enrich nitrogen-poor soils, thus supporting subsequent plant succession (Li et al. 2024). Likewise, mycorrhizal fungi enhance soil aggregation and water retention, which is crucial in arid and semi-arid degraded soils (Pauwels et al. 2023). Moreover, microbial communities degrade or transform contaminants such as heavy metals and hydrocarbons, thereby facilitating phytoremediation (Tang et al. 2024).

1.6 Future Perspectives and Challenges

Although individual abiotic and biotic stresses have been extensively investigated, the interactions between multiple simultaneous or sequential stresses remain largely unclear (Pehlivan et al. 2025). This complexity stems from the fact that combined stresses often trigger distinctive physiological and molecular responses that are not simply additive effects of individual stresses. For instance, when drought and heat occur simultaneously in the field, stomatal responses are antagonistically regulated under heat, stomata open to cool leaves, but under drought, they close to conserve water. This conflicting regulation presents a major challenge for maintaining optimal plant productivity (Bjerring Jensen et al. 2024). Addressing knowledge gaps requires the integration of high-throughput omics (transcriptomics, epigenomics, metabolomics), systems biology, advanced phenotyping under realistic stress combinations, and predictive modeling approaches.

Advances in real-time monitoring technologies are transforming the way plant responses to environmental stressors are studied. By capturing detailed physiological, biochemical, and environmental data streams, these tools allow for timely and precise characterization of plant–environment interactions. For example, hyperspectral and multispectral imaging from satellites, drones, and ground platforms provide non-destructive monitoring of plant health by recording reflectance across multiple wavelengths. Vegetation indices, such as NDVI, can detect changes in canopy vigor. Similarly, thermal infrared imaging enables the measurement of leaf temperature differentials, which reflect stomatal closure and changes in transpiration. These measurements serve as sensitive indicators of drought stress and facilitate proactive irrigation management (Stoll and Jones 2007). In addition, chlorophyll fluorescence and PAM fluorometry allow for the assessment of photosystem II activity, helping to differentiate between nutrient deficiencies, biotic stress, and abiotic stress (Lysenko et al. 2022). Portable electrochemical sensors and volatile organic compound (VOC) detectors further enhance early diagnostics by capturing stress-induced changes in plant emissions, allowing rapid identification of biotic stress interactions (Nejad et al. 2024).

Moreover, wearable and miniaturized sensors attached to plant organs can continuously track real-time changes in growth, turgor, humidity, and leaf movements. Inertial measurement units, for instance, detect subtle shifts in stomatal activity or leaf posture associated with early abiotic stress, while flexible adhesive sensors enable continuous in situ monitoring of water content, electrolyte leakage, and stress hormone levels (Guo 2023; Yan et al. 2024). Internet of Things (IoT) networks, combined with artificial intelligence, integrate multiple environmental and plant sensor inputs with cloud-based processing, supporting real-time decision-making for irrigation, fertilization, and pest management. Additionally, machine learning models trained on multi-parameter datasets, can identify early stress signatures and recommend adaptive interventions to optimize yields and resource use efficiency (Mansoor et al. 2025).

Furthermore, geophysical and non-invasive imaging techniques, such as ground-penetrating radar, electrical resistivity tomography, and electromagnetic induction, allow assessment of root architecture, soil moisture, and drought effects on subsurface plant parts, extending monitoring beyond visible tissues (Rocha et al. 2024). Practical applications of these technologies include vineyard management, where drone-based hyperspectral and thermal imaging, combined with AI analytics, enable early detection of water stress and disease, improving irrigation and pesticide efficiency (Walsh et al. 2024).

Despite their transformative potential, widespread adoption of these advanced monitoring technologies faces challenges such as high costs, calibration requirements, environmental variability, and integration into existing breeding and farming workflows. Effective data management and interpretation demand interdisciplinary collaboration across plant biology, computer science, and engineering. Therefore, future research should focus on developing affordable and robust sensors suitable for diverse crop types and environments, integrating multi-scale data from molecular to ecosystem levels to build comprehensive plant stress models (Otieno et al. 2025).

CRISPR/Cas gene editing is transforming crop improvement by enabling precise, heritable modifications that enhance tolerance to stresses like drought, heat, salinity, and pathogens. This technology allows direct manipulation of stress-responsive genes without foreign DNA insertion, accelerating breeding for climate resilience (Chen et al. 2024). For example, CRISPR/Cas9 editing of the OsProDH gene in rice increased heat tolerance by boosting proline and antioxidant activity without yield loss (Chakraborty and Wylie 2025). Multiplexed editing targets multiple stress pathways simultaneously, improving overall resilience. Challenges include minimizing off-target effects and gaining regulatory approval, yet advancements continue to expand its sustainable agriculture potential (Jin and Ding 2025).

Artificial intelligence models analyze multi-omics, phenomics, and environmental data to predict stress responses, identify key genes, and enhance diagnostics. These predictive tools guide breeding and cultivation decisions, improving resilience and sustainability (Ali et al. 2024). CRISPR, HTP, and AI provide an integrated and powerful toolkit to address complex plant–environment challenges and drive future crop improvement (Li et al. 2025).

1.7 Conclusion

Understanding and optimizing plant–environment interactions to enhance resilience against multifactorial stresses remains a dynamic and evolving frontier in plant science. Despite substantial progress, several critical knowledge gaps and technological challenges still limit our ability to fully harness plant adaptive potential and develop robust, sustainable agricultural and ecological systems. Therefore, addressing these gaps through innovative interdisciplinary approaches, integrative technologies, and collaborative research is essential to unlock plant adaptive potential and advance sustainable agricultural and ecological systems.

References

Abdulraheem MI, Xiong Y, Moshood AY, Cadenas-Pliego G, Zhang H, Hu J (2024) Mechanisms of plant epigenetic regulation in response to plant stress: recent discoveries and implications. Plants 13(2):163

Adeniji A, Fadiji AE, Ofoe ST (2025) Redefining root exudation with proteins at the forefront of rhizosphere function. J Agric Food Chem 73:17387

Ahmad HM, Fiaz S, Hafeez S, Zahra S, Shah AN, Gul B, Aziz O, Fakhar A, Rafique M, Chen Y, Yang SH (2022) Plant growth-promoting rhizobacteria eliminate the effect of drought stress in plants: a review. Front Plant Sci 13:875774

Ahmed N, Li J, Li Y, Deng L, Deng L, Chachar M, Chachar Z, Chachar S, Hayat F, Raza A, Umrani JH (2025) Symbiotic synergy: how Arbuscular Mycorrhizal fungi enhance nutrient uptake, stress tolerance, and soil health through molecular mechanisms and hormonal regulation. IMA fungus 16:e144989

Ait-El-Mokhtar M, Meddich A, Baslam M (2023) Plant-microbiome interactions under drought—insights from the molecular machinist's toolbox. Front Sustainable Food Syst 7:1253735

Ali T, Rehman SU, Ali S, Mahmood K, Obregon SA, Iglesias RC, Khurshaid T, Ashraf I (2024) Smart agriculture: utilizing machine learning and deep learning for drought stress identification in crops. Sci Rep 14(1):30062

Banerjee A, Chakraborty N, Acharya K, Chakraborty N (2023) Role of endophytes, plant growth promoting Rhizobacteria, and Arbuscular Mycorrhizal fungi in stress tolerance. In: Biology and biotechnology of environmental stress tolerance in plants. Apple Academic Press, pp 57–96

Behr JH, Kampouris ID, Babin D, Sommermann L, Francioli D, Kuhl-Nagel T, Chowdhury SP, Geistlinger J, Smalla K, Neumann G, Grosch R (2023) Beneficial microbial consortium improves winter rye performance by modulating bacterial communities in the rhizosphere and enhancing plant nutrient acquisition. Front Plant Sci 14:1232288

Bjerring Jensen N, Vrobel O, Akula Nageshbabu N, De Diego N, Tarkowski P, Ottosen CO, Zhou R (2024) Stomatal effects and ABA metabolism mediate differential regulation of leaf and flower cooling in tomato cultivars exposed to heat and drought stress. J Exp Bot 75(7):2156–2175
Bouremani N et al (2024) Osmotolerant plant growth-promoting bacteria mitigate drought stress in wheat by modulating osmolyte levels and antioxidant enzyme activities. Agron Sustain Dev 44(2):27–41
Branco S, Schauster A, Liao HL, Ruytinx J (2022) Mechanisms of stress tolerance and their effects on the ecology and evolution of mycorrhizal fungi. New Phytol 235(6):2158–2175
Brear EM, Day DA, Smith PM (2013) Iron: an essential micronutrient for the legume-rhizobium symbiosis. Front Plant Sci 4:359
Cargill RI, Shimizu TS, Kiers ET, Kokkoris V (2025) Cellular anatomy of arbuscular mycorrhizal fungi. Curr Biol 35(11):R545–R562
Chakraborty A, Wylie SJ (2025) CRISPR/Cas9 for heat stress tolerance in Rice: a review. In: Plant molecular biology reporter, pp 1–10
Chen K, Li GJ, Bressan RA, Song CP, Zhu JK, Zhao Y (2020) Abscisic acid dynamics, signaling, and functions in plants. J Integr Plant Biol 62(1):25–54
Chen F, Chen L, Yan Z, Xu J, Feng L, He N, Guo M, Zhao J, Chen Z, Chen H, Yao G (2024) Recent advances of CRISPR-based genome editing for enhancing staple crops. Front Plant Sci 15:1478398
Chieb M, Gachomo EW (2023) The role of plant growth promoting rhizobacteria in plant drought stress responses. BMC Plant Biol 23(1):407
Chilakala AR, Mali KV, Irulappan V, Patil BS, Pandey P, Rangappa K, Ramegowda V, Kumar MN, Puli COR, Mohan-Raju B, Senthil-Kumar M (2022) Combined drought and heat stress influences the root water relation and determine the dry root rot disease development under field conditions: a study using contrasting chickpea genotypes. Front Plant Sci 13:890551
Concha C, Doerner P (2020) The impact of the rhizobia–legume symbiosis on host root system architecture. J Exp Bot 71(13):3902–3921
Cruz-Paredes C, Tájmel D, Rousk J (2023) Variation in temperature dependences across Europe reveals the climate sensitivity of soil microbial decomposers. Appl Environ Microbiol 89(5):e0209022
Dalling JW, Willey HI (2025) Mycorrhizal barriers to the upslope migration of tropical trees. Biotropica 57(5):e70100
Devi J, Sagar V, Mishra GP, Jha PK, Gupta N, Dubey RK, Singh PM, Behera TK, Prasad PV (2023) Heat stress tolerance in peas (Pisum sativum L.): current status and way forward. Front Plant Sci 13:1108276
Du Y, Yang Y, Wu S, Gao X, He X, Dong S (2025) Core microbes regulate plant-soil resilience by maintaining network resilience during long-term restoration of alpine grasslands. Nat Commun 16(1):3116
Dvořák P, Krasylenko Y, Zeiner A, Šamaj J, Takáč T (2021) Signalingtoward reactive oxygen species-scavenging enzymes in plants. Front Plant Sci 11:618835
Ejaz U, Khan SM, Khalid N, Ahmad Z, Jehangir S, Fatima Rizvi Z, Lho LH, Han H, Raposo A (2023) Detoxifying the heavy metals: a multipronged study of tolerance strategies against heavy metals toxicity in plants. Front Plant Sci 14:1154571
Fodor J, Nagy JK, Király L, Mészáros K, Bányai J, Cséplő MK, Schwarczinger I, Künstler A (2024) Heat treatments at varying ambient temperatures and durations differentially affect plant defense to Blumeria hordei in a resistant and a susceptible Hordeumvulgare line. Phytopathology 114(2):418–426
Fu H, Yang Y (2023) How plants tolerate salt stress. Curr Issues Mol Biol 45(7):5914–5934
Fu H, Zhao J, Ren Z, Yang K, Wang C, Zhang X, Elesawi IE, Zhang X, Xia J, Chen C, Lu P (2022) Interfered chromosome pairing at high temperature promotes meiotic instability in autotetraploid Arabidopsis. Plant Physiol 188(2):1210–1228
Gao T, Li Y, Yang N, Xiong W, Liang X, Wang Y (2025) Plant-rhizosphere microbe interactions and their roles in nitrogen cycles under periodic flooding: from cooperation mechanisms to ecological responses. Crit Rev Environ Sci Technol 55(17):1358–1382

Goncharuk EA, Zagoskina NV (2023) Heavy metals, their phytotoxicity, and the role of phenolic antioxidants in plant stress responses with focus on cadmium. Molecules 28(9):3921

Guo Y (2023) Wearable sensors to monitor plant health. Nat Food 4(5):350–350

Hanif MS, Tayyab M, Baillo EH, Islam MM, Islam W, Li X (2024) Plant microbiome technology for sustainable agriculture. Front Microbiol 15:1500260

Hönig M, Roeber VM, Schmülling T, Cortleven A (2023) Chemical priming of plant defense responses to pathogen attacks. Front Plant Sci 14:1146577

Hou R, Yang L, Wuyun T, Chen S, Zhang L (2023) Genes related to osmoregulation and antioxidation play important roles in the response of Trolliuschinensis seedlings to saline-alkali stress. Front Plant Sci 14:1080504

Jin M, Ding D (2025) Multiplex CRISPR-Cas9 editing of yield-related genes in Rice. Rice Genomics Genet 16:140

Jin Z, Jiang F, Wang L, Declerck S, Feng G, Zhang L (2024) Arbuscular mycorrhizal fungi and Streptomyces: brothers in arms to shape the structure and function of the hyphosphere microbiome in the early stage of interaction. Microbiome 12(1):83

Justamante MS, Larriba E, Luque A, Nicolás-Albujer M, Pérez-Pérez JM (2025) A systematic review to identify target genes that modulate root system architecture in response to abiotic stress. Sci Rep 15(1):13219

Kafle A, Timilsina A, Gautam A, Adhikari K, Bhattarai A, Aryal N (2022) Phytoremediation: mechanisms, plant selection and enhancement by natural and synthetic agents. Environ Adv 8:100203

Kamran M, Imran QM, Ahmed MB, Falak N, Khatoon A, Yun BW (2022) Endophyte-mediated stress tolerance in plants: a sustainable strategy to enhance resilience and assist crop improvement. Cells 11(20):3292

Kou X, Han W, Kang J (2022) Responses of root system architecture to water stress at multiple levels: A meta-analysis of trials under controlled conditions. Front Plant Sci 13:1085409

KrzesBowska M, Rabęda I, Lewandowski M, Samardakiewicz S, Basińska A, Napieralska A, Mellerowicz EJ, Wozny A (2013) Pb induces plant cell wall modifications-in particular-the increase of pectins able to bind metal ions level. In: E3S web of conferences, vol 1. EDP Sciences, p 26008

Kumar M et al (2023) Application of CRISPR/Cas9-mediated gene editing for improving abiotic stress tolerance in crop plants. Front Plant Sci 14:1157678. https://doi.org/10.3389/fpls.2023.1157678

Lahlali R, Taoussi M, Laasli SE, Gachara G, Ezzouggari R, Belabess Z, Aberkani K, Assouguem A, Meddich A, El Jarroudi M, Ait Barka E (2024) Effects of climate change on plant pathogens and host-pathogen interactions. Crop Environ 3(3):159–170

Leng J, Wei X, Jin X, Wang L, Fan K, Zou K, Zheng Z, Saridis G, Zhao N, Zhou D, Duanmu D (2023) ARBUSCULAR MYCORRHIZA-INDUCED KINASES AMK8 and AMK24 associate with the receptor-like KINASE KINASE3 to regulate arbuscular mycorrhizal symbiosis in Lotus japonicus. Plant Cell 35(6):2006–2026

Li F, He C, Chang Z, Ma C, Yu J, Liu L, Zhang Y, Hao L (2023) Effects of elevated carbon dioxide on plant growth and leaf photosynthesis of annual ryegrass along a phosphorus deficiency gradient. Front Plant Sci 14:1271262

Li X, Zheng X, Yadav N, Saha S, Salama ES, Li X, Wang L, Jeon BH (2024) Rational management of the plant microbiome for the second green revolution. Plant Commun 5(4):100812

Li G, An L, Yang W, Yang L, Wei T, Shi J, Wang J, Doonan JH, Xie K, Fernie AR, Lagudah ES (2025) Integ rated biotechnological and AI innovations for crop improvement. Nature 643(8073):925–937

Lin Q, Li M, Wang Y, Xu Z, Li L (2022) Root exudates and chemotactic strains mediate bacterial community assembly in the rhizosphere soil of casuarina equisetifoliaL. Front Plant Sci 13:988442

Liu W, Thapa P, Park SW (2023) RD29A and RD29B rearrange genetic and epigenetic markers in priming systemic defense responses against drought and salinity. Plant Sci 337:111895

Liu S, Wu J, Cheng Z, Wang H, Jin Z, Zhang X, Zhang D, Xie J (2025) Microbe-mediated stress resistance in plants: the roles played by core and stress-specific microbiota. Microbiome 13(1):111

Lucaciu R, Pelikan C, Gerner SM, Zioutis C, Köstlbacher S, Marx H, Herbold CW, Schmidt H, Rattei T (2019) A bioinformatics guide to plant microbiome analysis. Front Plant Sci 10:1313

Lysenko V, Rajput VD, Kumar Singh R, Guo Y, Kosolapov A, Usova E, Varduny T, Chalenko E, Yadronova O, Dmitriev P, Zaruba T (2022) Chlorophyll fluorometry in evaluating photosynthetic performance: key limitations, possibilities, perspectives and alternatives. Physiol Mol Biol Plants 28(11):2041–2056

Lyu D, Zajonc J, Pagé A, Tanney CAS, Shah A, Monjezi N, Msimbira LA, Antar M, Nazari M, Backer R (2021) Plant holobiont theory: the phytomicrobiome plays a central role in evolution and success. Microorganisms 2021(9):675

Ma W, Tang S, Dengzeng Z, Zhang D, Zhang T, Ma X (2022) Root exudates contribute to belowground ecosystem hotspots: a review. Front Microbiol 13:937940

Mansoor S, Iqbal S, Popescu SM, Kim SL, Chung YS, Baek JH (2025) Integration of smart sensors and IOT in precision agriculture: trends, challenges and future prospectives. Front Plant Sci 16:1587869

Mo L, Pang Z, Tan Z, Liu Q, Jia Y, Xiao Y, Yuan Z (2025) Transcriptome and WGCNA reveal the key genes of Arbuscular Mycorrhizal fungi in regulating sugarcane growth and nutrient absorption. Food Sci Nutr 13(7):e70508

Mohanavel V, Muthu V, Kambale R, Palaniswamy R, Seeli P, Ayyenar B, Rajagopalan V, Manickam S, Rajasekaran R, Rahman H, Nallathambi J (2024) Marker-assisted breeding accelerates the development of multiple-stress-tolerant rice genotypes adapted to wider environments. Front Plant Sci 15:1402368

Muhammad A, Kong X, Zheng S, Bai N, Li L, Khan MHU, Fiaz S, Zhang Z (2024) Exploring plant-microbe interactions in adapting to abiotic stress under climate change: a review. Front Plant Sci 15:1482739

Mukherjee A, Dwivedi S, Bhagavatula L, Datta S (2023) Integration of light and ABA signaling pathways to combat drought stress in plants. Plant Cell Rep 42(5):829–841

Murali M, Gowtham HG, Singh SB, Shilpa N, Aiyaz M, Niranjana SR, Amruthesh K (2021) Bioprospecting of ACC deaminase producing Rhizobacteria towards sustainable agriculture: a special emphasis on abiotic stress in plants. Appl Soil Ecol 168:104142

Mus F, Crook MB, Garcia K, Garcia Costas A, Geddes BA, Kouri ED, Paramasivan P, Ryu MH, Oldroyd GE, Poole PS, Udvardi MK (2016) Symbiotic nitrogen fixation and the challenges to its extension to nonlegumes. Appl Environ Microbiol 82(13):3698–3710

Nawaz M, Sun J, Shabbir S, Khattak WA, Ren G, Nie X, Bo Y, Javed Q, Du D, Sonne C (2023) A review of plants strategies to resist biotic and abiotic environmental stressors. Sci Total Environ 900:165832

Nejad SK, Ma H, Al-Shami A, Soleimani A, Mohamed MA, Dankwah P, Lee HJ, Mousavi MP (2024) Sustainable agriculture with LEAFS: a low-cost electrochemical analyzer of foliage stress. Sensors Diagn 3(3):400–411

Nowroz F, Hasanuzzaman M, Siddika A, Parvin K, Caparros PG, Nahar K, Prasad PV (2024) Elevated tropospheric ozone and crop production: potential negative effects and plant defense mechanisms. Front Plant Sci 14:1244515

Otieno M, Karpati Z, Peters MK, Duque L, Schmitt T, Steffan-Dewenter I (2023) Elevated ozone and carbon dioxide affects the composition of volatile organic compounds emitted by Vicia faba (L.) and visitation by European orchard bee (Osmia cornuta). PLoS One 18(4):e0283480

Otieno JA, Omia DO, Amwata DA (2025) Vertical gardening undergirds household food security: evidence from Nairobi's kibera informal settlements. Front Sustainable Food Syst 9:1654777

Pattnaik D, Dash D, Mishra A, Padhiary AK, Dey P, Dash GK (2021) Emerging roles of osmoprotectants in alleviating abiotic stress response under changing climatic conditions. In: Climate impacts on sustainable natural resource management, pp 303–324

Pauwels R, Graefe J, Bitterlich M (2023) An arbuscular mycorrhizal fungus alters soil water retention and hydraulic conductivity in a soil texture specific way. Mycorrhiza 33(3):165–179

Pedroncelli L, Carter-House D, Nghacd B (2019) The consequences of drought on plant pathology. J Sci Policy Governance 15(1):1–6
Pehlivan N, Altaf MT, Emamverdian A, Ghorbani A (2025) Beyond the lab: future-proofing agriculture for climate resilience and stress management. Front Plant Sci 16:1565850
Peng P, Li R, Chen ZH, Wang Y (2022) Stomata at the crossroad of molecular interaction between biotic and abiotic stress responses in plants. Front Plant Sci 13:1031891
Petrillo E (2023) Do not panic: an intron-centric guide to alternative splicing. Plant Cell 35(6):1752–1761
Qi S, Wang J, Wan L, Dai Z, da Silva Matos DM, Du D, Egan S, Bonser SP, Thomas T, Moles AT (2022) Arbuscular mycorrhizal fungi contribute to phosphorusuptake and allocation strategies of Solidago canadensis in a phosphorus-deficientenvironment. Front Plant Sci 13:831654
Qian Z, He L, Li F (2024) Understanding cold stress response mechanisms in plants: an overview. Front Plant Sci 15:1443317
Ramakrishnan M, Zhang Z, Mullasseri S, Kalendar R, Ahmad Z, Sharma A, Liu G, Zhou M, Wei Q (2022) Epigenetic stress memory: a new approach to study cold and heat stress responses in plants. Front Plant Sci 13:1075279
Raupp P, Carrillo Y, Nielsen UN (2024) Soil health to enhance ecological restoration and conservation. Journal of Sustainable Agriculture and Environment 3(4):e70022
Rocha AA, Borges WR, Von Huelsen MG, de Oliveira e Sousa FRFR, Maciel STR, Rocha JDA, Jacobson TKB (2024) Imaging tree root systems using ground penetrating radar (GPR) data in Brazil. Front Earth Sci 12. https://doi.org/10.3389/feart.2024.1353572
Saad MM, Eida AA, Hirt H (2020) Tailoring plant-associated microbial inoculants in agriculture: a roadmap for successful application. J Exp Bot 71(13):3878–3901
Saikia B, S R, Debbarma J, Maharana J, Sastry GN, Chikkaputtaiah C (2024) CRISPR/Cas9-based genome editing and functional analysis of SlHyPRP1 and SlDEA1 genes of Solanum lycopersicum L. in imparting genetic tolerance to multiple stress factors. Front Plant Sci 15:1304381
Sharma JK, Kumar N, Singh NP, Santal AR (2023) Phytoremediation technologies and their mechanism for removal of heavy metal from contaminated soil: an approach for a sustainable environment. Front Plant Sci 14:1076876
Shepherd DA, Vos N, Reid SM, Godler DE, Guzys A, Moreno-Betancur M, Amor DJ (2020) Growth trajectories in genetic subtypes of Prader–Willi syndrome. Genes 11(7):736
Singh P, Choudhary KK, Chaudhary N, Gupta S, Sahu M, Tejaswini B, Sarkar S (2022) Salt stress resilience in plants mediated through osmolyte accumulation and its crosstalk mechanism with phytohormones. Front Plant Sci 13. https://doi.org/10.3389/FPLS.2022.1006617/FULL
Singh BK, Delgado-Baquerizo M, Egidi E, Guirado E, Leach JE, Liu H, Trivedi P (2023) Climate change impacts on plant pathogens, food security and paths forward. Nat Rev Microbiol 21(10):640–656
Soth S, Hampton JG, Alizadeh H, Wakelin SA, Mendoza-Mendoza A (2025) Microbiomes in action: multifaceted benefits and challenges across academic disciplines. Front Microbiol 16:1550749
Stoll M, Jones HG (2007) Thermal imaging as a viable tool for monitoring plant stress. OENO One 41(2):77–84
Szczepaniec A, Finke D (2019) Plant-vector-pathogen interactions in the context of drought stress. Front Ecol Evol 7:262
Tan YS, Zhang RK, Liu ZH, Li BZ, Yuan YJ (2022) Microbial adaptation to enhance stress tolerance. Front Microbiol 13:888746
Tang H, Xiang G, Xiao W, Yang Z, Zhao B (2024) Microbial mediated remediation of heavy metals toxicity: mechanisms and future prospects. Front Plant Sci 15:1420408
Teshome DT, Zharare GE, Naidoo S (2020) The threat of the combined effect of biotic and abiotic stress factors in forestry under a changing climate. Front Plant Sci 11:601009
Traubenik S, Charon C, Blein T (2024) From environmental responses to adaptation: the roles of plant lncRNAs. Plant Physiol 195(1):232–244
Voolstra CR, Ziegler M (2020) Adapting with microbial help: microbiome flexibility facilitates rapid responses to environmental change. BioEssays 42(7):2000004

Wahab A, Muhammad M, Munir A, Abdi G, Zaman W, Ayaz A, Khizar C, Reddy SPP (2023) Role of arbuscular mycorrhizal fungi in regulating growth, enhancing productivity, and potentially influencing ecosystems under abiotic and biotic stresses. Plants 12(17):3102
Walsh JJ, Mangina E, Negrão S (2024) Advancements in imaging sensors and AI for plant stress detection: a systematic literature review. Plant Phenomics 6:0153
Woods LC, Gorrell RJ, Taylor F, Connallon T, Kwok T, McDonald MJ (2020) Horizontal gene transfer potentiates adaptation by reducing selective constraints on the spread of genetic variation. Proc Natl Acad Sci 117(43):26868–26875
Wu L, Weston LA, Zhu S, Zhou X (2023) Rhizosphere interactions: root exudates and the rhizosphere microbiome. Front Plant Sci 14:1281010
Xu H, Liang X, Lloyd JR, Chen Y (2024) Visualizing calcium-dependent signaling networks in plants. Trends Plant Sci 29(2):117–119
Xu M, Zuo H, He M, Yang Y, Zhang L, Zhai X, Hu D, Lü H, Chu S Wang J, Zhang D (n.d.) Functional analysis of Gmmate transporters in soybean phosphorus homeostasis and abiotic stress resilience. Available at SSRN 5229646
Xue Y, Kang H, Cui Y, Lu S, Yang H, Zhu J, Fu Z, Yan C, Wang D (2022) Consistent plant and microbe nutrient limitation patterns during natural vegetation restoration. Front Plant Sci 13:885984
Yan A, Wang Y, Tan SN, Mohd Yusof ML, Ghosh S, Chen Z (2020) Phytoremediation: a promising approach for revegetation of heavy metal-polluted land. Front Plant Sci 11:513099
Yan B, Zhang F, Wang M, Zhang Y, Fu S (2024) Flexible wearable sensors for crop monitoring: a review. Front Plant Sci 15:1406074
Yavas I, Jamal MA, Ul Din K, Ali S, Hussain S, Farooq M (2024) Drought-induced changes in leaf morphology and anatomy: overview, implications and perspectives. Pol J Environ Stud 33(2). https://doi.org/10.15244/pjoes/174476
Yu MH, Liao WC, Wu K (2025) Histone methylation in plant responses to abiotic stresses. J Exp Bot 76:eraf058
Zandi A, Hosseinirad S, Kashani Zadeh H, Tavakolian K, Cho BK, Vasefi F, Kim MS, Tavakolian P (2025) A systematic review of multi-mode analytics for enhanced plant stress evaluation. Front Plant Sci 16:1545025
Zhang Y, Xu J, Li R, Ge Y, Li Y, Li R (2023) Plants' response to abiotic stress: mechanisms and strategies. Int J Mol Sci 24(13):10915
Zhang D, Ma S, Yang X, Huang E, Yang C, Dong K, Liu L, Zhu B, Tang Z, Ji C, Zhu J (2025) Mycorrhizal association shapes responses of plant biomass but not soil carbon to nitrogen addition in global forests. Forest Ecol Manage 586:122685
Zhou Y et al (2024a) Salt stress affects the bacterial communities in rhizosphere soil of rice. Front Microbiol 15:1505368
Zhou Y, Feng C, Wang Y, Yun C, Zou X, Cheng N, Zhang W, Jing Y, Li H (2024b) Understanding of plant salt tolerance mechanisms and application to molecular breeding. Int J Mol Sci 25(20):10940
Zhou Y, Liu D, Li F, Dong Y, Jin Z, Liao Y, Li X, Peng S, Delgado-Baquerizo M, Li X (2024c) Superiority of native soil core microbiomes in supporting plant growth. Nat Commun 15(1):6599

Climate Change and Plant Physiology

2

Hem C. Joshi, Neeraj Joshi, Saurabh Gangola, S. K. Guru, and Narendra Singh Bhandari

Abstract

Climate change is emerging as one of the most pressing global challenges, profoundly influencing the physiological processes of plants. Rising atmospheric carbon dioxide (CO_2) levels, increased temperatures, altered precipitation patterns, and more frequent extreme weather events are disrupting plant growth, development, and productivity. These environmental shifts impact critical physiological functions such as photosynthesis, transpiration, respiration, nutrient uptake, and water-use efficiency. Elevated CO_2 concentrations can stimulate photosynthesis in C_3 plants, a phenomenon known as the CO_2 fertilization effect; however, this benefit is often offset by heat stress and nutrient limitations. Increased temperatures accelerate respiration and may shorten growth periods, leading to reduced biomass accumulation and yield. Water availability, significantly influenced by climate change, affects stomatal behavior, leading to either excessive water loss or limited carbon assimilation. Furthermore, climate-induced oxidative stress elevates the production of reactive oxygen species

H. C. Joshi (✉)
Department of Botany, D.S.B. Campus, Kumaun University, Nainital, Uttarakhand, India

N. Joshi
L'institut Agro Montpellier, Montpellier, France

S. Gangola
Department of Microbiology, Graphic Era (Deemed to be) University, Dehradun, Uttarakhand, India

S. K. Guru (✉)
Department of Plant Physiology, College of Basic Sciences & Humanities, G.B. Pant University of Agriculture & Technology, Pantnagar, Uttarakhand, India
e-mail: skguru.bpy@gbpuat.ac.in

N. S. Bhandari
Shree Guru Gobind Singh Tricentenary University, Gurugram, Haryana, India

H. C. Joshi et al. (eds.), *Climate-Responsive Plant Biology*,
https://doi.org/10.1007/978-981-95-7354-7_2

(ROS), damaging cellular components and impairing plant metabolism unless mitigated by antioxidant defense systems. The physiological responses of plants to climate change are species-specific and can vary even within cultivars, highlighting the need for a detailed understanding at both cellular and whole-plant levels. Insights into these responses are crucial for developing climate-resilient crops through traditional breeding and biotechnological interventions. Adopting sustainable agricultural practices and leveraging stress-tolerant genotypes can help mitigate the adverse impacts of climate change on plant physiology and secure food systems in the future. This chapter explores the multifaceted effects of climate change on plant physiological mechanisms and emphasizes adaptive strategies that enhance plant resilience under changing climatic conditions.

Keywords

Climate change · Plant physiology · Carbon dioxide fertilization · Oxidative stress · Reactive oxygen species (ROS) · Climate-resilient crops · Water use efficiency

2.1 Introduction

Climate change is reshaping the global environment in unprecedented ways, with direct and far-reaching consequences for plant life. The major drivers of this phenomenon include increasing concentrations of greenhouse gases such as carbon dioxide (CO_2), methane (CH_4), and nitrous oxide (N_2O), rising global mean temperatures, shifts in precipitation patterns, and the growing frequency of extreme climatic events. According to the Intergovernmental Panel on Climate Change (IPCC 2023), atmospheric CO_2 concentrations have risen from pre-industrial levels of ~280 ppm to more than 410 ppm today, a trajectory strongly linked to fossil fuel use and deforestation. Global surface temperatures have already warmed by approximately 1.1–1.2 °C since the late nineteenth century, and the most recent climate data indicate that the 12-month period through April 2025 was about 1.58 °C warmer than the pre-industrial baseline (Financial Times 2025). Such changes are not occurring in isolation; instead, they interact in complex ways to alter the very conditions that underpin plant growth and survival. For plants, which are immobile and must withstand the environments in which they grow, even subtle changes in climate can trigger profound physiological consequences.

Elevated CO_2 levels, for example, may stimulate photosynthetic activity in C_3 plants, yet the concurrent stresses of heat and drought often offset these potential gains. Rising temperatures alter enzymatic activity, respiration rates, and reproductive success, while irregular precipitation patterns affect soil moisture availability, leading to droughts in some regions and flooding in others (IPCC 2023). Furthermore, extreme weather events such as heat waves, cold snaps, and cyclones can inflict acute damage during sensitive phenological stages such as flowering or grain filling. These physiological disruptions affect fundamental processes like photosynthesis, stomatal regulation, transpiration, and nutrient uptake, ultimately influencing plant productivity, reproductive fitness, and survival.

The importance of understanding these physiological responses becomes evident when considering global food security. Agriculture is highly sensitive to climatic fluctuations, and the livelihoods of billions depend on stable crop yields. Modeling studies in India project that rainfed rice yields may decline by nearly 20% by 2050 and up to 47% by 2080, while wheat yields could decrease by 19% by mid-century and up to 40% by the late twenty-first century under current emission trajectories (Organiser 2024; Livemint 2024). Similarly, yields of Kharif maize are expected to decline by 18–23% over the same time horizons (Livemint 2024). Globally, it has been estimated that warming has already reduced the yields of barley, maize, and wheat by 4–13% relative to a no-warming baseline (Stanford Woods Institute 2023). Beyond quantity, climate stress also affects the nutritional quality of crops, reducing protein and micronutrient concentrations in staple foods. Such declines directly threaten food security in regions where agricultural systems are predominantly rain-fed, and farmers have limited capacity for adaptation.

In addition to threatening crop yields, climate-induced shifts in plant physiology disrupt ecosystem stability and services. Plants are at the heart of carbon and water cycles, biodiversity maintenance, and soil fertility. Physiological stress can reduce net primary productivity, alter species composition, and lead to mismatches in phenological interactions such as those between flowering plants and pollinators. Evidence from long-term monitoring indicates that such disruptions have cascading effects on entire ecosystems, diminishing resilience to further climatic shocks (IPCC 2023). Regions that are already ecologically fragile or socio-economically vulnerable, such as many rainfed agricultural zones in South Asia and Sub-Saharan Africa, are expected to experience the most severe consequences unless targeted interventions are introduced (Organiser 2024).

Given this context, the study of plant physiological responses to climate change is not only of academic interest but also of urgent practical importance. Understanding these processes provides the foundation for developing adaptation and mitigation strategies, including climate-smart agriculture, crop diversification, genetic improvement, and biotechnological innovations aimed at enhancing tolerance to heat, drought, and salinity. A recent study published in *Nature* projected that even with adaptation measures, yield losses in major breadbasket regions could average 41% for staple crops under high-emission scenarios by the end of the century (Down to Earth 2025). This underscores the need for integrated approaches that link plant physiology with genomics, systems biology, ecological modeling, and agricultural policy.

The present chapter, therefore, aims to examine the intricate relationship between climate change and plant physiology. It will explore the impacts of key climatic drivers such as rising CO_2, temperature extremes, and altered water regimes on core physiological processes, discuss the adaptive and acclimation mechanisms plants employ, and evaluate the broader implications for crop productivity and ecosystem sustainability. By synthesizing physiological knowledge with emerging mitigation and adaptation strategies, this chapter seeks to provide insights into how agriculture and ecosystems can build resilience to the uncertainties of a rapidly changing world.

2.2 Climate Change Drivers Impacting Plant Physiology

Climate change impacts plants through multiple interacting drivers: elevated atmospheric CO_2, rising temperature, altered precipitation and water regimes, and more frequent extreme weather events. These drivers do not act in isolation; their combinations can amplify or mitigate effects depending on species, developmental stage, and environmental context. Below, we consider recent empirical findings for each major driver and how they influence key physiological responses.

2.2.1 Rising Atmospheric CO_2 Concentration

Atmospheric CO_2 has increased from ~280 ppm in the pre-industrial era to over 410 ppm today. Elevated CO_2 often enhances photosynthetic carboxylation in C_3 plants and reduces photorespiration, leading to increased biomass, growth rate, and yield under otherwise suboptimal CO_2 levels. However, C_4 plants (e.g., maize, sorghum) that have carbon concentrating mechanisms derive less benefit from further increases in CO_2. Similarly, CAM plants can respond variably depending on water availability and thermal stress.

A recent meta-analysis ("Stomatal responses of terrestrial plants to global change") found that elevated CO_2 generally reduces stomatal conductance across many species, increasing intrinsic water use efficiency (iWUE), particularly in C_3 plants (Liang et al. 2023). Another study in wetlands (SMARTX experiment) showed that warming + elevated CO_2 caused declines in stomatal conductance and chlorophyll fluorescence in both C_3 sedge and C_4 grass species; the C_3 sedge suffered more under warming (Sendall et al. 2024). Experiments that combine drought with elevated CO_2 provide interesting insight: C_3 species often show stronger improvements in WUE under elevated CO_2 than C_4 species, due to reduced stomatal conductance and enhanced photosynthesis, but increases in leaf area or biomass may offset water savings under severe drought (Cao et al. 2022).

2.2.2 Global Temperature Rise

Rising temperatures influence enzymatic activities, respiration, phenology, and reproductive success. Even moderate warming accelerates growth and shortens developmental cycles in many temperate plants, but excessive heat causes metabolic imbalances and reduces net carbon gain due to increased respiration.

Heat stress during reproductive stages is especially damaging. Several recent studies indicate that high temperature exposure around anthesis drastically reduces pollen viability, fertilization, and seed set in crop species (Qian et al. 2025). For example, pollen development under high temperature involves disruption of molecular and hormonal regulation, raising reactive oxygen species (ROS), destabilizing cellular membranes, and impairing protein folding (heat shock, unfolded protein response) (Fartyal et al. 2025; Chaturvedi et al. 2021).

Moreover, a global review ("Short- and long-term warming events on photosynthetic traits") reports that increased temperature reduces photosynthetic capacity in many crop species, with damage more severe under longer or more intense heat episodes, with yield penalties even in elevated CO_2 environments (Bernacchi et al. 2023).

2.2.3 Altered Precipitation and Water Availability

Changes in precipitation patterns—including more frequent or prolonged droughts, shifts in seasonal rainfall, and increased risk of flooding—are already altering plant water availability. Under drought, plants respond with physiological changes: stomatal closure to reduce water loss (which reduces CO_2 uptake), accumulation of abscisic acid (ABA), osmolytes like proline, and protective proteins. Severe or prolonged drought can lead to cavitation in xylem conduits, reducing hydraulic conductivity and risking irreversible damage.

Flooding or waterlogging causes hypoxic or even anoxic soil conditions, impeding root respiration, disturbing nutrient uptake, and accumulation of toxic metabolites (e.g., ethanol). Some species develop adaptive features such as aerenchyma in roots to facilitate oxygen diffusion. Recent meta-analyses show C_3 plants often have more plastic responses to drought under elevated CO_2 than C_4, but limits exist: prolonged drought can overwhelm these adaptations (Du et al. 2025).

2.2.4 Extreme Weather Events

The frequency, severity, and unpredictability of extreme weather events—such as heat waves, cold snaps, and frosts—are increasing under climate change. These events often occur during sensitive physiological/developmental windows and impose disproportionate damage.

For instance, exposure to extreme heat (e.g., ~37.5 °C for short periods) during bud swelling or pollen development in highbush blueberry reduced pollen protein content and amino acid profiles, lowered fruit set significantly, and had effects on pollinator health when heat-stressed pollen was consumed (Walters et al. 2025). Similarly, studies show that heat waves heighten pollen limitation via reductions in both pollen quantity and quality (male gametophyte development) (Rosenberger et al. 2024). Cold spells or frost events can disrupt membrane integrity, freeze intracellular fluid, and cause cellular damage, especially if occurring unexpectedly after warm periods. Although fewer recent studies focus on cold extremes compared to heat, the evidence supports that both ends of the thermal extremes are impactful.

2.3 Physiological Processes Affected by Climate Change

Rising atmospheric CO_2 has enhanced terrestrial photosynthesis relative to pre-industrial times, with global analyses showing that a large share of increased gross primary productivity is attributable to the so-called "CO_2 fertilization effect" (Keenan et al. 2023). At the leaf level, C_3 plants such as rice, wheat, and soybean benefit most because higher CO_2 suppresses photorespiration and boosts carboxylation efficiency. By contrast, C_4 plants like maize show limited gains, while long-term exposure may down-regulate Rubisco and photosynthetic capacity when sinks or nutrients are constrained (Mondal et al. 2023). Heat and drought interact strongly with these effects: drought restricts stomatal conductance and CO_2 entry, but elevated CO_2 improves intrinsic water-use efficiency, providing some buffering under mild water stress (Qiao et al. 2024). Warming, however, accelerates respiration, which reduces net carbon gain. Under combined heat and drought, excessive reactive oxygen species (ROS) accumulate, overwhelming antioxidant defenses and causing oxidative damage (Mondal et al. 2023) (Table 2.1).

Water balance is another critical process sensitive to climate change. Plants typically respond to drought by closing stomata, conserving water, but limiting photosynthesis. Stomatal behavior is tightly regulated by abscisic acid signaling and hydraulic cues (He et al. 2024). Osmotic adjustment-accumulation of solutes such as proline and sugars—helps maintain turgor, supporting metabolism during water deficit, though at the cost of growth. Severe drought often induces cavitation in xylem conduits, causing embolism and reducing hydraulic conductivity; such hydraulic failure can persist beyond the drought event (Fickle et al. 2025). Conversely, excess rainfall and flooding reduce oxygen diffusion to roots, lowering respiration and ATP supply while impairing ion uptake. Some species develop aerenchyma or radial oxygen loss barriers as adaptive traits, but prolonged waterlogging generally restricts growth. In parallel, rising vapor pressure deficit (VPD) increases atmospheric demand, exacerbating transpirational water loss and hydraulic stress (Oishy et al. 2025).

Climate change also alters nutrient dynamics. Warming accelerates decomposition and mineralization, which can increase nutrient supply temporarily but also enhance leaching and gaseous losses. Elevated CO_2 tends to dilute nutrients in plant tissues, raising C:N ratios and lowering protein content in crops (Li et al. 2024). Drought restricts nitrogen uptake by limiting mass flow and root activity, while heat stress impairs enzymes such as nitrate reductase and glutamine synthetase (Zhang et al. 2023). Phosphorus dynamics are equally climate sensitive: flooding can mobilize P in some soils but may convert it to unavailable forms under anoxic conditions. Together, these changes reshape nutrient stoichiometry and limitation patterns, forcing crops to reallocate metabolic energy under stress (Oishy et al. 2025).

Shifts in climate also disrupt plant growth and development. Warming has advanced phenological events such as flowering and fruiting in many regions (Zu et al. 2025). While earlier flowering may extend the growing season in some systems, it also increases risks of frost damage and mismatches with pollinator activity (Solakis-Tena et al. 2025). Under stress, plants adjust biomass partitioning, often

Table 2.1 Overview of key plant physiological processes influenced by various climate change drivers and their associated responses

Physiological process	Climate change driver(s)	Plant response/Mechanism	Key references
Photosynthesis and carbon assimilation	Elevated CO_2 concentration	Enhanced photosynthesis in C_3 plants (e.g., rice, wheat, soybean) due to reduced photorespiration and higher carboxylation efficiency. Limited gain in C_4 plants like maize. Long-term CO_2 enrichment may cause down-regulation of Rubisco and photosynthetic capacity under nutrient/sink limitation.	Keenan et al. (2023), Mondal et al. (2023)
Interaction of heat and drought with CO_2	Combined heat and water stress	Elevated CO_2 improves intrinsic water-use efficiency. Drought limits stomatal conductance and CO_2 diffusion. Warming accelerates respiration, reducing net carbon gain. Combined stress enhances ROS generation, causing oxidative damage.	Qiao et al. (2024), Mondal et al. (2023)
Water relations and hydraulic function	Drought, flooding, and vapor pressure deficit (VPD)	Stomatal closure reduces water loss but restricts photosynthesis. Abscisic acid (ABA) regulates stomatal behavior under drought. Osmotic adjustment via proline and sugar accumulation maintains turgor. Drought can cause xylem cavitation and embolism, leading to hydraulic failure. Flooding impairs root respiration and ion uptake; adaptive traits like aerenchyma formation may occur. Rising VPD increases transpiration demand and hydraulic stress.	He et al. (2024), Fickle et al. (2025), Oishy et al. (2025)
Nutrient dynamics	Elevated CO_2, temperature rise, drought, and flooding	Warming enhances decomposition and mineralization but increases leaching/gaseous loss.Elevated CO_2 dilutes nutrients, increasing the C:N ratio and lowering protein content.Drought limits N uptake and root activity; heat impairs nitrate reductase and glutamine synthetase. Flooding alters P availability-mobilization or conversion to unavailable forms. Altered stoichiometry affects metabolic energy allocation.	Li et al. (2024), Zhang et al. (2023), Oishy et al. (2025)

(continued)

Table 2.1 (continued)

Physiological process	Climate change driver(s)	Plant response/Mechanism	Key references
Growth, development, and phenology	Warming, drought, CO_2 enrichment	Warming advances flowering and fruiting times, potentially lengthening or shortening growing seasons. Early flowering risks frost damage and pollinator mismatch. Stress-induced reallocation to roots improves nutrient/water access. Chronic stress reduces biomass and alters yield quality, affecting food security.	Zu et al. (2025), Solakis-Tena et al. (2025), Li et al. (2024)

investing more in root systems to improve water and nutrient access, or reallocating resources between vegetative and reproductive sinks depending on stress timing (Li et al. 2024). Chronic stress usually reduces overall biomass and alters yield quality, with consequences for food security and ecosystem productivity.

2.4 Stress Physiology and Climate Change

Plants under climate change are increasingly challenged by abiotic stresses-heat, cold/frost, drought, salinity, and oxidative stress. Each of these stresses triggers specific physiological and molecular responses. The interaction of multiple stresses (e.g., heat + drought) often exacerbates damage. Heat stress, particularly during sensitive stages like flowering or seed set, destabilizes proteins, disrupts membrane fluidity, and can denature enzymes. To cope, plants synthesize heat shock proteins (HSPs) that act as molecular chaperones, preserving proteins from misfolding and aggregation. A recent study by David et al. (2023a) showed that in tomato, overexpression of HSP70 increased pollen viability under sustained heat of 38 °C, improving set and yield. Membrane stability under heat is mediated by adjustments in lipid composition-more saturated lipids can increase the melting point, but at the cost of flexibility. Heat also increases lipid peroxidation; plants with higher baseline antioxidant capacity or induced expression of lipid-repair enzymes show better membrane integrity (Chen et al. 2024).

Cold stress (including frost, freezing nights) causes ice formation in the apoplast or within cells, leading to cell dehydration, mechanical damage, and metabolic disruptions. Protective mechanisms include cryoprotection via accumulation of soluble solutes (e.g., sugars like sucrose, trehalose; compatible solutes; proline) that help reduce freezing point and protect membranes. Antifreeze proteins (AFPs) are another defense: these bind to ice crystals and limit their growth. Research in alpine grasses (Lee et al. 2024a) found that certain AFP isoforms were upregulated during night frost, correlating with reduced tissue damage. Cold acclimation (gradual exposure to low but non-freezing temperatures) induces changes in membrane lipid unsaturation, expression of cold-responsive (COR) genes, and increased antifreeze protein expression (Müller and Smith 2023).

Drought and salinity often produce overlapping stress responses because both cause osmotic stress and water deficit. Abscisic acid (ABA) is a central hormone: elevated ABA levels under soil drying trigger stomatal closure, reduce transpiration, induce expression of stress-responsive genes, and downstream protective mechanisms. Osmolytes (proline, glycine betaine, trehalose, etc.) accumulate to help maintain cell turgor, stabilize proteins and membranes, and scavenge ROS. Under salinity, ion homeostasis becomes critical: high external Na^+ or Cl^- disrupts potassium (K^+) uptake, and plants with better Na^+ exclusion or compartmentalization into vacuoles fare better. Recent work (David et al. 2023a, b, c; Patel and Rao 2024) showed that salt-tolerant rice varieties express high levels of SOS pathway (Salt Overly Sensitive) genes, maintain favorable K^+/Na^+ ratios, and accumulate osmolytes better than sensitive ones.

Many stresses (heat, cold, drought, salinity, high light) lead to overproduction of reactive oxygen species (ROS) such as superoxide (O_2^-), hydrogen peroxide (H_2O_2), hydroxyl radical (·OH). Without detoxification, ROS damage proteins, lipids, DNA. The antioxidant defense system includes enzymatic components (superoxide dismutase, catalase, peroxidases, glutathione reductase) and non-enzymatic antioxidants (ascorbate, glutathione, carotenoids, tocopherols). Recent meta-analysis (Lee and Kim 2024) found that plants under combined heat + drought stress increase expression and activities of SOD and CAT by ~20–40% compared to single-stress conditions. Furthermore, overexpression of specific antioxidant genes (e.g., APX - ascorbate peroxidase) in Arabidopsis and rice has conferred better tolerance to ROS bursts under heat or salinity (Chen et al. 2024).

2.5 Adaptive and Acclimation Mechanisms

Plants exhibit a wide range of adaptive and acclimation strategies that enable survival under changing climatic conditions. These mechanisms operate at morphological, physiological, biochemical, and molecular levels, and often interact synergistically to maintain growth, reproduction, and productivity.

2.5.1 Morphological Adaptations

Morphological plasticity is one of the first visible responses to environmental stress. Leaf rolling in cereals such as rice and wheat reduces exposed surface area, thereby minimizing transpiration under drought conditions. Enhanced root system architecture, including deeper taproots and increased lateral root branching, improves soil water and nutrient acquisition, a trait strongly selected in drought-resilient cultivars (Singh et al. 2023). Additionally, deposition of epicuticular waxes on leaf surfaces enhances cuticular resistance, reduces non-stomatal water loss, and provides photoprotection under high irradiance and temperature (Zhang et al. 2024b). Together, these morphological traits confer resilience by directly mitigating stress exposure.

2.5.2 Physiological Acclimation

Plants fine-tune their physiology under climatic stresses through altered stomatal sensitivity, hormonal signaling, and photosynthetic adjustments. Stomata may become more responsive to abscisic acid (ABA), closing rapidly under soil moisture deficits to conserve water while still balancing carbon uptake. Hormones such as salicylic acid, jasmonic acid, and ethylene interact with ABA to orchestrate stress responses, often with cross-talk mechanisms that allow plants to respond to multiple stresses simultaneously (Kaur and Sharma 2023). Physiological acclimation also includes adjustments in photosynthetic electron transport, alternative respiratory pathways, and maintenance of photochemical efficiency during high-temperature or drought episodes (David et al. 2023a, b, c).

2.5.3 Biochemical Adjustments

At the biochemical level, plants accumulate a variety of osmoprotectants and stress-related metabolites. Solutes such as proline, glycine betaine, and trehalose act as osmolytes to stabilize proteins, membranes, and enzymes under water or salt stress. Secondary metabolites such as flavonoids, phenolics, and terpenoids function as antioxidants, scavenging reactive oxygen species (ROS) and reducing oxidative damage (Patel et al. 2024a). Stress-induced synthesis of heat shock proteins (HSPs) and late embryogenesis abundant (LEA) proteins further helps preserve structural integrity under temperature extremes. Collectively, these biochemical adjustments allow plants to maintain cellular homeostasis during stress exposure.

2.5.4 Epigenetic and Genetic Regulation

Recent studies have highlighted the critical role of epigenetic modifications and genetic reprogramming in climate stress adaptation. Stress memory, mediated by DNA methylation, histone modification, and small RNAs, enables plants to "remember" prior stress episodes and mount a faster or stronger response upon re-exposure (Müller and Singh 2024b). Transcriptomic changes under heat, drought, or salinity stress involve activation of transcription factors such as DREB (Dehydration Responsive Element Binding), NAC, and WRKY families, which regulate large sets of stress-responsive genes. Advances in genome editing, particularly CRISPR/Cas systems, are now being applied to enhance crop stress resilience by directly modifying such key regulatory genes (Lee et al. 2024a, b). These molecular strategies highlight the dynamic nature of plant acclimation beyond immediate physiological responses.

2.6 Mitigation and Adaptation Strategies

Addressing the multifaceted impacts of climate change on plant physiology requires integrated strategies that combine agronomic practices, advanced breeding and biotechnology, targeted physiological improvements, and supportive policy frameworks. These approaches aim not only to sustain crop productivity but also to promote resilience, environmental sustainability, and global food security (Fig. 2.1).

2.6.1 Agronomic Approaches

At the field level, climate-smart agriculture (CSA) has emerged as a central framework for adaptation. CSA emphasizes practices that enhance productivity, build resilience, and reduce greenhouse gas emissions. Efficient irrigation methods such as drip and sprinkler systems significantly improve water-use efficiency (WUE) compared with conventional flood irrigation (Shah et al. 2023). Similarly, conservation agriculture practices-including minimum tillage, residue retention, and cover cropping—improve soil structure and moisture retention while enhancing carbon sequestration. Crop diversification and intercropping reduce risks associated with climate variability, as different species respond differently to stress (Patel et al. 2024a, b). Integrating agroforestry into farming systems further contributes to microclimate regulation and biodiversity preservation.

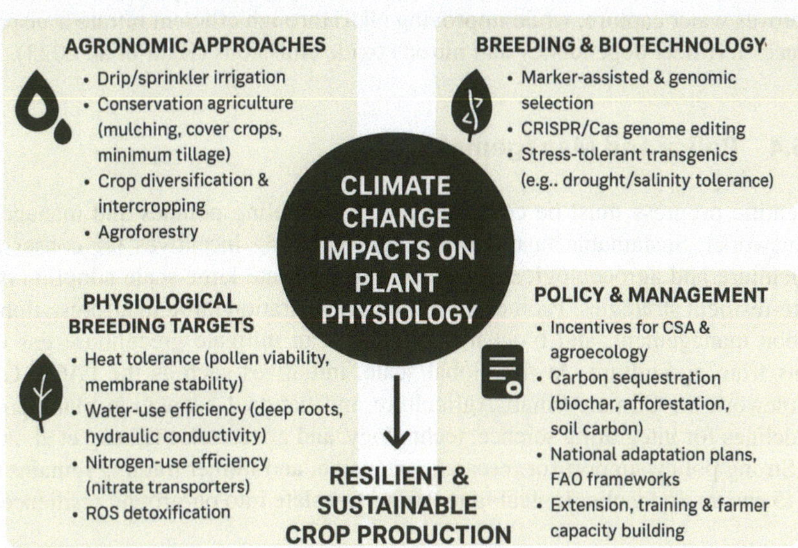

Fig. 2.1 Integrated strategies for mitigating climate change impacts on plant physiology

2.6.2 Breeding and Biotechnology

Traditional breeding has contributed substantially to the development of stress-tolerant cultivars, but recent advances in molecular breeding and biotechnology have accelerated progress. Marker-assisted selection (MAS) and genomic selection enable precise identification and deployment of stress-resilient alleles. The use of CRISPR/Cas genome editing has opened new opportunities for targeted modifications, such as enhancing drought tolerance by editing genes regulating ABA sensitivity, or improving salinity tolerance through modification of ion transporters (Lee et al. 2024a, b). Transgenic approaches, though controversial, have produced crops with improved stress tolerance—for example, rice engineered with drought-responsive transcription factors (NAC and DREB families) shows higher yield stability under water-limited conditions (Müller and Singh 2024a, b).

2.6.3 Physiological Breeding Targets

Future crop improvement requires prioritizing physiological traits linked to climate resilience. Heat tolerance is an increasingly important target, as high temperatures during flowering can significantly reduce grain yield in cereals. Breeding for traits such as membrane thermostability, pollen viability, and efficient ROS detoxification has shown promise (David et al. 2023a, b, c). Similarly, water-use efficiency (WUE) and nitrogen-use efficiency (NUE) are key targets to maintain productivity under resource-limited conditions. Enhancing root depth and hydraulic conductivity improves water capture, while improving NUE through efficient nitrate transporters reduces fertilizer dependency and nitrous oxide emissions (Kaur et al. 2023).

2.6.4 Policy and Management

Scientific progress must be complemented by enabling policies and management frameworks. Sustainable land-use policies, including incentives for conservation agriculture and agroecological practices, can facilitate large-scale adoption of climate-resilient strategies. Promoting carbon sequestration through afforestation, soil carbon management, and biochar application can mitigate greenhouse gas emissions from agriculture. At the global scale, initiatives such as the FAO's Global Framework on Climate-Smart Agriculture and national adaptation plans provide guidelines for integrating science, technology, and governance (Zhang et al. 2024a, b). Strong policy support for research, extension, and farmer training remains critical to ensure that technological innovations translate into on-ground resilience.

2.7 Conclusion

Climate change poses profound challenges to plant physiology, altering fundamental processes such as photosynthesis, respiration, water relations, nutrient uptake, and growth dynamics. Rising atmospheric CO_2, global temperature increases, shifts in precipitation, and the intensification of extreme weather events interact in complex ways, often amplifying stress effects on plants. These physiological disruptions translate into yield instability, nutritional deficiencies, and increased vulnerability of agricultural and natural ecosystems. The urgency of addressing these challenges cannot be overstated. While individual plant responses—such as morphological adaptations, biochemical adjustments, and stress memory—provide some level of resilience, they are insufficient in isolation. Sustainable agricultural productivity in the face of climate change demands integrated approaches that combine climate-smart agronomy, advanced breeding and biotechnological tools, physiological trait-based selection, and robust policy interventions. Looking forward, building resilience must remain the cornerstone of agricultural development strategies. Resilient cropping systems, supported by scientific innovation and sound policy frameworks, will be essential for safeguarding global food security and ecosystem stability. By integrating physiological insights with technological, ecological, and socioeconomic solutions, agriculture can adapt to the evolving climate while ensuring productivity, sustainability, and resilience for future generations.

References

Bernacchi CJ, Bagley JE, Ruiz-Vera UM (2023) Modelling the impacts of climate change on crop photosynthesis and productivity. New Phytologist 238(2):456–470

Cao Y, Luo Q, Tian Y, Meng F (2022) Physiological and molecular responses of plants to drought stress. International Journal of Molecular Sciences 23(4):1922

Chaturvedi P, Mishra A, Tiwari RK (2021) Omics approaches in abiotic stress tolerance in plants: current status and future prospects. Plant Physiology and Biochemistry 160:258–273

Chen X, Zhao L, Feng Y et al (2024) Membrane lipid remodeling and peroxidation across heat-stressed plant species: comparative analyses. Planta 259:82

David J, Kumar P, Singh R et al (2023a) Overexpression of HSP70 improves pollen viability and yield under heat stress in tomato. Plant Sci 305:111242

David J, Patel M, Rao S et al (2023b) Physiological acclimation strategies in crops exposed to combined heat and drought stress. J Exp Bot 74(11):3290–3303

David J, Kumar V, Rao S et al (2023c) Breeding physiological traits for heat tolerance in cereals under climate change. J Exp Bot 74(12):3561–3574

Down to Earth (2025) Despite farmers adapting to climate change, yield losses—especially in India's wheat growing regions-will be severe: study. Down to Earth

Du H, Zhang W, Li R (2025) Carbon–water interactions in plants under future climate scenarios. Forthcoming

Fartyal D, Agarwal A, Kumar S (2025) Physiological and biochemical strategies for climate resilience in crops. Forthcoming

Fickle J, Brown P, Turner N (2025) Interactive effects of heat and drought on plant productivity. Forthcoming

Financial Times (2025) World has already warmed 1.58°C in past 12 months, EU scientists say. Financial Times

He Z, Sun X, Li Y (2024) Molecular and physiological regulation of stress tolerance in plants. Forthcoming

Intergovernmental Panel on Climate Change (IPCC) (2023) Climate change 2023: synthesis report. In: Contribution of Working Groups I, II, and III to the Sixth Assessment Report of the Intergovernmental Panel on Climate Change. IPCC, Geneva

Kaur J, Sharma P (2023) Hormonal cross-talk in abiotic stress responses: ABA, jasmonates and salicylic acid signaling networks. Front Plant Sci 14:1178942

Kaur J, Verma S, Gupta R (2023) Physiological and molecular targets for improving nitrogen-use efficiency in major crops. Front Plant Sci 14:1180123

Keenan TF, Prentice IC, Canadell JG (2023) Recent pause in the growth rate of atmospheric CO_2 due to enhanced terrestrial carbon uptake. Nature Climate Change 13(1):15–21

Lee M, Kim Y (2024) Combined drought and heat stress: oxidative stress metrics and antioxidant enzyme responses across plant functional types. Plant Physiol Biochem 191:59–68

Lee H, Park S, Kang H et al (2024a) Antifreeze protein isoform-specific induction and cryoprotection in alpine grasses under night frost. J Exp Bot 75(9):2931–2944

Lee H, Kang Y, Park S (2024b) Genome editing for abiotic stress resilience: CRISPR-based advances in crop improvement. Biotechnol Adv 65:108176

Li X, Wang Y, Chen Z (2024) Climate change impacts on plant physiological processes. In press

Liang J, Zhou Z, Huo C (2023) Root system architecture and drought adaptation in plants. Journal of Experimental Botany 74(9):2621–2635

Livemint (2024) Govt study predicts significant decline in yields of rice, wheat and maize by 2080 due to climate change. Mint

Mondal S, Rutkoski JE, Velu G (2023) Harnessing drought tolerance in crops through physiological breeding. Trends in Plant Science 28(5):462–476

Müller S, Singh A (2024a) Epigenetic stress memory and transcriptional reprogramming in plants under climate extremes. Trends Plant Sci 29(3):221–234

Müller S, Singh A (2024b) Transcription factor-based transgenic crops for climate resilience: current progress and future prospects. Plant Biotechnol J 22(4):512–528

Oishy R, Rahman M, Hossain M (2025) Climate-driven physiological responses of crops. Forthcoming

Organiser (2024) Impact of climate change on farmers, agriculture and food security in India. Organiser

Patel M, Rao S (2024) Differential expression of SOS pathway genes and ion homeostasis in salt-tolerant vs. sensitive rice under saline stress. Crop Sci 64(2):560–575

Patel D, Meena R, Choudhary S et al (2024a) Role of osmoprotectants and secondary metabolites in oxidative stress mitigation under salinity and drought. Plant Physiol Biochem 196:234–248

Patel R, Joshi V, Sharma N (2024b) Crop diversification and intercropping systems as adaptive measures under climate variability. Field Crop Res 305:108673

Qian H, Xu Z, Chen G (2025) Climate change effects on plant growth, productivity, and resilience. Forthcoming

Qiao Y, Huang J, Deng X (2024) Crop physiological adaptation to climate change and implications for food security. Forthcoming

Rosenberger K, Müller C, Smith P (2024) Ecosystem-level consequences of climate-induced plant stress. Forthcoming

Sendall KM, Reich PB, Zhao C (2024) Acclimation of photosynthesis to warming across plant functional types. Global Change Biology 30(3):e17012

Shah D, Mehta P, Raval S et al (2023) Climate-smart irrigation strategies for improving water-use efficiency in arid and semi-arid regions. Agric Water Manag 280:108145

Singh R, Kumar V, Joshi H et al (2023) Root architectural plasticity and drought resilience in cereal crops under climate change. Plant Cell Environ 46(5):1432–1445

Solakis-Tena K, Martínez A, Gómez R (2025) Plant acclimation to combined climatic stressors. Forthcoming

Stanford Woods Institute for the Environment (2023) Heat and drought are quietly hurting crop yields. Stanford Woods Institute

Müller S, Smith A (2023) Cold acclimation responses in temperate perennials: COR gene upregulation, lipid unsaturation, and resilience to frost. Front Plant Sci 14:1020100

Zhang Y, Li P, Chen X et al (2024a) Epicuticular wax deposition improves water-use efficiency and heat tolerance in wheat and rice cultivars. Environ Exp Bot 214:105621

Walters M, Green J, Lopez R (2025) Climate extremes and plant resilience: physiological perspectives. Forthcoming

Zhang Y, Chen L, Wang H (2024b) Policy frameworks for climate-smart agriculture: integrating carbon sequestration, land use, and food security. Clim Pol 24(2):229–246

Zu Y, Liu Q, Zhang H (2025) Regulation of photosynthesis under drought and heat stress. Forthcoming

Zhang L, Chen J, Wang S (2023) Plant physiological responses to climate variability and extremes. Forthcoming

Adaptation Strategies in Plants

3

Prem Prakash and Deep Pandey

Abstract

In recent years, climate change has gained significant attention from all fields of study due to the impacts it has laid upon all living organisms from plants to animals and even microorganisms. Especially, environmental stresses resulting from changes in climate have affected the growth and development of floras. These abiotic stressors refer to the adverse effects of any environmental factor on organisms. The stresses which include extreme temperatures, salinity, drought, nutrient deficiency, heavy metals, and many more have jeopardized the survival and reproduction of flora. They severely impose negative effects on plants on multiple levels such as physiological, phenological, biochemical and molecular. Therefore, to cope with changing climate and the stresses laid upon by it, plants have adopted different adaptation measures to ensure their survival. The different alterations a plant adapt to sustain in unfavourable and changing climatic conditions includes regulating water loss, shifts in flowering and fruiting patterns, undergoing seed dormancy, altering dispersal mechanisms and other modifications involving changes in its morphology. These adaptation strategies help plants to ensure resources for their subsistence and provide them with reproductive success in changing climatic conditions. Several measures involving physiological, behavioral, phenological changes, migration and production of certain phytohormones and antioxidants help plants to manage with the stresses imposed on them.

By understanding the role of adaptation strategies in plants, it will become easier to monitor the responses of plants with respect to climate change, which eventually will become crucial for developing conservation strategies for ecosystems. This will also help plant breeders to selectively adopt those adaptations

P. Prakash (✉) · D. Pandey
Department of Botany, M.B. Government P.G. College, Haldwani, Uttarakhand, India

H. C. Joshi et al. (eds.), *Climate-Responsive Plant Biology*,
https://doi.org/10.1007/978-981-95-7354-7_3

that help plants survive during harsh circumstances and develop new varieties that are stress-resistant in nature.

Keywords

Climate change · Environmental stresses · Adaptation strategies · Phytohormones · Stress-resistant

3.1 Introduction

Climate change has always been a source of concern for environmentalists, scientists, biologists, and policymakers. It is making headlines as a global concern affecting every organism on Earth. The shifts in temperature and weather patterns leading to other sets of problems, such as global warming, melting of glaciers, intense rainfall and floods, and droughts have greatly affected our ecosystem and the organisms living in it.

Anthropogenic actions such as deforestation, fossil fuels burning, industrial pollution, urbanization and many more has led to a severe change in climate over the past years, affecting biodiversity, causing habitat loss and degradation. Human interference has led to an increase in the level of carbon dioxide since the pre-industrial period. Notably, other greenhouse gases have also increased in the atmosphere, causing an increase in temperature. Similarly, salinity, droughts and floods have resulted in the downside of our environment and therefore organism try to adopt various measures to thrive in such ever-changing climatic conditions.

Organisms, through a variety of mechanisms such as migration, evolution at a genetic level, plasticity and other such measures try to adapt to climate change allowing them to survive, flourish and evolve. As animals migrate when abiotic stresses are imposed on them; plants, being immobile organisms adopt various adaptation methods to survive in changing climatic conditions.

Being one of the megadiverse countries of the world, India encompasses a wide variety of habitats from tropical forests to deserts to coastal and marine ecosystems, leading to a rich array of floras. The key phytogeographical areas in India includes tropical rainforests, temperate deciduous, coniferous forests, deserts and grasslands. Due to this diverse geographical distribution, abiotic factors involving temperature, water, and edaphic factors throughout the country vary, thereby shaping the plant distribution patterns. These diverse environments where plants thrive are referred to as habitats.

Habitats are the different places where plants grow well. There are two main types of habitats: land and water. Henceforth, plants can be classified according to the type of habitat they grow best in. Xerophytes are plants that live in deserts, mesophytes are plants that live on land, and hydrophytes are plants that live in water. Plants adapt to their specific environments by developing special traits called "adaptations." These traits help them thrive in their own environments. For a plant to live and thrive in its environment, a number of elements are necessary, including soil, water, light, etc. However, these variables are observed to deviate from their typical range and are needed by the plants as a result of climate change. For instance, changes in temperature lead to variations in light intensity, variations in water level result in floods and droughts, and soil nutrient deficiencies. The natural habitat of a plant is altered by these stressors.

Due to this change in climate, plants are always in a state of stress. These stresses, which include temperature fluctuations, droughts, floods, salinity, nutrient deficiencies in soil, pollution, etc., affect plants' growth and development. As stated by Zhu (2016), plant growth and geographic distribution are severely limited by different abiotic stresses.

So, to survive in these extreme conditions, plants adopt various strategies or methods known as "adaptation strategies." These adaptation strategies could be modifications in physiological, morphological, and reproductive features of plants often also involving migration and symbiotic associations for increasing resource utilization.

Plants have various adaptation measures, but this chapter undertakes a detailed discussion of the environmental stressors that plants face and the ways in which they have adapted to these stresses with respect to climate change.

3.2 Environmental Stresses

The change in climate has led to a severe downfall of our ecosystem with plant migration, resource scarcity, species competition and conflicts, decrease in productivity, exacerbation in food insecurity, thereby impairing plants and animals alike. Climate change also intensifies weather events which increases the percentage of many environmental stresses, for instance, droughts, floods, salinity, temperature fluctuations and others, which eventually affects the ecosystem, causing widespread damage (Fig. 3.1). Let us understand some of these abiotic stresses brought on by shifting weather patterns.

3.2.1 Extreme Temperatures

Temperature fluctuations due to climate change have become one of the most challenging abiotic stresses, impairing the normal physiological processes of plants. Global warming causes an increase in heatwaves, which becomes detrimental for organisms on Earth. These arid and warm environments are vulnerable to wildfires, which significantly harm the local flora and fauna. As rising temperatures cause heat waves, similarly, a drop in ambient temperatures results in cold waves, causing chilling stress in plants.

Extreme temperatures, involving both heat and cold waves, have shown to exacerbate plant's physiological, biochemical, and molecular impairments, lowering photosynthesis; ultimately affecting crop yield.

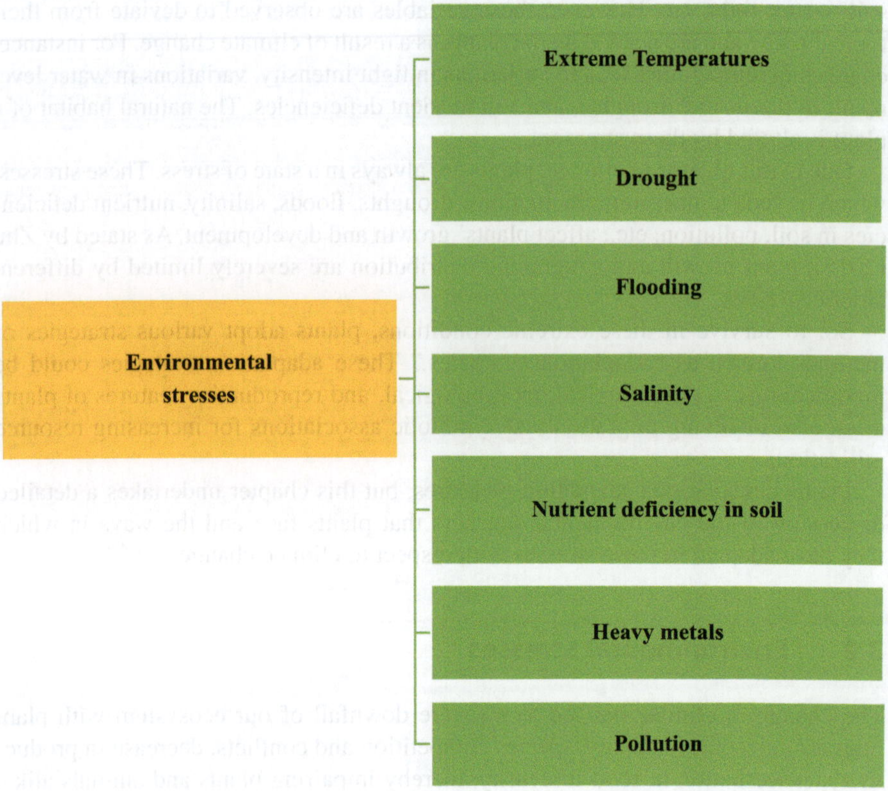

Fig. 3.1 Environmental stresses imposed on plants due to changing climatic conditions

3.2.2 Drought

Droughts have increased in frequency and severity due to climate change, which has an annual effect on agricultural productivity. Due to changes in precipitation patterns brought about by climate change, some regions endure protracted droughts while others receive heavy rainfall. A prolonged drought puts agricultural productivity at risk because rising temperatures accelerate evaporation rates, which causes the soils to dry out, leading to water scarcity.

As a result of drought, hydrological imbalances are created due to prolonged dry weather, which eventually affects plant development (Gray et al. 2016). Regions already prone to droughts face more dangers due to greenhouse gas emissions. These gases are known to increase aridity in such areas, which might enhance the rate of desertification.

3.2.3 Flooding

Melting of glaciers combined with thermal expansion of water significantly contributes towards rising sea levels, particularly threatening coastal areas, causing floods. A warm atmosphere due to global warming causes an increase in evaporation rate, resulting in more moisture accumulation in the atmosphere, thereby leading to heavier downpours, severe storms, and an increase in the cloud burst frequencies.

Due to the detrimental effects floods provoke on terrestrial plants, flooding has become a natural disturbance, thereby affecting crop production (Colmer and Voesenek 2009). The increment in the intensity and frequency of floods has affected plants by damaging their root systems, causing reduced growth. Waterlogged soils lead to oxygen deprivation; with such reduced oxygen levels, plants are unable to uptake oxygen, disrupting their cellular components.

3.2.4 Salinity

Plants experience salinity stress when soil salt concentrations are higher. Due to overuse of fertilizers, poor irrigation practices, and alterations in precipitation patterns, there has been an increase in the accumulation of salt content in the soil.

An obscure dimension of climate change is global drying which refers to the increase in the evaporative demand (Barkhordarian et al. 2019) that is a decrease in the moisture content of Earth's mainland. As a consequence of global warming, temperature has been creating new records every year. This rise in temperature causes excessive evaporation from soil and a surge in transpiration rates in the plants. As a result of this, salt is drawn upwards and gets concentrated in the root zone area (Rengasamy et al. 2003).

Insufficient leaching of ions from the soil due to reduced precipitation leads to salt accumulation resulting in soil salinity (Blaylock 1994) affecting plants by causing osmotic stress and nutrient imbalance. Similarly, rising sea levels through saltwater intrusion in coastal areas lead to groundwater salinity (Werner and Simmons 2009). The rise in the amounts of sodium and chloride ions in soil causes ionic toxicity, impacting the physiological processes of the plant by drawing water out of plant roots. This disrupts the normal equilibrium of water and ions in plant cells, affecting overall plant growth and development.

3.2.5 Nutrient Deficiency in Soil

Numerous climatic stressors are thought to be the cause of nutrient deficiencies, which lead to imbalances in the soil nutrient content. Prolonged dry spells reduce the amount of water available in soil leading to a shift in soil microbial population. This affects soil microbial processes like nitrogen fixation and ultimately interferes with plants' ability to absorb nutrients.

Nutrient leaching has been observed in areas receiving recurring and intense rainfall (Rashmi et al. 2017). Essential macronutrients such as nitrogen and phosphorus are washed away from the soil due to heavy and severe rainfalls becoming more frequent with a change in climate. These changes reduce the amount of nutrients in soil that are crucial for plant health, consequently, a reduction in essential nutrients in crops exacerbating food insecurity.

3.2.6 Heavy Metals

Heavy metal as stated by Wang et al. (2023) refers to any heavy metal element with relatively high density that is toxic at low concentrations. Recent shifts in precipitation trends have caused an increase in soil erosion rates and the frequency of floods. Such events carry heavy metals from industrial and polluted areas transferring them to water bodies and agricultural lands contaminating them. The increase in cumulation of heavy metals in soil is becoming a matter of distress in agricultural productivity (Yang et al. 2019) as it disrupts soil quality potentially harming flora and fauna.

Melting of glaciers is another factor responsible for the release of heavy metals (Xiao et al. 2024), such as lead and mercury into our ecosystem risking the environment and the organisms residing in it. This mobilization of heavy metals negatively affects soil microbes and their functions impacting soil health. Elevated levels of heavy metals in soil enters our food chain when up taken by plants posing risks to animals and humans.

3.2.7 Pollution

One of the greatest threats to our planet Earth is the deterioration of air, water, and soil quality due to change in climate. Anthropogenic actions such as burning fossil fuels, deforestation and urbanization release greenhouse gases, for example, methane, carbon dioxide, CFCs and nitrous oxide. These gases are responsible for warming the planet ultimately enhancing the rate of pollution.

Due to change in climate, the atmosphere is getting warmer with warming rates being predicted to be rapid about 0.3 °C per decade (Change 1990). Such dry and warm conditions increase the frequency of forest fires (Yáñez et al. 2025) which release harmful smoke and other particulate matter such as soot in air affecting the air quality.

In a similar vein, rising global temperatures make it easier for algae to grow, lower the amount of oxygen in water, and contaminate freshwater bodies with saltwater intrusion. Floods worsen water quality by increasing the amount of sediments and nutrients that runoff into water bodies. Floods cause chemicals from industrial sites to run off into underground water, making it toxic (Aldardasawi and Eren 2021).

Unpredictable precipitation patterns and temperature fluctuations are known to increase soil pollution. According to Fayinminnu et al. (2024), climate change affects the movement of both organic and inorganic soil pollutants that enter the soil from various sources. Changes in soil microbial community, increase in erosion and intrusion of heavy metals in soil linked with climate shifts have led to a degradation in soil quality.

3.3 Adaptation Strategies in Plants

The structural, physiological, and behavioral traits that improve survival and reproduction in particular environmental circumstances are referred to as adaptations in plants. Since plants are sessile and unable to flee from stress as animals can, they have evolved special coping mechanisms to deal with variations in light, temperature, water, soil, and interactions with other living things (Fig. 3.2).

- Adaptation: A heritable characteristic that has been cultivated over many generations and enhances survival (e.g., cactus spines to reduce water loss).
- Acclimatization: A brief, reversible physiological response to environmental stress is called acclimation (e.g., sun vs. shade leaves in the same tree).
- Evolution: Natural selection-driven long-term changes in plant populations.

Note The terms adaptation and acclimatization are often confused with each other as both terms refer to responses of organisms to environmental changes, but the terms carry significantly different meanings. Let us understand the differences between them through the following Table 3.1.

The term "adaptation" describes a plant's capacity to endure, develop, and procreate under challenging or less-than-ideal circumstances. As scientists and crop breeders search for plants that could thrive and adapt better to the changing climate, it has emerged as a major topic of discussion among biologists. To deal with stressors like temperature changes, nutrient shortages, drought, and so forth, plants use a variety of adaptation strategies that involve changes at the physiological, structural, and reproductive levels. For instance, plants respond to temperature stress through long- and short-term responses comprising of phenological adaptations and increased transpiration, respectively (Baidya et al. 2022). These adaptation strategies help plants to ensure reproductive success, access enough resources for their survival, and thrive in challenging conditions.

Adaptations could be changes in the physiology of a plant involving regulating water loss, nutrient uptake methods; phenological adaptations such as shifts in flowering and fruiting, seed dormancy; or structural modifications involving changes in leaf, stem, and root systems or changes in seed dispersal mechanisms. Often, forming symbiotic associations with other organisms also helps plants to cope with harsh conditions making them more easily accessible to water and nutrients.

Let us understand the different adaptation strategies adopted by plants in detail.

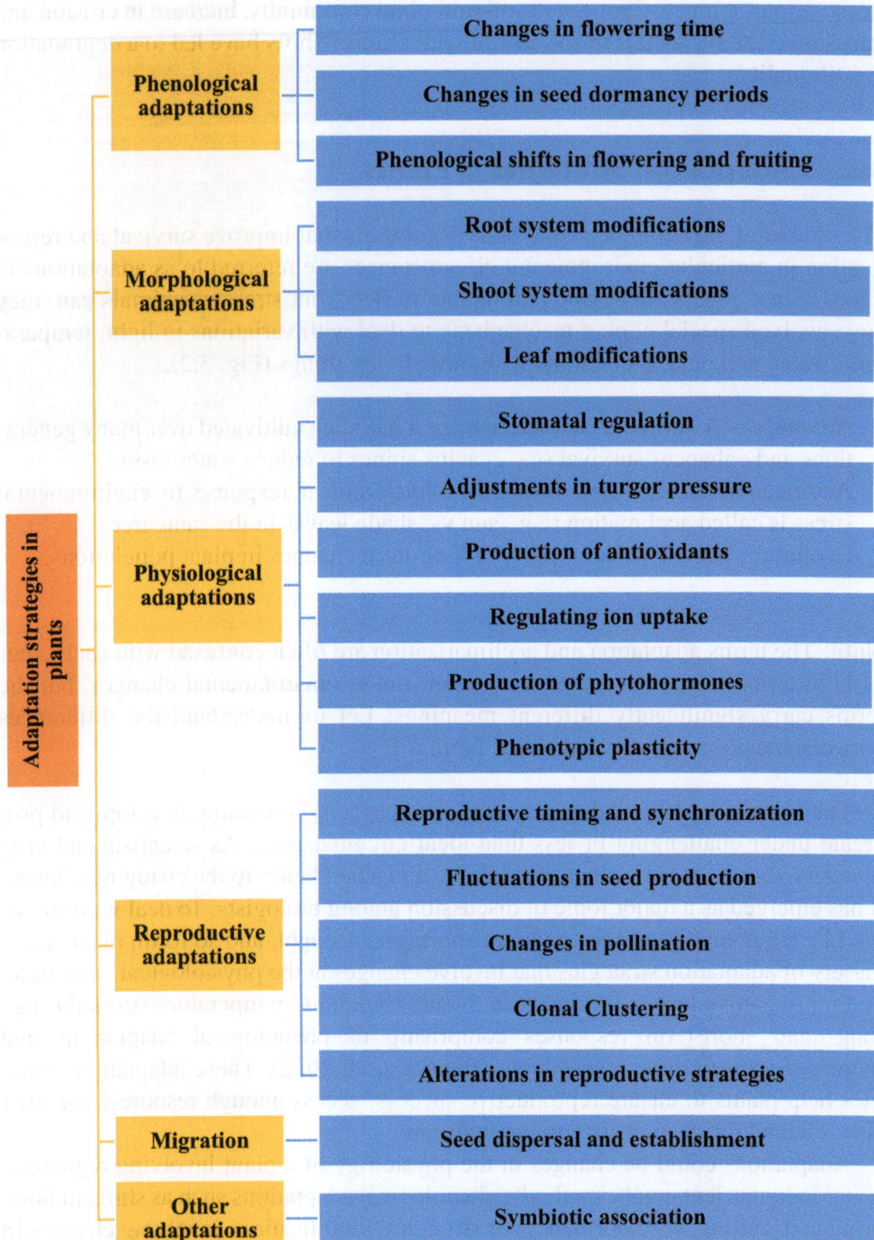

Fig. 3.2 Adaptation strategies in plants

Table 3.1 Difference between adaptation and acclimatization

Adaptation	Acclimatization
Adaptation is a long-lasting change	Acclimatization is a short-term change
Heritable in nature	Non-heritable
Passed down over generations	Occurs in an individual life
Permanent and gradual process	Temporary changes
Part of evolutionary process	Not concerned with evolution
Changes occur in organism both physically and chemically	Changes take place in organism only physically

3.3.1 Phenological Adaptations

Phenological adaptations are those in which plants modify the timing of their life cycle stages, particularly in response to environmental factors. For plants to reproduce successfully and to flourish in unfavorable environments, the crucial stages of their life cycle—flowering and fruiting—need to be modified. To adapt to or avoid these changed conditions, plants use phenological changes to help them deal with droughts, floods, and extreme temperatures.

3.3.1.1 Changes in Flowering Time

To survive during severe periods of stress such as temperature fluctuations, plants change their flowering time by either flowering earlier or later when temperature no more becomes a limiting factor. This adaptation ensures successful pollination for plants. In regions with less rainfall, flowering is often delayed due to lack of water. Similarly, in areas where summer temperatures are unbearably high, some plants, such as *Tagetes* (marigolds) and *Vinca* might flower earlier to avoid periods of dryness. *Salvia* plants are known to be adapted to thrive in dry conditions (Caser et al. 2012).

3.3.1.2 Changes in Seed Dormancy Periods

For many species, the timing of seed germination is strongly affected by environmental conditions (Baskin and Baskin 2000). So, during unfavourable environmental ambience, plants tend to delay germination by extending the dormancy period of seeds until the stresses imposed upon them are reduced. This helps them survive harsh, cold climates and prolonged dry periods.

Plants in such cases often link their germination timing with the onset of favourable conditions that is when the impacts of environmental stresses are lessened. One of the best examples is leguminous plants, well known for the physical dormancy of their seeds (Carruggio et al. 2020).

3.3.1.3 Phenological Shifts in Flowering and Fruiting

Abiotic stresses cause disruption in the normal temperature and weather patterns, thereby creating imbalances in the biological clock of plants. Because temperature variations cause changes in the timing of flowering, fruiting, and senescence, these alterations impair the overall interactions between plants and other organisms in the ecosystem, particularly pollinators. Plants adapt their flowering and fruiting schedules to coincide with favorable times in response to shifting environmental conditions. Studies have indicated that woody and spring-flowering species tend to advance their flowering time in response to warming environments (Calinger et al. 2013; Zeng et al. 2025).

3.3.2 Morphological Adaptations

Morphological or Structural adaptations in plants are defined as those specialized physical features that help plants survive in harsh climatic conditions. These include modifications in the root, stem, and leaves of the plant to ensure survival.

3.3.2.1 Root System Modifications

In recent years, rising temperatures, extreme weather patterns, and alterations in rainfall due to changes in climate have notably contributed to the formation of arid areas throughout the world. Scientists have even stated about the expansion of arid regions globally and the simultaneous conversion of semi-arid regions to arid regions. Huang et al. (2016) mentioned the long-term global warming trend that has particularly enhanced over semi-arid regions. Therefore, in arid regions or regions with less soil moisture, plants tend to adopt various changes in their root system to access water and other nutrients essential for their growth.

In arid regions, evaporation rates are high, and precipitation is low, thereby the environment is dry and scarce, barely sustaining any vegetation. Owing to climate change, these regions now receive intense solar radiations and even lesser rainfall limiting water resources. Plants in such ecosystem therefore develop deep root systems, which allow them to access water from deeper layers of soil and make up for the water shortage. For example, studies on *Prosopis juliflora* (Mesquite) indicate the development of deep roots in such shrubs (Phillips 1963).

A prolonged period of low precipitation leads to drought conditions that cause the soil to lose its moisture content, causing insufficient soil moisture. Agricultural areas are particularly exposed to excessive water evaporation leading to deficit soil moisture percentage. Plants adopt shallow root system modifications in such cases. Rather than penetrating deep into the soil, in a shallow root system, roots spread horizontally, absorbing nutrients and water from the upper layers, covering wider areas near the soil surface. A network of fine roots is formed that helps plants to absorb nutrients from the topsoil (Saraf et al. 2025).

3.3.2.2 Shoot System Modifications

Modifications in the stem include ability to store resources during harsh climatic conditions, changes in stem form, adaptations in stem growth patterns, spreading vegetatively, developing aerial roots, and others. These changes alter the structure and functioning of plants and help them better adapt in altered environmental conditions.

With more frequent changes in climate leading to longer periods of dryness, often causing resource scarcities for several plants, it becomes significant for plants to store resources such as water and minerals in case of even harsher conditions. For these, plants often modify their stem as rhizomes and tubers, storing nutrients in them. These underground modified stems help plants survive extreme periods of drought.

Excessive temperature rise leading to water shortage makes it impossible for plants to retain water for their physiological functions. As a result, some plants develop fleshy stems for water storage. Since the current habitats of some plants become unsuitable to live in due to climate change, plants such as strawberries, colonize to newer areas with the help of runners and stolons which are modified stems that help plants to grow vegetatively and spread rapidly to more suitable environments.

Climate change also contributes to increased temperatures leading to higher evaporation rates thereby causing areas to become drier and more arid. Once habitable areas are now being converted to drier lands due to desertification, affecting the growth and survival of flora. Henceforth, to minimize water loss in arid environments, stems of some plants are modified into phylloclades. Phylloclades are flattened modified stem structures that allow plants to minimize water loss in such environments; for example: Cactus.

Development of thorns and prickles in extremely dry conditions helps plants against water losses in an already drought environment brought upon by climate change. It is also discovered that some plants show modifications in their stem growth patterns with response to change in temperatures. In general, nutrients and temperature are recognized as limiting factors for growth (Linder and Axelsson 1982). This is one of the components found to be taking part in a potential mechanism for radial growth (Zweifel et al. 2006). For instance, under warmer temperatures, some tree species, *Juglans mandshurica* and *Fraxinus mandshurica* (Yuan et al. 2021), have shown increased radial growth, which allows them to adapt to a longer growing season and better utilize the resources (Fang et al. 2014).

3.3.2.3 Leaf Modifications

Climate change has disrupted the normal precipitation patterns causing intense rainfall in some areas while also leading to prolonged drought conditions in others. Plants in drier climates (xerophytes) therefore, develop waxy coating on their leaves to reduce this water loss. The smaller-sized leaves help in water retention and enhance plants resilience to changing climate. Example includes xerophytic plants such as *Yucca*, *Agave*, Oleander, etc. Needle-like leaves of conifers also help them retain water in dry environments.

During prolonged drought periods, some plants develop thick, fleshy leaves to store water and survive periods of water scarcity. For example: *Aloe vera*, *Sedum* (Stonecrops).

3.3.3 Physiological Adaptations

Those adaptations in plants that involve modification in functions and internal changes that help plants cope up with environmental stresses are referred to as physiological adaptations. Plants modify their physiology to enhance survival and thrive in harsh climates. These adaptations involve stomatal regulation, adjustments in turgor pressure, production of antioxidants and phytohormones, and many more. Let us take a look at some of the physiological adaptations.

3.3.3.1 Stomatal Regulation

Stomata on leaves regulate water loss by opening and closing the stomatal pores, regulating transpiration rates, and maintaining the level of water for plants to survive during dry conditions. In drought-prone areas, plants close their stomata or reduce their pore size to limit transpiration and conserve water. Cacti, succulents, and crops such as maize and wheat are some examples of stomatal regulation as an adaptation strategy in plants.

Generally, when plants face extremely high temperatures, they tend to close their stomata, preventing water loss. But some desert plants, during high temperatures, increase their leaf transpiration rates by opening stomata more readily. This helps in cooling the leaf surface and prevents overheating.

3.3.3.2 Adjustments in Turgor Pressure

The internal water pressure exerted by cell sap from inside on the cell wall is referred to as turgor pressure, or we can simply define it as the internal water pressure within a plant cell. It is essential for maintaining the plant cell structure and rigidity. Under drought and flooding conditions, plants make adjustments in their turgor pressure as a modification under physiological adaptation. For instance, under prolonged drought periods, as in desert plants, a lack of water within the plant cells can potentially lead to dehydration, eventually cell damage. When plants lose the turgor pressure, they become flaccid and wilt. Therefore, under such conditions, plants adjust the concentration of solutes (Morgan 1977) in their cells to maintain this pressure, which prevents cell damage, helps plants stand upright, providing structural support, and prevents wilting.

3.3.3.3 Production of Antioxidants

During stress conditions, for instance, heavy metal toxicity, salinity, droughts, and temperature fluctuations, Reactive oxygen species (ROS) are produced in various plant cellular compartments, including peroxisomes (Sandalio and Romero-Puertas 2015) and chloroplasts (Dietz et al. 2016). Since plants produce ROS as a signalling molecule under both biotic and abiotic stress conditions, ROS are referred to as a by-product of stress. However, too much ROS damages plants' cells and membranes. It may result in DNA mutations and cell death. In order to scavenge and neutralize reactive oxygen species and thereby lessen their detrimental effects, plants produce several antioxidants. For example: superoxide dismutase (SOD), catalase, and peroxidase (POD). This shield cells from oxidative damage due to abiotic stresses such as extreme temperatures (Sharma et al. 2019). This is an antioxidant defence mechanism adopted by plants during stress conditions.

3.3.3.4 Regulating Ion Uptake

Melting of glaciers due to climate change, altered precipitation patterns and thermal expansion leads to rise in sea levels, which causes saltwater intrusion into agricultural lands; in turn, affecting crop yield. Similarly, increased intensity of droughts has led to higher evapotranspiration, leading to salt accumulation in the soil. Excessive rainfall causes flooding, due to which saltwater moves underground into freshwater aquifers. These processes eventually increase the levels of salt in soil, causing salinity stress. Plants in such environments develop mechanisms to regulate ion uptake. For instance, halophytes (Duarte et al. 2014) adapt to saline environments by reducing sodium ion influx. This prevents salt toxicity in plant cells and maintains their normal cellular functions.

3.3.3.5 Production of Phytohormones

When environmental stresses are imposed on plants, physiological responses by plants lead to the production of phytohormones, which act as signaling molecules helping plants to adapt to these stresses. These hormones interact with each other enhancing plant defense mechanism and ensuring survival. Biologists have found the production of plant hormones such as Strigolactones (SLs) and Abscisic Acid (ABA) during stress conditions in plants (Roychoudhury and Tripathi 2020).

Strigolactones (SLs): When exposed to stress conditions, SLs are being produced by many plants for improving their coping mechanism. It is known as a multifunctional molecule with significant role in growth and development specially produced in the course of deficient and stress conditions in plants.

Functions of Strigolactones

- SLs play regulatory roles in plant defense responses under abiotic conditions (Shafi et al. 2022).
- Plants produce strigolactones during nutrient stress conditions to complex plant responses, for instance, during the regulation of root system development by the promotion of primary roots (Waters et al. 2017).
- Ling et al. (2020) have stated about the responses of SLs in drought, salinity, and chilling conditions.
- SLs are also known to work in collaboration with other phytohormones such as auxin, cytokinin and ethylene in supplementing plant tolerance against environmental stressors.

Abscisic Acid (ABA) Referred to as a stress hormone, Abscisic acid is known to trigger stomatal closure to reduce water loss during droughts. This helps plants to conserve water in already dry environments. ABA also stimulates a variety of physiological processes during stress conditions, such as root system modulation, metabolic alterations, organization of soil microbial communities (Muhammad Aslam et al. 2022), and many more. Fluctuations in temperature also stimulate the production of ABA in plants (Saraf et al. 2025). During cxtremely cold temperatures, plants such as *Momordica charantia* (Bitter gourd), *Cucumis melo* (Melon) activate the production of abscisic acid, which helps them adapt to cold stress.

3.3.3.6 Phenotypic Plasticity

Another type of physiological adaptation that occurs in plants during stress is phenotypic plasticity. It refers to the ability of plants to express different phenotypes without any change in the underlying genotype, particularly in response to environmental fluctuations. Phenotypic plasticity helps plants to make adjustments in their traits to adapt in changing climate, which otherwise could jeopardize their survival and reproduction. It involves changes in the internal processes of plants, such as altering the rates of photosynthesis, better utilization of nutrients, and water uptake during stress conditions. Experiments on *Arabidopsis thaliana* indicate the significance of phenotypic plasticity in helping plants adapt to changing climatic conditions, with plasticity becoming a possible adaptive trait in natural populations of such species (Scheepens et al. 2018).

Developing different root structures according to the water availability is another example of phenotypic plasticity in plants. Rice, one of the cereal crops being cultivated under a wide range of soil hydrologic conditions, exhibits root plasticity as an adaptation to water stresses (Lucob-Agustin et al. 2021).

Plants also respond to temperature fluctuations with the help of plasticity. Experiments on *Aechmea aquilega* (Tank bromeliad) by Lafont Rapnouil et al. (2023) stated how bromeliads growing along a light intensity gradient show significant differences in morphological and functional traits, with plants exposed to higher light intensities having smaller and vertical leaves.

3.3.4 Reproductive Adaptations

Those adaptations that enhance the chance of successful reproduction in plants are referred to as reproductive adaptations. These involve alterations in reproductive strategies, pollination mechanisms seasonal reproduction, and others. Let us understand these adaptations.

3.3.4.1 Reproductive Timing and Synchronization

For plants to survive in diverse habitats such as tropical rainforests to temperate to arid deserts, plants have adapted to reproduce at specific time of the year to increase the chances of fertilization. This synchronization with favourable conditions during the year optimizes seedling survival. For instance, temperate plants such as *Calendula* (Pot marigolds) and *Primula* (Primroses) are known to flower in the spring (Tooke and Battey 2010) as conditions here are generally in favour for seedling growth and survival.

Certain plants are also known to use environmental cues to time their reproduction (Karban 2008), ensuring favourable conditions are met for optimum seed production. For example, *Arabidopsis*.

3.3.4.2 Fluctuations in Seed Production

Climate change particularly has affected pollen viability and seed formation, eventually leading to reduced seed and fruit set. In the face of climate change, plant persistence and migration depend on successful reproduction by seeds (Luo et al. 2022). During harsh environmental conditions such as droughts, plants tend to reduce their seed production rates by prioritizing their resource allocation to other processes, which ensures their survival. When conditions become favourable that is when stresses are alleviated, plants exhibit higher seed production. For example, chickpea (Pang et al. 2017), maize (Sah et al. 2020), soybean (Poudel et al. 2023) reduce their seed production during stressful conditions such as extreme droughts by limiting the number of flowers and pods.

3.3.4.3 Changes in Pollination

With changing climate conditions, it becomes significant for plants to alter their pollination strategies in an effort to disperse better and potentially attain reproductive success. For this, plants change their pollinating behaviour and adopt those pollination mechanisms that ensure attracting better pollinators, flowering at that exact time when their pollinators are most active, relying on wind dispersal mechanisms for long distances travel and others, therefore, accordingly adapting to environmental stresses. For example:

- Some night-blooming flowers, such as *Ipomoea alba* (moonflower), *Oenothera biennis* (Evening primrose), attract nocturnal pollinators such as moths for successful pollination.
- It is also known that some pollinator species are particularly sensitive to fluctuations in temperature and precipitation patterns (Whipple and Bowser 2023), thereby a decline in pollinators has been observed with a change in climate. For instance, the climate thermal limit might exceed the physiological thermal limit of bumble bees resulting in a decline in their population (Soroye et al. 2020).

Henceforth, compared to diurnal pollinator species, nocturnal pollinators have become significant alternatives for attaining pollination in such conditions.

3.3.4.4 Clonal Clustering/Clonal Reproduction

To survive in harsh climatic conditions, some plants are known to form multiple individuals that are genetically identical, referred to as Clonal Clustering. It also seems to be of particular importance in ecosystems with strong climatic constraints and/or resource-deprived regions (Couteron et al. 2014). Therefore, clonal reproduction helps plants to thrive during stresses or competitive environments by spreading out resources and minimizing competition among clones. For instance, Anthelme et al. (2008) have stated about clonal propagation of *Olea europaea* in the Saharan mountains of Niger.

3.3.4.5 Alterations in Reproductive Strategies/Shift to Vegetative Reproduction

Rather than entirely focusing on seed production, due to environmental stresses, plants tend to shift their reproductive strategies towards vegetative growth (Raihan et al. 2021), which helps them counteract the negative effects of climate change. At times, when resource availability is reduced due to abiotic stressors, some plants tend to alter their reproductive strategies towards vegetative growth, ensuring their persistence and survival. For example, vegetative propagation through tubers, runners, allow plants to colonize new areas with comparatively stable environmental factors.

3.3.5 Migration (Seed Dispersal and Establishment)

Plants, being immobile organisms cannot move from one place to another unlike other organisms during unfavourable conditions. So, they rely on seed dispersal mechanism as a means of migration. Hoffmann and Parsons (1997) have stated how organisms when exposed to variable environments incline towards exhibiting higher dispersal rates taking into account a conservation perspective.

This could be understood as due to climate change; when conditions become unsuitable for a plant population, the seeds of such plants disperse to newer, more favourable habitats. This change results in the shifting of the plant population to a suitable environment. Ecologists identify this migration as a need of plants to stay within their limited range of tolerance to climatic variables known as the 'climate envelope'.

Each species has a limited range of climatic conditions within which it can thrive, survive, and reproduce. But due to a change in climate, this 'envelope' moves across the surface of the Earth (Corlett and Westcott 2013). If populations do not shift with this movement of 'climate envelope', they might not be able to survive this ever-changing climate and its impacts. To survive, plant populations must move to follow their 'climate envelope'.

As a result, plants, though not being able to move physically, count on seed dispersal for migration. Seeds get dispersed and establish themselves in habitable environments where abiotic factors are more suitable for their growth and survival. Especially, the wind-dispersal mechanism is advantageous for plants as seeds are carried to long distances.

Following are the examples of seed dispersal and establishment as adaptation measures taken up by plants.

- Ejecting seeds far away from the unsuitable environment is one example of a seed dispersal adaptation strategy adopted by plants to ensure dispersal over wide and favourable areas. Examples: *Impatiens* (Touch-me-not), *Viola* (Violets), *Lupinus* (Lupine).
- Due to rise in temperature, wind velocity changes, which enhances wind-dispersal of seeds to longer distances. Seeds can spread to wider areas, reaching suitable environments. For example: *Acer*, *Taraxacum*, *Typha*.

- Some plants in flood-prone regions, such as water lilies develop buoyant seeds that can float, thereby travelling long distances in water. In long-flooded wetland regions, high buoyancy along with inhibition of germination for the period the seeds float on water is definitely one of the most successful adaptation measures taken by plants for high dispersal (van den Broek et al. 2005).
- Studies have indicated uphill migration of plants as climate change causes an increase in temperature, leading plants to shift to higher altitudes where temperatures are still favourable. For instance, west Asian apple species *Malus orientalis* Uglitzk. (Walas et al. 2024) have shown potential shift due to changing climatic conditions.
- Some plant species move from the equator to the poles, expanding their ranges northward or towards higher elevations. As winters become warmer, mangroves are moving northward in some regions of the world, shifting their habitable range to new areas (Cavanaugh et al. 2014).
- Alpine areas, which are now facing extreme weather events are habitat to many endangered and endemic species (Dirnböck et al. 2011). Therefore, as lower slopes become more and more warm due to climate change, alpine plants are shifting to higher altitudes (Auld et al. 2022).

3.3.6 Other Adaptations

3.3.6.1 Symbiotic Association

Symbiotic association is a type of species interaction where one species is benefited and the other (different) species could be positively, negatively, or neutrally affected. Plants develop symbiotic relations with organisms such as fungi and bacteria. It is known that symbiosis helps plants in mitigating the negative impacts of environmental stresses by increasing resilience through communication with other organisms, enhancing stress tolerance, improving nutrient uptake and other resources, thereby aiding plants' adaptation to climate change (Table 3.2).

- Association with organisms such as fungi becomes crucial in improving nutrient absorption in areas with nutrient-poor soil.
- Bacterial associations increase the plant's tolerance against various stresses.

Table 3.2 Symbiotic associations of plants with other organisms and its potential benefits with respect to environmental stresses

Symbiotic association	Symbionts	Benefits to the host plant
Mycorrhiza	Fungi	Enhancement in nutrient uptake in nutrient-deficient soils Water acquisition in soil water-deficit regions
Bacterial symbiosis	Bacteria	Promotes stress tolerance Resilience towards temperature fluctuations
Plants-endophytes	Microbes	Drought-resistant management Production of essential phytohormones during stresses

- Rhizobial and mycorrhizal associations help plants in water acquisition during droughts and nutrient uptake in nutrient-deficient soils.

Let us understand through various examples how symbiotic associations help plants withstand multiple abiotic stresses associated with climate change.

1. Higher plants and fungal associations forming mycorrhiza play crucial role in drought-prone regions. As soil moisture content is expected to reduce in some areas due to increased temperatures (Le Houérou 1996); compared to plants with no mycorrhizal relations, ecto- and ericoid mycorrhizal associations have brought enhancement in nutrient uptake and water acquisition particularly during drought periods.
   - Plants may even replace the type of mycorrhiza colonizing their host when exposed to longer durations of drought (Querejeta et al. 2009). These fungi form symbiotic relations with plant roots, accessing water and nutrients by extending deeper into the soil. This becomes significant when soil is deficient in nitrogen and phosphorus and is soil water-deficit. Particularly, in dry regions, plant water uptake can be enhanced through such associations.
   - Due to greenhouse gas emissions, carbon dioxide levels elevate in the atmosphere, impacting plants. Various studies have found that plants with Arbuscular Mycorrhizal associations thrive better compared to plants with no symbiotic relations during such stress conditions. Compant et al. (2010) have stated that elevated atmospheric CO_2 levels usually escalate the colonization behaviour of Arbuscular Mycorrhizal Fungi (AMF). This becomes essential, taking into view the fact that AMF colonization is known to improve nutrient uptake (Smith and Read 2010).
   - According to some studies, mycorrhizal fungi potentially induce the expression of stress-responsive genes (Hadian-Deljou et al. 2020) helping plants tolerate abiotic stressors.
2. Plants that have a symbiotic association with bacteria are known to better adapt to climatic changes. As studies by Zhang et al. (2024) have shown, genetic factors take part in the adaptability of *Rhizobium* by tuning the expression levels of genes that promote stress tolerance.
   - Abiotic stress due to temperature fluctuations severely affects plant growth and development. Both extremely low and high temperatures disrupt the normal plant processes such as respiration, photosynthesis and germination, consequently impairing growth and threatening their survival. However, plants with symbiotic associations have shown greater tolerance towards such fluctuations. Nitrogen-fixing plants and their associated strains of N_2-fixing bacteria are known to exhibit an extraordinary resilience to temperature regimes of their natural environment and are found in both cold and warm climates (Dixon and Wheeler 1983). Development of nodules deeper in the soil system

as an adaptation to higher soil temperatures has also been noted (Munns et al. 1977).

- Plants such as legumes form symbiotic relations with N_2-fixing bacteria, which reside in the nodules of roots and fix nitrogen by converting atmospheric nitrogen into ammonia (NH_3), which is used by plants for their growth. Plants in areas exposed to higher greenhouse gas emissions make use of nitrogen-fixing bacteria. As a single factor, Thomas et al. (2007) have stated how increased carbon dioxide levels in the atmosphere stimulate biomass production of Nitrogen-fixing plants by increasing nodulation.

3. The microbes that reside inside the plant tissues are referred to as endophytes. These endosymbiotic groups of microorganisms are known to play crucial role in plant development and management during droughts by producing various phytohormones (Singh et al. 2017).

    - As documented by Clay and Holah (1999), endophytic fungi offer a competitive advantage to their host plant by augmenting their resistance to environmental stresses.
    - Similarly, endophytic bacteria are known to produce certain compounds that enhance a plant's ability against multiple stresses.
    - Endophytic bacteria and plant interaction lead to the production of plant hormones such as auxin and compounds such as acetylene, which benefit the host (plant) in reducing stresses (Hardoim et al. 2008). Glick (2012) has stated about the potential of endophytic bacteria in fixing nitrogen, nutrient acquisition, and production of plant growth hormones.

3.4 Conclusion and Future Perspectives

Climate change is one of the major constraints that limits plant growth and development throughout the world, resulting in alterations in the environmental conditions such as temperature, intensity of rainfall and concentration of carbon dioxide (Chaudhry and Sidhu 2022).

Although plants show adaptations to the stresses inflicted upon them due to climate change, we cannot ignore the fact that all species have limits to their capacity for adaptive responses to changing environments (Williams et al. 2008). This becomes a concerning factor taking in view the fact that global temperature is still rising, extreme weather conditions like floods and droughts are worsening, greenhouse gas emissions are setting records with the overall trend in climate change barely showing any signs of slowdown.

What needs to be done, on one hand, is shifting our focus on investing more in adopting measures that would protect our planet from the aggravating effects of global climate change. Biodiversity conservation and management, mitigation of greenhouse gases, setting targets to reduce the rise in temperature and afforestation practices need to be highlighted when climate change policies are implemented.

Similarly, on the other hand, there is a need to understand the gaps that have to be filled when relation between the environment and plants is considered; especially in this scenario of climate change, where plants tend to adapt to changing conditions, but the climate crisis is worsening to the point where there might be no going back. With a long-term goal to mitigate the effects of climate change, the focus also needs to be shifted towards understanding how plants adopt different strategies to survive such conditions and if, any new adaptation measures by a particular plant can be focused upon and used as a tackling measure for other plants too. Using advanced methods to monitor plants that have shown better adaptation measures, primarily in harsh environments, should be the focal point for plant breeders in further developing such varieties for optimum growth and yield of crops.

In the face of climatic shift, it becomes crucial for crop breeders to select climate-resilient crops with drought-, salinity-tolerant, and heat-resistant features to ensure food security. Understanding the adaptation strategies of plants becomes significant in such a scenario where climate change has impacted every other organism.

With a changing climate, interaction patterns between plants and ecosystems are bound to change with plants adopting measures to adapt to such conditions to ensure survival. Therefore, it becomes necessary to consider the interactions between environmental conditions, crop growth, and management practices to determine appropriate crop management adaptation strategies (Müller et al. 2014). By assessing how these adaptations help plants enhance their resilience, biologists can predict how plant communities will respond to climate change. This will help in understanding the vulnerabilities of different ecosystems with respect to climate change and then, accordingly, developing conservation strategies for plants and organisms living in them. Simultaneously, agronomic practices such as soil management, crop diversification, along with selective breeding of plants for stress tolerance, and genetic modifications for increasing resilience can also be focused upon.

References

Aldardasawi AM, Eren B (2021) Floods and their impact on the environment. Acad Perspect Procedia 4(2):42–49

Anthelme F, Abdoulkader A, Besnard G (2008) Distribution, shape and clonal growth of the rare endemic tree Olea europaea subsp. laperrinei (Oleaceae) in the Saharan mountains of Niger. Plant Ecol 198(1):73–87

Auld J, Everingham SE, Hemmings FA, Moles AT (2022) Alpine plants are on the move: quantifying distribution shifts of Australian alpine plants through time. Divers Distrib 28(5):943–955

Baidya A, Ali MA, Atta K (2022) Climate change and abiotic. In: Advances in plant defense mechanisms, p 351

Barkhordarian A, Saatchi SS, Behrangi A, Loikith PC, Mechoso CR (2019) A recent systematic increase in vapor pressure deficit over tropical South America. Sci Rep 9(1):15331

Baskin CC, Baskin JM (2000) Seeds: ecology, biogeography, and, evolution of dormancy and germination. Academic

Blaylock AD (1994) Soil salinity, salt tolerance, and growth potential of horticultural and landscape plants. University of Wyoming, Cooperative Extension Service, Department of Plant, Soil, and Insect Sciences, College of Agriculture, Laramie

Calinger KM, Queenborough S, Curtis PS (2013) Herbarium specimens reveal the footprint of climate change on flowering trends across north-central North America. Ecol Lett 16(8):1037–1044

Carruggio F, Onofri A, Impelluso C, Giusso del Galdo G, Scopece G, Cristaudo A (2020) Seed dormancy breaking and germination in Bituminaria basaltica and B. bituminosa (Fabaceae). Plants 9(9):1110

Caser M, Ruffoni B, Scariot V (2012) Screening for drought tolerance in Salvia spp. and Helichrysum petiolare: a way to select low maintenance ornamental plants. In: XXIV international Eucarpia symposium section ornamentals: ornamental breeding worldwide, vol 953, pp 239–246

Cavanaugh KC, Kellner JR, Forde AJ, Gruner DS, Parker JD, Rodriguez W, Feller IC (2014) Poleward expansion of mangroves is a threshold response to decreased frequency of extreme cold events. Proc Natl Acad Sci 111(2):723–727

Change C (1990) The IPCC scientific assessment. In: Contribution of Working Group I to the first assessment report of the intergovernmental panel on climate change, p 365

Chaudhry S, Sidhu GPS (2022) Climate change regulated abiotic stress mechanisms in plants: a comprehensive review. Plant Cell Rep 41(1):1–31

Clay K, Holah J (1999) Fungal endophyte symbiosis and plant diversity in successional fields. Science 285(5434):1742–1744

Colmer TD, Voesenek LACJ (2009) Flooding tolerance: suites of plant traits in variable environments. Funct Plant Biol 36(8):665–681

Compant S, Van Der Heijden MG, Sessitsch A (2010) Climate change effects on beneficial plant–microorganism interactions. FEMS Microbiol Ecol 73(2):197–214

Corlett RT, Westcott DA (2013) Will plant movements keep up with climate change? Trends Ecol Evol 28(8):482–488

Couteron P, Anthelme F, Clerc M, Escaff D, Fernandez-Oto C, Tlidi M (2014) Plant clonal morphologies and spatial patterns as self-organized responses to resource-limited environments. Philos Trans R Soc A Math Phys Eng Sci 372(2027):20140102

Dietz KJ, Turkan I, Krieger-Liszkay A (2016) Redox-and reactive oxygen species-dependent signaling into and out of the photosynthesizing chloroplast. Plant Physiol 171(3):1541–1550

Dirnböck T, Essl F, Rabitsch W (2011) Disproportional risk for habitat loss of high-altitude endemic species under climate change. Glob Chang Biol 17(2):990–996

Dixon RO, Wheeler CT (1983) Biochemical, physiological and environmental aspects of symbiotic nitrogen fixation. In: Biological nitrogen fixation in forest ecosystems: foundations and applications. Springer Netherlands, Dordrecht, pp 107–171

Duarte B, Sleimi N, Caçador I (2014) Biophysical and biochemical constraints imposed by salt stress: learning from halophytes. Front Plant Sci 5:746

Fang J, Kato T, Guo Z, Yang Y, Hu H, Shen H, Zhao X, Kishimoto-Mo AW, Tang Y, Houghton RA (2014) Evidence for environmentally enhanced forest growth. Proc Natl Acad Sci 111(26):9527–9532

Fayinminnu OO, Onitayo FA, Ogunkunle FA, Daodu BJ (2024) Soil pollution and climate change. In: Environmental pollution and public health. Elsevier, pp 289–302

Glick BR (2012) Plant growth-promoting bacteria: mechanisms and applications. Scientifica 2012(1):963401

Gray SB, Dermody O, Klein SP, Locke AM, McGrath JM, Paul RE, Rosenthal DM, Ruiz-Vera UM, Siebers MH, Strellner R, Ainsworth EA (2016) Intensifying drought eliminates the expected benefits of elevated carbon dioxide for soybean. Nat Plants 2(9):1–8

Hadian-Deljou M, Esna-Ashari M, Mirzaie-Asl A (2020) Alleviation of salt stress and expression of stress-responsive gene through the symbiosis of arbuscular mycorrhizal fungi with sour orange seedlings. Sci Hortic 268:109373

Hardoim PR, van Overbeek LS, van Elsas JD (2008) Properties of bacterial endophytes and their proposed role in plant growth. Trends Microbiol 16(10):463–471

Hoffmann AA, Parsons PA (1997) Extreme environmental change and evolution. Cambridge University Press

Huang J, Ji M, Xie Y, Wang S, He Y, Ran J (2016) Global semi-arid climate change over last 60 years. Clim Dyn 46(3):1131–1150

Karban R (2008) Plant behaviour and communication. Ecol Lett 11(7):727–739

Lafont Rapnouil T, Gallant Canguilhem M, Julien F, Céréghino R, Leroy C (2023) Light intensity mediates phenotypic plasticity and leaf trait regionalization in a tank bromeliad. Ann Bot 132(3):443–454

Le Houérou HN (1996) Climate change, drought and desertification. J Arid Environ 34(2):133–185

Linder S, Axelsson B (1982) Changes in carbon uptake and allocation patterns as a result of irrigation and fertilization in a young Pinus sylvestris stand. In: Carbon uptake and allocation in subalpine ecosystems as a key to management, pp 38–44

Ling F, Su Q, Jiang H, Cui J, He X, Wu Z, Zhang Z, Liu J, Zhao Y (2020) Effects of strigolactone on photosynthetic and physiological characteristics in salt-stressed rice seedlings. Sci Rep 10(1):6183

Lucob-Agustin N, Kawai T, Kano-Nakata M, Suralta RR, Niones JM, Hasegawa T, Inari-Ikeda M, Yamauchi A, Inukai Y (2021) Morpho-physiological and molecular mechanisms of phenotypic root plasticity for rice adaptation to water stress conditions. Breed Sci 71(1):20–29

Luo W, Griffin-Nolan RJ, Felton AJ, Yu Q, Wang H, Zhang H, Wang Z, Han X, Collins SL, Knapp AK (2022) Drought has inconsistent effects on seed trait composition despite their strong association with ecosystem drought sensitivity. Funct Ecol 36(11):2690–2700

Morgan JM (1977) Differences in osmoregulation between wheat genotypes. Nature 270(5634):234–235

Muhammad Aslam M, Waseem M, Jakada BH, Okal EJ, Lei Z, Saqib HSA, Yuan W, Xu W, Zhang Q (2022) Mechanisms of abscisic acid-mediated drought stress responses in plants. Int J Mol Sci 23(3):1084

Müller C, Waha K, Bondeau A, Heinke J (2014) Hotspots of climate change impacts in sub-Saharan Africa and implications for adaptation and development. Glob Chang Biol 20(8):2505–2517

Munns DN, Fogle VW, Hallock BG (1977) Alfalfa root nodule distribution and inhibition of nitrogen fixation by heat 1. Agron J 69(3):377–380

Pang J, Turner NC, Du YL, Colmer TD, Siddique KH (2017) Pattern of water use and seed yield under terminal drought in chickpea genotypes. Front Plant Sci 8:1375

Phillips WS (1963) Depth of roots in soil. Ecology 44:424–429

Poudel S, Vennam RR, Shrestha A, Reddy KR, Wijewardane NK, Reddy KN, Bheemanahalli R (2023) Resilience of soybean cultivars to drought stress during flowering and early-seed setting stages. Sci Rep 13(1):1277

Querejeta J, Egerton-Warburton LM, Allen MF (2009) Topographic position modulates the mycorrhizal response of oak trees to interannual rainfall variability. Ecology 90(3):649–662

Raihan T, Geneve RL, Perry SE, Rodriguez Lopez CM (2021) The regulation of plant vegetative phase transition and rejuvenation: miRNAs, a key regulator. Epigenomes 5(4):24

Rashmi I, Shirale A, Kartikha KS, Shinogi KC, Meena BP, Kala S (2017) Leaching of plant nutrients from agricultural lands. In: Essential plant nutrients: uptake, use efficiency, and management. Springer International Publishing, Cham, pp 465–489

Rengasamy P, Chittleborough D, Helyar K (2003) Root-zone constraints and plant-based solutions for dryland salinity. Plant Soil 257(2):249–260

Roychoudhury A, Tripathi DK (2020) Protective chemical agents in the amelioration of plant abiotic stress. Wiley

Sah RP, Chakraborty M, Prasad K, Pandit M, Tudu VK, Chakravarty MK, Narayan SC, Rana M, Moharana D (2020) Impact of water deficit stress in maize: phenology and yield components. Sci Rep 10(1):2944

Sandalio LM, Romero-Puertas MC (2015) Peroxisomes sense and respond to environmental cues by regulating ROS and RNS signalling networks. Ann Bot 116(4):475–485

Saraf M, Goswami D, Maheshwari DK (eds) (2025) Climate Change and soil microorganisms for environmental sustainability. Springer

Scheepens JF, Deng Y, Bossdorf O (2018) Phenotypic plasticity in response to temperature fluctuations is genetically variable, and relates to climatic variability of origin, in Arabidopsis thaliana. AoB Plants 10(4):ply043

Shafi A, Hassan F, Khanday FA (2022) Strigolactones: regulation of biosynthesis, hormonal cross-talk, and its role in abiotic stress adaptation. In: Emerging plant growth regulators in agriculture. Academic, pp 287–302

Sharma P, Jha AB, Dubey RS (2019) Oxidative stress and antioxidative defense system in plants growing under abiotic stresses. In: Handbook of plant and crop stress, 4th edn. CRC press, pp 93–136

Singh M, Kumar A, Singh R, Pandey KD (2017) Endophytic bacteria: a new source of bioactive compounds. 3 Biotech 7(5):315

Smith SE, Read DJ (2010) Mycorrhizal symbiosis. Academic

Soroye P, Newbold T, Kerr J (2020) Climate change contributes to widespread declines among bumble bees across continents. Science 367(6478):685–688

Thomas RB, Van Bloem SJ, Schlesinger WH (2007) Climate change and symbiotic nitrogen fixation in agroecosystems. In: Agroecosystems in a changing climate, pp 85–116

Tooke F, Battey NH (2010) Temperate flowering phenology. J Exp Bot 61(11):2853–2862

van den Broek T, van Diggelen R, Bobbink R (2005) Variation in seed buoyancy of species in wetland ecosystems with different flooding dynamics. J Veg Sci 16(5):579–586

Walas Ł, Alipour S, Haq SM, Alamri S (2024) The potential range of west Asian apple species Malus orientalis Uglitzk. under climate change. BMC Plant Biol 24(1):381

Wang M, Yan L, Dou S, Yang L, Zhang Y, Huang W, Li S, Lu P, Guo Y (2023) Blood multiple heavy metals exposure and lung function in young adults: a prospective Cohort study in China. J Hazard Mater 459:132064

Waters MT, Gutjahr C, Bennett T, Nelson DC (2017) Strigolactone signaling and evolution. Annu Rev Plant Biol 68(1):291–322

Werner AD, Simmons CT (2009) Impact of sea-level rise on sea water intrusion in coastal aquifers. Groundwater 47(2):197–204

Whipple S, Bowser G (2023) The buzz around biodiversity decline: detecting pollinator shifts using a systematic review. IScience 26(11):108101

Williams SE, Shoo LP, Isaac JL, Hoffmann AA, Langham G (2008) Towards an integrated framework for assessing the vulnerability of species to climate change. PLoS Biol 6(12):e325

Xiao W, Zhang Y, Chen X, Sha A, Xiong Z, Luo Y, Peng L, Zou L, Zhao C, Li Q (2024) The easily overlooked effect of global warming: diffusion of heavy metals. Toxics 12(6):400

Yáñez MA, Garrido P, Blanco J, Gajardo J, Osborne D, Osborne T (2025) Effects of burn severity on the diversity and relative abundance of birds 3-years postfire in a coastal forest of Nothofagus glauca (Phil.) Krasser in Central Chile. J Nat Conserv 86:126965

Yang S, Zhao J, Chang SX, Collins C, Xu J, Liu X (2019) Status assessment and probabilistic health risk modeling of metals accumulation in agriculture soils across China: a synthesis. Environ Int 128:165–174

Yuan D, Zhu L, Cherubini P, Li Z, Zhang Y, Wang X (2021) Species-specific indication of 13 tree species growth on climate warming in temperate forest community of Northeast China. Ecol Indic 133:108389

Zeng K, Sentinella AT, Armitage C, Moles AT (2025) Species that require long-day conditions to flower are not advancing their flowering phenology as fast as species without photoperiod requirements. Ann Bot 135(1–2):113–124

Zhang Y, Ku YS, Cheung TY, Cheng SS, Xin D, Gombeau K, Cai Y, Lam HM, Chan TF (2024) Challenges to rhizobial adaptability in a changing climate: genetic engineering solutions for stress tolerance. Microbiol Res 288:127886

Zhu JK (2016) Abiotic stress signaling and responses in plants. Cell 167(2):313–324

Zweifel R, Zimmermann L, Zeugin F, Newbery DM (2006) Intra-annual radial growth and water relations of trees: implications towards a growth mechanism. J Exp Bot 57(6):1445–1459

Plant Growth and Development in the Changing Environment

4

Munmun Kothari, Tribhuwan Singh, Diksha Sharma, Niharika Pandey, and S. K. Guru

Abstract

Plant growth and development are highly sensitive to environmental fluctuations, which have become more intense and unpredictable due to changes in the climatic conditions. The different physiological and developmental processes such as photosynthesis, transpiration, assimilation of nutrient and reproductive growth are severely affected by abiotic stresses like drought and heat. To cope with these stresses, plants have evolved various adaptive mechanisms at morphological, physiological, biochemical, and molecular levels. With the increase in the plant hormone abscisic acid (ABA) under drought stress, closure of stomata and the activation of stress-responsive genes like *RD29A*, *DREB1A*, and *NCED3* is observed. The expression of heat shock proteins like HSP70 and HSP101, acting as molecular chaperones are upregulated under heat stress to stabilize proteins and protect cells from thermal denaturation. The compatible solutes such as proline, glycine betaine, etc., released by plants under stressed conditions maintain osmotic balance. The ion transporters like *NHX1* and *SOS1* control Na^+ exclusion under salinity stress. Exposure to UV-B triggers the activation of UV RESISTANCE LOCUS 8 (UVR8), enhancing flavonoid biosynthesis for photoprotection. Moreover, different transcription factors like *MYB, WRKY*, *bZIP,* and

M. Kothari · D. Sharma · S. K. Guru (✉)
Department of Plant Physiology, College of Basic Sciences & Humanities,
G.B. Pant University of Agriculture & Technology, Pantnagar, Uttarakhand, India
e-mail: skguru.bpy@gbpuat.ac.in

T. Singh
Centre for Oilseeds Research, S.D. Agricultural University, Gujarat, India

N. Pandey
Department of Molecular Biology and Genetic Engineering, College of Basic Sciences & Humanities, G.B. Pant University of Agriculture & Technology,
Pantnagar, Uttarakhand, India

H. C. Joshi et al. (eds.), *Climate-Responsive Plant Biology*,
https://doi.org/10.1007/978-981-95-7354-7_4

NAC play a critical role in the signal transduction pathways and regulation of downstream defense genes under stressed conditions. The advances in functional genomics, transcriptomics, and genome-editing technologies, including CRISPR/Cas9, have enabled precise manipulation of these regulatory pathways. These insights offer valuable opportunities for engineering climate-resilient plant species capable of withstanding complex environmental challenges. This chapter also highlights the integrative mechanisms by which plants perceive and respond to environmental stressors and discusses innovative strategies for promoting sustainable plant growth under climate change.

Keywords

Climate change · Abiotic stress responses · Heat shock proteins · Stress-responsive genes · Compatible solutes · Photoprotection

4.1 Introduction

Plants form the foundation of life on Earth, providing food, oxygen, and ecological balance for all living organisms. Their growth and development are genetically governed, yet these processes are never isolated from the surrounding environment. Light, temperature, water availability, nutrient supply, and atmospheric composition function not only as resources but also as regulatory signals that shape plant physiology, morphology, and reproductive processes. While moderate fluctuations in these cues can promote adaptive growth, extreme or prolonged deviations often impair development, productivity, and survival (Singh 2020; Singh et al. 2024a, b).

In the current era, global environmental change is occurring at a pace faster than natural evolutionary adaptation. Rising atmospheric carbon dioxide, increasing mean temperatures, erratic precipitation, salinity intrusion, flooding, and drought are redefining the ecological and agricultural aspects of plant growth (Jha and Dev 2024). Unlike past centuries, when slow climatic shifts allowed for gradual genetic adaptation, modern plants must cope with rapid transformations within decades. This urgency is particularly significant for global agriculture, as staple crops like rice, wheat, maize, barley, sorghum, and millets feed billions of people worldwide but continue to be highly vulnerable to environmental stresses (Shaw and Etterson 2012; Muitire et al. 2021). Rice (*Oryza sativa*), the major staple food for over half of the global population, is particularly susceptible to elevated temperatures during its flowering stage, which can lead to spikelet sterility and significant reductions in yield (Shrestha et al. 2022; Kothari et al. 2024). The submergence-tolerant rice varieties such as 'Swarna-Sub1' have been developed to withstand prolonged flooding through the activation of *Sub1A* genes that regulate quiescence under low oxygen (Haque et al. 2023; Kuanar et al. 2023). Wheat (*Triticum aestivum*) shows dramatic reductions in grain filling under terminal heat and drought, but deeper-rooting and heat-tolerant varieties exhibit improved resilience (Mao et al. 2023; Sine et al. 2024). Maize (*Zea mays*), which is sensitive to water deficits during tasseling and silking, adapts through altered anthesis silking intervals, and newer hybrids with early flowering have been introduced to shorten the exposure time to drought (Danilevskaya et al. 2019).

Other cereals also demonstrate important adaptive strategies. Barley (*Hordeum vulgare*), often cultivated in marginal environments, exhibits tolerance to both salinity and drought by accumulating protective metabolites and regulating cellular water balance (Isayenkov et al. 2020). Sorghum (*Sorghum bicolor*), a traditional crop of arid regions, uses stay-green traits to sustain photosynthetic activity during drought (Abreha et al. 2022). Some millets, including pearl millet (*Pennisetum glaucum*) and finger millet (*Eleusine coracana*), are naturally adapted to poor soils and high temperature conditions. They have efficient root systems and effective water-use traits, which enable them to face harsh environments (Tadele 2016).

The diversity of plant adaptations highlights the remarkable resilience of developmental processes. Plants have evolved mechanisms to detect and interpret stress signals and regulate the changes in gene expression aimed at maintaining cellular homeostasis, such as the adjustments in stomatal density, which can regulate water-use efficiency under drought (Bertolino et al. 2019; Raza et al. 2020), while changes in membrane lipid composition stabilize the cellular integrity under temperature stress. Antioxidant systems and heat shock proteins further enhance cellular machinery against oxidative and heat damage (Szyller and Bil-Lula 2021). The osmolyte accumulation, like proline and sugars, contributes to the continuity of cellular metabolism under salinity and drought (Singh et al. 2015). Early flowering in wheat or altered vernalization requirements in barley show phenological shifts and represent an effective strategy of developmental adaptation to climatic variability (Hyles et al. 2020).

These adaptive strategies can help in stress tolerance, but it also reduces biomass accumulation or yield potential under unfavorable conditions. These factors indicate the importance of understanding between plant and environment interactions across scales from molecular networks to ecosystem functioning, so that resilience traits can be optimized without compromising productivity (Quinn et al. 2015; Fahad et al. 2017). Genomics, physiology, ecology, and breeding can be integrated to discover the underlying mechanisms of resilience (Amin et al. 2025).

4.2 Environmental Factors Influencing Plant Growth and Development

4.2.1 Light Intensity, Quality, and Photoperiodism

Light serves as the key determinant in the regulation of growth and yield in cereal crops, in which intensity, spectral quality, and photoperiod interact with genetic programs to influence productivity. In wheat, recent studies showed that moderate increases in light intensity (around 600–800 μmol m^{-2} s^{-1}) significantly enhanced grain filling rates and protein accumulation, while excessive irradiance caused oxidative stress and reduced photosystem efficiency (Kharshiing et al. 2022; Li et al.

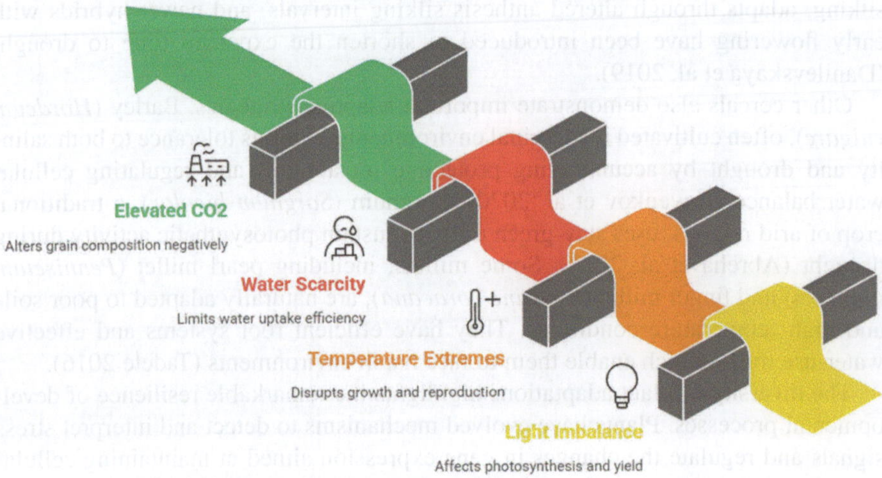

Fig. 4.1 Major environmental stresses affecting plant growth

2023). The spectral quality experiments in rice demonstrated that supplementing red and blue LED light in controlled conditions increases chlorophyll content, tillering, and grain yield by up to 18% as compared with white light alone (Monostori et al. 2018). Similarly, maize exposed to altered red:far-red ratios displayed pronounced shade avoidance responses, like increased stem elongation and reduced leaf expansion, which lowered yield in high-density planting systems (Liu et al. 2025a, b) (Fig. 4.1).

Photoperiodism further governs flowering in cereals, with rice serving as a classical short-day crop, photoperiod-sensitive varieties flower earlier under decreasing daylengths, while insensitivity bred into modern cultivars allows flexibility in multiple cropping systems (Hill and Li 2016). Wheat and barley are predominantly long-day plants, and recent genome-wide studies confirm CONSTANS and FLOWERING LOCUS T orthologs as key regulators of heading time, directly affecting adaptation to higher latitudes (Benaouda et al. 2022).

4.2.2 Temperature

Different crop species exhibit varying responses to temperature at different stages of their life cycle. Each species has a specific minimum and maximum temperature range that supports growth, along with an optimum temperature where growth occurs most rapidly (Wood et al. 2012). When temperatures exceed this optimum, growth rates gradually decline, and plant development stops entirely at the maximum temperature threshold. Vegetative growth, like rate of leaf emergence and node formation typically accelerates as temperatures approach the species' optimum (Bahuguna and Jagadish 2015). Rice, maize, barley, and wheat are highly vulnerable to temperature extremes, both low and high. Low temperatures cause slow germination, weaken seedlings, and disrupt reproduction, as rice struggles to survive below 15 °C (Bhattacharya 2022), and maize exposed to chilling during tasseling shows reduced pollen viability by up to 40% (Hayashi et al. 2015). Barley can germinate at around 4–5 °C, but prolonged cold hampers both the growth of the root and the uptake of nutrients (Bhattacharya 2022). Rising temperatures can be detrimental during flowering and grain filling period. In rice, spikelet fertility drops sharply above 38 °C, causing yield losses (Shrestha et al. 2022; Kothari et al. 2024), while maize exposed to temperatures above 35 °C during pollination experiences reduced silk receptivity and fewer kernels (Wang et al. 2021). Barley and wheat also suffer under heat, with shortened duration of grain filling and decreased grain weight; the yield of wheat can decline upto 10% for each 1 °C rise in mean temperatures (Shirdelmoghanloo et al. 2022). The early growth and reproductive development is delayed under cold stress, whereas heat stress accelerates development but reduces fertility and final yield.

4.2.3 Water Availability and Drought Response

Plants have evolved sophisticated strategies to survive under limited water conditions with the help of different morphological, physiological, biochemical, and molecular adaptations. Morphologically, plants enhance root depth and density to optimize water uptake, while reducing leaf area and modifying stomatal density to minimize transpiration (Haghpanah et al. 2024; Zahedi et al. 2025). Physiologically, drought stress induces stomatal closure, osmotic adjustment via accumulation of compatible solutes as well as activation of antioxidant systems to scavenge reactive oxygen species (ROS), thereby maintaining photosynthetic performance even under limited water availability (Dikilitas et al. 2020; Haghpanah et al. 2024). At the molecular level, both ABA-dependent and ABA-independent signaling pathways regulate the expression of drought-responsive genes. Key transcription factors, including *DREB, NAC,* and *MYB,* regulate osmoprotection, cell wall remodeling, and stress-induced metabolic adjustments (Baldoni et al. 2015; Hussain et al. 2023). The epigenetic mechanisms (DNA methylation, histone modifications, and small RNAs) help plants to retain a memory of drought, allowing them to react more efficiently when confronted with recurring water stress (Rahman et al. 2025; Bashir and Setyawati 2025). Furthermore, integrative "omics" approaches combining transcriptomics, proteomics, metabolomics, and phenomics are uncovering previously unrecognized drought tolerance mechanisms and identifying promising candidate genes for crop improvement (Saleem et al. 2025). Recent advances also emphasize the role of hydraulic signaling and aquaporins in controlling water flux at the cellular and tissue levels, improving water use efficiency without compromising the growth of plants (Cardoso et al. 2020).

4.2.4 Elevated Carbon Dioxide

Rising atmospheric CO_2 levels significantly influence plant physiology by enhancing photosynthetic carbon assimilation while suppressing photorespiration, primarily through increased carboxylation activity of Rubisco (Dusenge et al. 2019; Busch 2020). This stimulatory effect is highly pronounced in the C3 crops, like wheat and rice, whereas C4 species, including maize, millet, and sorghum, show a limited response due to their CO_2-concentrating mechanisms that saturate Rubisco activity (Leegood 2017). Elevated CO_2 also reduces stomatal conductance, improving leaf transpiration efficiency and reducing water loss (Li et al. 2017). However, at the canopy scale, increased leaf area and biomass in C3 crops can result in higher total water use despite improved water-use efficiency per leaf (Leakey et al. 2019). Additionally, elevated CO_2 can alter grain composition by increasing non-structural carbohydrate accumulation and reducing protein content, with potential implications for nutritional quality (Wang et al. 2021).

4.3 Physiological, Biochemical, and Molecular Mechanisms of Abiotic Stress Adaptation

Abiotic stresses have significant consequences on the growth and productivity of plants, not only reducing overall yield but also affecting the nutritional quality of crops (Yadav et al. 2020; Kumari et al. 2022). This issue is particularly critical for global food security, where rising populations demand more stable and higher-quality crop yields (Rasheed et al. 2023). While conventional breeding and agronomic practices have offered some relief, but their impact remains limited (Alatawi et al. 2023). As a result, there is growing emphasis on advanced physiological and molecular approaches to enhance plant resilience to environmental stresses (Sarraf et al. 2022; Ali et al. 2022a, b). At the biochemical and physiological levels, plants respond to stress through mechanisms, including osmotic adjustment, enhanced antioxidant activity, modulation of growth patterns, and strategic allocation of resources to prioritize survival under stress conditions (Ghadirnezhad Shiade et al. 2023). On a molecular scale, exposure to stress triggers intricate changes in gene expression, activating stress-responsive genes and modulating signaling pathways that regulate coordinated responses to the environmental extremes (Sahil et al. 2021). Cutting-edge genomic and molecular studies have identified key transcription factors and regulatory molecules providing opportunities for the development of genetically engineered crops with improved stress tolerance (Saleem et al. 2025) (Fig. 4.2).

Integrating these physiological and molecular insights provides a promising path to develop crops that are prepared to sustain productivity under climate change conditions. Such innovations are essential to building resilient agricultural systems and safeguarding global food security (Noor et al. 2022; Saleem 2024).

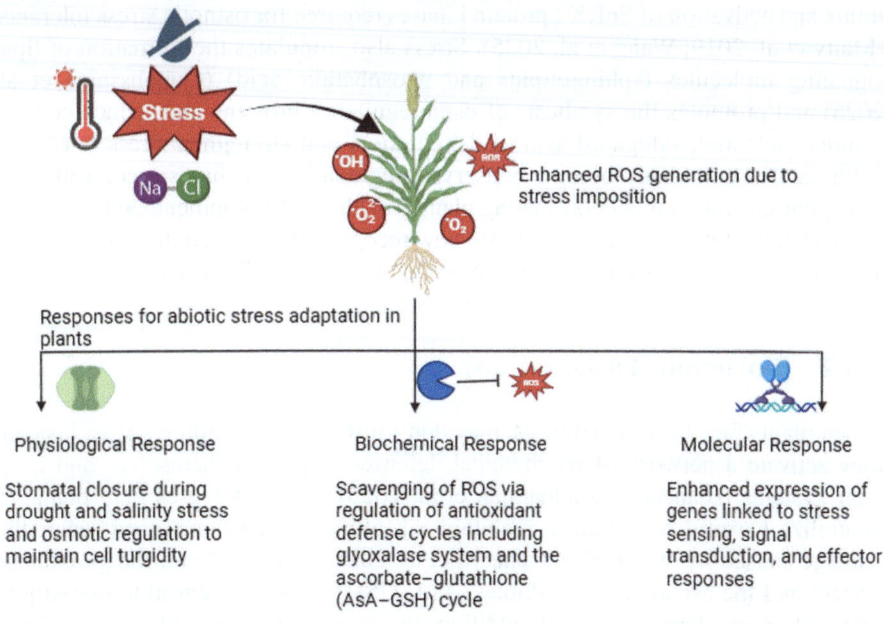

Fig. 4.2 Plant responses to abiotic stress. Stress conditions trigger enhanced reactive oxygen species (ROS) generation, which is mitigated through physiological (stomatal regulation), biochemical (antioxidant defense cycles), and molecular (stress-related gene expression) responses to maintain cellular homeostasis and enhance stress tolerance

4.3.1 Physiological Responses

Stomatal regulation represents an essential physiological strategy that supports crop to tolerate abiotic stresses. Stomata, tiny pores on the leaf surface, mediate CO_2 uptake and water loss, with their opening and closure tightly controlled by guard cells when exposed to the changing environments (Ozfidan-Konakci et al. 2020; Saleem et al. 2023). Abscisic acid (ABA) produced in the roots triggers guard cell signaling in the leaves which causes stomatal closure to conserve water under drought stress and salt stress (Abeed et al. 2023; Akram et al. 2023).

Under hyperosmotic conditions arising from salinity and drought plants maintain osmoregulation, a process where solute concentrations increase to lower osmotic potential and enhance water retention (Gong et al. 2020) This involves increasing intracellular organic solute concentrations and inorganic ions to reduce osmotic potential and mitigate dehydration stress (Saleem et al. 2024; Jeddi et al. 2025). In drought-tolerant species like naked fruitwood accumulate proline, SSs, and soluble proteins (SPs) to maintain their turgor and improve drought resistance (Carpenter et al. 1990). As soil moisture declines further, proline level rises but then plateaus, SP levels initially decrease and later increase below a critical water threshold (31.38%), and SS levels continue to rise (Kumar et al. 2025). Osmotic stress activates the opening of *OSCA* ion channels in rice (*Oryza sativa*) to facilitate Ca^{2+}

influx and activation of SnRK2 protein kinases required for osmotic stress tolerance (Maity et al. 2019; Wang et al. 2025). Stress also stimulates the formation of lipid signaling molecules (sphingolipids and phosphatidic acid) (Oubohssaine et al. 2025) and promotes the synthesis of osmoregulatory proteins that reduce cellular osmotic potential, safeguard against dehydration, and strengthen stress tolerance. Additionally, autophagy being a conserved degradation mechanism controlled by *ATG* genes, is integral in coordinating plant growth, its development, and responses to environmental stress. *ATG* genes are now recognized as key contributors to plant stress resilience across multiple species (Haq et al. 2025) (Table 4.1).

4.3.2 Biochemical Adaptations

When plants face harsh conditions, they don't just passively endure stress; instead, they activate a network of biochemical defenses to protect themselves and keep their systems balanced (Varadharajan et al. 2025). These defenses help plants to neutralize harmful compounds, maintain a healthy redox balance, and repair the damage caused by stress (Bhattacharya et al. 2025). Among them, the glyoxalase system and the ascorbate–glutathione (AsA–GSH) cycle are central to detoxification and antioxidant defense. In addition, the secondary metabolites and antioxidants are synthesized to provide further protection by stabilizing cell structures. Together, these biochemical processes form a coordinated defense network that enhances plant resilience under stress (Cai and Gao 2020; Wahab et al. 2023).

4.3.2.1 The Glyoxalase System

The glyoxalase system represents one of the most ubiquitous and conserved detoxification mechanisms in plants. It functions as a critical safeguard against methylglyoxal (MG) which is a highly reactive and cytotoxic by-product of glycolysis that tends to accumulate under abiotic stress conditions (Berdowska et al. 2025). Excessive MG can severely impair cellular metabolism by modifying proteins, nucleic acids, and other essential biomolecules, thereby disrupting normal physiological processes. This two-step enzymatic pathway is mediated primarily by glyoxalase I (Gly I) and glyoxalase II (Gly II). In the presence of reduced glutathione (GSH), Gly I catalyzes the conversion of MG into S-D-lactoylglutathione, which is subsequently hydrolyzed by Gly II to produce D-lactate, thereby completing the detoxification cycle (Chen et al. 2024). The efficiency of the glyoxalase system is tightly linked to intracellular GSH levels (Zheng et al. 2024). Recent studies have demonstrated that enhancing the glyoxalase pathway through genetic engineering, molecular breeding, or exogenous application of GSH precursors can markedly improve plant tolerance to multiple abiotic stresses (Charagh et al. 2024).

4.3.2.2 Ascorbate–Glutathione Cycle

The ascorbate–glutathione (AsA–GSH) cycle is a central component of the plant antioxidant defense network, which safeguards the cells from reactive oxygen species (ROS) generated during stress caused by changing environments (Rao et al.

Table 4.1 Comprehensive overview of physiological, biochemical, and molecular mechanisms underlying plant adaptation to abiotic stress

Category	Mechanism	Key components/ processes	Role in stress adaptation
Physiological responses	Stomatal regulation	Abscisic acid, guard cell signaling	Triggers stomatal closure to conserve water under drought and salt stress
	Osmoregulation	Proline, soluble sugars, soluble proteins, *OSCA* ion channels, SnRK2 protein kinases	Lowers osmotic potential, enhances water retention, and maintains turgor under drought and salinity stress
	Lipid signaling	Sphingolipids, phosphatidic acid	Reduces cellular osmotic potential, strengthens stress tolerance
	Autophagy	*ATG* genes	Coordinates growth, development, and stress responses, enhancing resilience
Biochemical adaptation	Glyoxalase system	Glyoxalase I (Gly I), Glyoxalase II (Gly II), glutathione (GSH)	Detoxifies methylglyoxal (MG), preventing cellular damage under stress
	Ascorbate-Glutathione cycle	Ascorbate, glutathione, APX, DHAR, GR, NADPH	Neutralizes ROS, maintains redox balance
	Secondary metabolites and antioxidants	Phenolics, flavonoids, alkaloids, terpenoids, SOD, CAT, peroxidases	Scavenges ROS, stabilizes membranes, modulates stress-responsive gene expression
Molecular responses	Gene expression changes	Aquaporins, ion transporters, HSPs, cold-regulated proteins CORs	Regulates water movement, ion homeostasis, and stabilizes cellular structures under stress
	Transcription factors	*DREB* transcription factors	Activates stress-responsive genes, enhances drought and salinity resistance
	Genetic manipulation	Overexpression of osmolyte biosynthesis genes, RNA interference	Improves stress tolerance, delays senescence, extends cellular survival
Ca^{2+}-dependent signal transduction	*LEA* genes activation	Calcium-dependent protein kinases (CDPKs), LEA proteins	Produces anti-desiccation proteins, stabilizes membranes and macromolecules under stress
	SOS signaling	*SOS1*, SOS2–SOS3 kinase complex, *NHX1*, Ca^{2+}-binding proteins	Maintains $Na^+/K^+/H^+$ homeostasis, enhances salt tolerance through Na^+ extrusion

2025). Functioning across multiple cellular compartments, including the chloroplasts, mitochondria, peroxisomes, and cytosol, this cycle operates as an enzymatic system designed to detoxify ROS and maintain intracellular redox equilibrium (Foyer and Kunert 2024). APX utilizes ascorbate (AsA) as an electron donor to convert harmful hydrogen peroxide into water, while DHAR regenerates AsA from its oxidized form, dehydroascorbate, by utilizing reduced glutathione (GSH). GR, in turn, restores reduced glutathione (GSH) from its oxidized state (GSSG) using NADPH as an electron donor (Mishra and Sharma 2024). This continuous regeneration of ascorbate and glutathione sustains cellular antioxidant capacity, stabilizes the redox environment, and activates downstream defense responses, making the AsA–GSH pathway a strategic method for enhancing plant stress tolerance through genetic and agronomic interventions (Kumar 2024).

4.3.2.3 Secondary Metabolites and Antioxidants

Secondary metabolites and antioxidants play a central role in plant defense against abiotic stresses. It functions not only as protectants but also as a regulator of stress signaling and adaptation (Khan et al. 2025). Plants accumulate bioactive compounds, including phenolics, flavonoids, alkaloids, and terpenoids. These molecules contribute to stress tolerance by scavenging reactive oxygen species (ROS), chelating toxic metal ions, stabilizing cellular membranes, and modulating the expression of stress-responsive genes (Rao and Zheng 2025). Plant antioxidant defenses operate on two complementary levels: (i) Enzymatic antioxidants (superoxide dismutase (SOD), catalase (CAT), and various peroxidases), which catalyze the conversion of ROS into less reactive and harmless molecules. (ii) Non-enzymatic antioxidants, including ascorbic acid, glutathione, carotenoids, tocopherols, and phenolic compounds, directly neutralize ROS or act as cofactors to enhance enzymatic activity (Chinmayi et al. 2025).

4.3.3 Molecular Responses to Abiotic Stress

Plants adapt to non-living environmental stressors through complex molecular mechanisms that involve large-scale changes in the expression of genes (Chaffai et al. 2024). The advances in molecular biology have revealed how plants reprogram their transcriptomes to survive under adverse conditions. Next-generation sequencing and transcriptomic studies have identified numerous genes against stress-response involved in osmoprotection, neutralization of harmful compounds (ROS), and recovery of stress-induced damage (Singh et al. 2024a, b). These include genes linked to stress perception, signal transduction, transcriptional regulation, and effector responses (Rai et al. 2024). The aquaporin genes are upregulated during drought stress and regulate cellular water movement, while salinity stress activates ion transporters and compartmentalization pathways to maintain ion homeostasis and minimise toxicity (Mishra et al. 2024a, b). The elevated temperatures trigger heat shock proteins (HSPs), and while decreased temperature beyond critical level stimulate cold-regulated proteins (CORs), which help stabilize and protect cellular structures under chilling stress (Zhou et al. 2025).

The expression of these stress-responsive genes is largely controlled by transcription factors binding to stress-responsive promoter elements. The *DREB* transcription factors stimulate drought-related genetic expression (Wei et al. 2024). Genetic manipulation strategies such as overexpressing transcription factors, osmolyte biosynthesis genes (e.g., proline, glycine betaine), or silencing negative regulators via RNA interference have produced effective results in enhancing the tolerance of crops against stress (Chen et al. 2025). The *DREB* overexpression improves drought and salinity resistance, while silencing genes involved in stress-causing production of ethylene. It can also delay deleterious stress responses, reduce premature senescence, and extend cellular survival under harsh environments (Hou et al. 2022).

4.4 Ca^{2+} Dependent Signal Transduction Pathways

4.4.1 LEA Genes Activation

Abiotic stress stimulates the rapid influx of calcium ions (Ca^{2+}) into the cytosol, where these specialized channels act as primary sensors of stress signals (Dong et al. 2022). These ions activate calcium-dependent protein kinases (CDPKs), large multigene families whose expression is spatially and temporally regulated during plant development. CDPK signaling induces the expression of late embryogenesis abundant (*LEA*) genes, which in turn produces anti-desiccation proteins that function separately from osmolyte synthesis pathways (Mu et al. 2024). Seeds naturally accumulate high levels of LEA proteins during maturation to reduce desiccation damage at germination (Matilla 2021). Similarly, drought, low temperatures, high salinity, and osmotic stress induce LEA protein accumulation in crops. These proteins stabilize membranes and proteins, prevent denaturation, sequester ions, and protect critical macromolecules such as enzymes, lipids, and mRNA (Mohanty and Hembram 2025). Along with osmolytes, LEA proteins help to maintain hydration and structural stability under fluctuating water conditions, highlighting their role in plant resilience.

4.4.2 SOS Signaling

Plants exposed to salinity stress rely on a coordinated signaling network to maintain their ionic and osmotic balance, detoxify reactive compounds, and regulate growth under adverse conditions (Singhal et al. 2021). A central component of this network is the regulation of ion transporters to maintain cellular Na^+, K^+, and H^+ homeostasis. Elevated extracellular or intracellular Na^+ concentrations trigger cytosolic calcium (Ca^{2+}) signals. It modulates the expression and activity of key transporters, including *NHX1* (vacuolar Na^+/H^+ exchanger) and *SOS1* (plasma membrane Na^+/H^+

antiporter) (Joshi et al. 2022). Salt stress activates the *SOS1* gene in arabidopsis, which is regulated by the SOS2–SOS3 kinase complex through direct phosphorylation. It enhances the antiporter's activity and enables efficient Na^+ extrusion from the cytosol (Ali et al. 2023). Functional analyses using wild-type and mutant lines have confirmed that SOS1 encodes a plasma membrane Na^+/H^+ exchanger, essential for salt tolerance (Xie et al. 2022). Because the SOS pathway responds specifically to ionic stress, homologs of SOS2 and SOS3 are believed to participate in other stress and hormonal signaling pathways as well (Ali et al. 2023). Transient elevations in cytosolic Ca^{2+} are detected by Ca^{2+}-binding proteins, including Ca^{2+}-dependent protein kinases (CDPKs) and SOS3-type sensors, which translate calcium fluctuations into targeted phosphorylation cascades. These cascades activate protective proteins and stress-responsive transcription factors. Calcium signaling is therefore central to salt tolerance in plants (Ren et al. 2023). In addition to this, secondary messenger (like inositol phosphates and reactive oxygen species) also modulates the intracellular Ca^{2+} concentration, initiating phosphorylation cascades to activate protective proteins or transcription factors (Ravi et al. 2023).

4.5 Different Approaches to Enhance Plant Growth Under a Changing Environment

4.5.1 Natural Genetic Variation

The recent progress in quantitative trait loci (QTL) mapping and genome-wide association studies (GWAS) has significantly advanced our understanding of the genetic basis of abiotic stress tolerance in plants (Huang 2024). These approaches have uncovered numerous key regulators and allelic variants that serve as valuable targets for breeding and biotechnological interventions aimed at developing high-yielding, climate-resilient crops. The *HKT1* alleles present in different cereal crops, i.e., rice, wheat, and maize, regulate salt tolerance and enhance yield on saline soils (Venkataraman et al. 2021). Likewise, *Temperature Tolerance 1 (TT1)* governs heat tolerance and local adaptation in African rice (*Oryza glaberrima*) (Zhang et al. 2023). Natural variation also extends to Na^+/K^+ transporters as demonstrated by allelic differences in the tomato genes *SlHAK20* and *SOS1*, which affect salt tolerance (Egea et al. 2023). In maize, allelic diversity in *ZmVPP1* and *ZmNAC111* contributes to drought tolerance, with transgenic maize overexpressing *ZmVPP1* exhibiting deeper root systems, improved photosynthetic capacity, and higher water-use efficiency (Liu et al. 2022a, b). Heat stress triggers the induction of heat shock proteins (HSPs) (HSP70 and HSP101). They function as molecular chaperones to stabilize protein structures, prevent aggregation, and protect cellular machinery from thermal denaturation (Kumar et al. 2024). Similarly, plant responses to UV-B stress are coordinated by the photoreceptor UV RESISTANCE LOCUS 8 (UVR8). It exists as a homodimer under normal conditions but undergoes monomerization upon UV-B perception. The monomeric UVR8 interacts with

CONSTITUTIVELY PHOTOMORPHOGENIC 1 (COP1), leading to the stabilization of the transcription factor HY5. It, in turn, activates genes involved in flavonoid biosynthesis such as *chalcone synthase, chalcone isomerase*, and *flavonol synthase* (Laad et al. 2023). Accumulation of flavonoids in epidermal tissues forms a UV-absorbing and antioxidant barrier, mitigating DNA damage and oxidative stress, and thereby enhancing photoprotection (Ferreyra et al. 2021).

4.5.2 Chemical Modulation

Chemical priming is exposing plants to specific compounds before stress to activate innate protective mechanisms and confer multi-stress tolerance (Nguyen et al. 2018). This approach is versatile and applicable across plant species. Endogenous hormones such as abscissic acid (ABA), function as core components in these responses. Drought, salinity, and low temperature stress trigger ABA synthesis, which regulates stomatal closure, osmolyte accumulation, and dormancy (Chauhan et al. 2021). Exogenous ABA application can further enhance stress resistance, improving antioxidant defenses and reducing oxidative damage under low temperature or alkaline stress in rice and other cereal crops (Liu et al. 2023). ABA also promotes stomatal closure to reduce the loss of water, thereby improving drought tolerance (Waseem et al. 2025). Similarly, phytohormones, foliar nutrient application, chemical growth regulators, biostimulants, and exogenous protectants (Table 4.2) can improve the plant resilience by various mode of action under the changing environment.

4.5.3 Genetic Engineering

Genetic modification allows precise manipulation of stress-related regulators, including kinases, transcription factors, enzymes, and membrane transporters. The transgenic rice lines expressing *OsDREB2A* showed markedly improved dehydration tolerance by activating downstream osmoprotective genes (Usman et al. 2023), while overexpressing genes in the glyoxalase pathway enhances resistance to salinity, drought, and heat. Suppression of *miR166* in rice mimics natural drought adaptations, resulting in reduced leaf curling and narrower xylem vessels that improve water-use efficiency (Zhang et al. 2018). Microbial genes, such as *Bacillus subtilis* cold shock protein B, have been introduced into plants to enhance cold and osmotic stress tolerance. Overexpression of *SAPK1* and *SAPK2* (SnRK2 family) enhances salt tolerance in *Oryza* spp., whereas loss-of-function mutants of these genes generated via CRISPR–Cas show reduced tolerance (Lou et al. 2018). CRISPR–Cas gene editing enables precise insertion, substitution, or regulatory sequence modification in stress-responsive genes, offering a powerful tool for functional genomics and breeding (Verma et al. 2023; Chavhan et al. 2025).

Table 4.2 Exogenous modulators and their mode of action in regulating plant growth, development, and stress adaptation in dynamic environmental conditions

Exogenous modulator	Description	Mode of action	Role	References
Phytohormones	Auxins (IAA, NAA), Gibberellins (GA_3), Cytokinins (BAP, kinetin), Abscisic acid (ABA), Ethylene releasers (ethephon), Brassinosteroids	Regulate cell division, elongation, and differentiation; modulate stress signaling pathways	Control seed germination, flowering, fruit set, senescence; improve tolerance to drought, salinity, and heat	Shah et al. (2023), Feng et al. (2024) and Huang and Jin (2025)
Nutrient sprays (Foliar application)	Foliar-applied solutions containing macro- and micronutrients like N, P, K, Fe, Zn, and Ca	Rapid absorption through leaf cuticle and stomata, chelation enhances bioavailability, activates metabolic enzymes for nutrient assimilation	Provide quick nutrient supplementation under nutrient-deficient or stress conditions, improve photosynthetic efficiency and yield in drought/ salinity-affected soils	Ofori et al. (2022)
Chemical growth regulators	Synthetic analogs of hormones like paclobutrazol (GA inhibitor) or chlormequat (ethylene enhancer)	Inhibit key enzymes (e.g., ent-kaurene oxidase for GAs), altering endogenous hormone balance and gene expression	Control excessive vegetative growth under heat/ drought, enhance stress tolerance by redirecting resources to roots and reproductive structures	Rademacher (2015), Desta and Amare (2021) and Huang and Jin (2025)
Nanomaterials (exogenous nanomodulators)	Engineered nanoparticles like ZnO, SiO_2, TiO_2, and carbon nanotubes	Penetrate cell walls, catalyze ROS scavenging as nanozymes, modulates gene expression via epigenetic changes	Enhance nutrient delivery and ROS detoxification under salinity/ drought, improve membrane stability and photosynthetic performance in fluctuating environments	Hong et al. (2021), El-Saadony et al. (2022)

(continued)

Table 4.2 (continued)

Exogenous modulator	Description	Mode of action	Role	References
Biostimulants	Organic substances like seaweed extracts, protein hydrolysates, and humic acids	Stimulate hormone-like signaling, enhance root exudation, and activate stress responsive transcription factors	Promote nutrient uptake and microbiome diversity, alleviate abiotic stresses by boosting osmolyte production and antioxidant defenses in changing climate	Basu et al. (2021)
Exogenous protectants/ stress modulators	Compounds like osmoprotectants (proline, glycine betaine) and antioxidants (ascorbate, glutathione)	Stabilize proteins and membranes, scavenge ROS via enzymatic/ non-enzymatic pathways, modulates signaling cascades	Mitigate oxidative damage from combined biotic/ abiotic stresses, enhance acclimation to fluctuating temperatures and water availability	Hasanuzzaman et al. (2014), Ghosh et al. (2022)

4.6 Conclusion and Future Perspectives

Plants survive and reproduce under fluctuating and often harsh environmental conditions. Physiological, biochemical and molecular mechanisms ranging from stomatal regulation and osmotic adjustment to antioxidant defenses and stress-responsive gene networks allow crops to sense, respond, and adapt to stresses such as drought, salinity, heat, and elevated CO_2. These adaptations, being vital for survival, often have compromises, such as reduced biomass or yield under unfavorable conditions, highlighting the delicate balance between resilience and productivity. Recent advances in genomics, transcriptomics, and phenomics have revealed key regulators of stress tolerance, offering novel pathways to harness natural genetic variation for crop improvement. Chemical priming and exogenous hormone treatments further enhance stress resilience by activating innate defense pathways, while precise genetic engineering, including CRISPR–Cas mediated editing, enables targeted enhancement of stress-responsive genes without compromising their growth.

Looking ahead, integrating traditional breeding with modern molecular and agronomic tools will be essential for developing climate-resilient crops. Emphasis on understanding stress memory, epigenetic regulation, and plant–microbe interactions can reveal novel avenues for improving adaptive capacity. To secure the production of food globally, high-yielding as well as nutritionally enriched varieties would be needed in the face of a rapidly changing environment. A multidisciplinary, systems-level approach that combines molecular insights with field-level management will be the foundation of climate-smart agriculture, ensuring sustainable food security for future generations.

References

Abeed AH, Saleem MH, Asghar MA, Mumtaz S, Ameer A, Ali B, Alwahibi MS, Elshikh MS, Ercisli S, Elsharkawy MM, Ali S (2023) Ameliorative effects of exogenous potassium nitrate on antioxidant defense system and mineral nutrient uptake in radish (Raphanus sativus L.) under salinity stress. ACS Omega 8(25):22575–22588

Abreha KB, Enyew M, Carlsson AS, Vetukuri RR, Feyissa T, Motlhaodi T, Ng'uni D, Geleta M (2022) Sorghum in dryland: morphological, physiological, and molecular responses of sorghum under drought stress. Planta 255(1):20

Akram J, Siddique R, Shafiq M, Tabassum B, Manzoor MT, Javed MA, Anwar S, Nisa BU, Saleem MH, Javed B, Malik T (2023) Genome-wide identification of CCO gene family in cucumber (Cucumis sativus) and its comparative analysis with A. thaliana. BMC Plant Biol 23(1):640

Alatawi A, Mfarrej MF, Alshegaihi RM, Asghar MA, Mumtaz S, Yasin G, Marc RA, Fahad S, Elsharkawy MM, Javed S, Ali S (2023) Application of silicon and sodium hydrosulfide alleviates arsenic toxicity by regulating the physio-biochemical and molecular mechanisms of Zea mays. Environ Sci Pollut Res 30(31):76555–76574

Ali B, Wang X, Saleem MH, Azeem MA, Afridi MS, Nadeem M, Ghazal M, Batool T, Qayyum A, Alatawi A, Ali S (2022a) Bacillus mycoides PM35 reinforces photosynthetic efficiency, antioxidant defense, expression of stress-responsive genes, and ameliorates the effects of salinity stress in maize. Life 12(2):219

Ali M, Wang X, Haroon U, Chaudhary HJ, Kamal A, Ali Q, Saleem MH, Usman K, Alatawi A, Ali S, Munis MF (2022b) Antifungal activity of zinc nitrate derived nano Zno fungicide synthesized from Trachyspermum ammi to control fruit rot disease of grapefruit. Ecotoxicol Environ Saf 233:113311

Ali A, Petrov V, Yun DJ, Gechev T (2023) Revisiting plant salt tolerance: novel components of the SOS pathway. Trends Plant Sci 28(9):1060–1069

Amin A, Zaman W, Park S (2025) Harnessing multi-omics and predictive modeling for climate-resilient crop breeding: from genomes to fields. Genes 16(7):809

Bahuguna RN, Jagadish KS (2015) Temperature regulation of plant phenological development. Environ Exp Bot 111:83–90

Baldoni E, Genga A, Cominelli E (2015) Plant MYB transcription factors: their role in drought response mechanisms. Int J Mol Sci 16(7):15811–15851

Bashir H, Setyawati A (2025) Decoding the Mechanisms of plant epigenetic adaptations to stress. Rev Agric Sci 13(2):1–9

Basu A, Prasad P, Das SN, Kalam S, Sayyed RZ, Reddy MS, El Enshasy H (2021) Plant growth promoting rhizobacteria (PGPR) as green bioinoculants: recent developments, constraints, and prospects. Sustainability 13(3):1140

Benaouda S, Dadshani S, Koua P, Léon J, Ballvora A (2022) Identification of QTLs for wheat heading time across multiple-environments. Theor Appl Genet 135(8):2833–2848

Berdowska I, Matusiewicz M, Fecka I (2025) A comprehensive review of metabolic dysfunction-associated steatotic liver disease: its mechanistic development focusing on methylglyoxal and counterbalancing treatment strategies. Int J Mol Sci 26(6):2394

Bertolino LT, Caine RS, Gray JE (2019) Impact of stomatal density and morphology on water-use efficiency in a changing world. Front Plant Sci 10:225

Bhattacharya A (2022) Effect of low-temperature stress on germination, growth, and phenology of plants: a review. In: Physiological processes in plants under low temperature stress, pp 1–06

Bhattacharya S, Gupta N, Dutta A, Bonomo MG, Milella L, Sarker SD, Nahar L (2025) Mitigating ROS signalling pathway-mediated defence mechanism: a novel approach to counteract bacterial resistance using natural antioxidant-based antibiotics. Phytochem Rev:1–33. https://doi.org/10.1007/s11101-025-10179-6

Busch FA (2020) Photorespiration in the context of rubisco biochemistry, CO_2 diffusion and metabolism. Plant J 101(4):919–939

Cai ZQ, Gao Q (2020) Comparative physiological and biochemical mechanisms of salt tolerance in five contrasting highland quinoa cultivars. BMC Plant Biol 20(1):70

Cardoso AA, Gori A, Da-Silva CJ, Brunetti C (2020) Abscisic acid biosynthesis and signaling in plants: key targets to improve water use efficiency and drought tolerance. Appl Sci 10(18):6322

Carpenter JF, Crowe JH, Arakawa T (1990) Comparison of solute-induced protein stabilization in aqueous solution and in the frozen and dried states. J Dairy Sci 73(12):3627–3636

Chaffai R, Ganesan M, Cherif A (2024) Plant adaptation to abiotic stress: from signaling pathways and microbiomes to molecular mechanisms. Springer Nature

Charagh S, Hui S, Wang J, Raza A, Zhou L, Xu B, Zhang Y, Sheng Z, Tang S, Hu S, Hu P (2024) Unveiling innovative approaches to mitigate metals/metalloids toxicity for sustainable agriculture. Physiol Plant 176(2):e14226

Chauhan DK, Yadav V, Vaculík M, Gassmann W, Pike S, Arif N, Singh VP, Deshmukh R, Sahi S, Tripathi DK (2021) Aluminum toxicity and aluminum stress-induced physiological tolerance responses in higher plants. Crit Rev Biotechnol 41(5):715–730

Chavhan RL, Jaybhaye SG, Hinge VR, Deshmukh AS, Shaikh US, Jadhav PK, Kadam US, Hong JC (2025) Emerging applications of gene editing technologies for the development of climate-resilient crops. Front Genome Ed 7:1524767

Chen TH, Wang HC, Chang CJ, Lee SY (2024) Mitochondrial glutathione in cellular redox homeostasis and disease manifestation. Int J Mol Sci 25(2):1314

Chen J, Zhang J, Liu Y, Zhang K, Zhu F, Xie Y (2025) Advances in the biosynthetic regulation and functional Mechanisms of glycine betaine for enhancing plant stress resilience. Int J Mol Sci 26(16):7971

Chinmayi S, Yadav M, Nandhan NS, Barot M, Saikrishna M (2025) Enzymatic and nonenzymatic antioxidant defense in plants under drought stress. In: Role of antioxidants in mitigating plant stress. Academic, pp 75–85

Danilevskaya ON, Yu G, Meng X, Xu J, Stephenson E, Estrada S, Chilakamarri S, Zastrow-Hayes G, Thatcher S (2019) Developmental and transcriptional responses of maize to drought stress under field conditions. Plant Direct 3(5):e00129

Desta B, Amare G (2021) Paclobutrazol as a plant growth regulator. Chem Biol Technol Agric 8(1):1

Dikilitas M, Simsek E, Roychoudhury A (2020) Role of proline and glycine betaine in overcoming abiotic stresses. In: Protective chemical agents in the amelioration of plant abiotic stress: biochemical and molecular perspectives, pp 1–23

Dong Q, Wallrad L, Almutairi BO, Kudla J (2022) Ca2+ signaling in plant responses to abiotic stresses. J Integr Plant Biol 64(2):287–300

Dusenge ME, Duarte AG, Way DA (2019) Plant carbon metabolism and climate change: elevated CO_2 and temperature impacts on photosynthesis, photorespiration and respiration. New Phytol 221(1):32–49

Egea I, Estrada Y, Faura C, Egea-Fernández JM, Bolarin MC, Flores FB (2023) Salt-tolerant alternative crops as sources of quality food to mitigate the negative impact of salinity on agricultural production. Front Plant Sci 14:1092885

El-Saadony MT, Saad AM, Soliman SM, Salem HM, Desoky ES, Babalghith AO, El-Tahan AM, Ibrahim OM, Ebrahim AA, Abd El-Mageed TA, Elrys AS (2022) Role of nanoparticles in enhancing crop tolerance to abiotic stress: a comprehensive review. Front Plant Sci 13:946717

Fahad S, Bajwa AA, Nazir U, Anjum SA, Farooq A, Zohaib A, Sadia S, Nasim W, Adkins S, Saud S, Ihsan MZ (2017) Crop production under drought and heat stress: plant responses and management options. Front Plant Sci 8:1147

Feng D, Liu W, Chen K, Ning S, Gao Q, Chen J, Liu J, Sun X, Xu W (2024) Exogenous substances used to relieve plants from drought stress and their associated underlying mechanisms. Int J Mol Sci 25(17):9249

Ferreyra ML, Serra P, Casati P (2021) Recent advances on the roles of flavonoids as plant protective molecules after UV and high light exposure. Physiol Plant 173(3):736–749

Foyer CH, Kunert K (2024) The ascorbate–glutathione cycle coming of age. J Exp Bot 75(9):2682–2699

Ghadirnezhad Shiade SR, Fathi A, Taghavi Ghasemkheili F, Amiri E, Pessarakli M (2023) Plants' responses under drought stress conditions: effects of strategic management approaches—a review. J Plant Nutr 46(9):2198–2230

Ghosh UK, Islam MN, Siddiqui MN, Cao X, Khan MA (2022) Proline, a multifaceted signalling molecule in plant responses to abiotic stress: understanding the physiological mechanisms. Plant Biol 2:227–239

Gong Z, Xiong L, Shi H, Yang S, Herrera-Estrella LR, Xu G, Chao DY, Li J, Wang PY, Qin F, Li J (2020) Plant abiotic stress response and nutrient use efficiency. Sci China Life Sci 63(5):635–674

Haghpanah M, Hashemipetroudi S, Arzani A, Araniti F (2024) Drought tolerance in plants: physiological and molecular responses. Plants 13(21):2962

Haq SI, Tariq F, Sama NU, Jamal H, Mohamed HI (2025) Role of autophagy in plant growth and adaptation to salt stress. Planta 3:49

Haque MA, Rafii MY, Yusoff MM, Ali NS, Yusuff O, Arolu F, Anisuzzaman M (2023) Flooding tolerance in Rice: adaptive mechanism and marker-assisted selection breeding approaches. Mol Biol Rep 50(3):2795–2812

Hasanuzzaman M, Alam MM, Rahman A, Hasanuzzaman M, Nahar K, Fujita M (2014) Exogenous proline and glycine betaine mediated upregulation of antioxidant defense and glyoxalase systems provides better protection against salt-induced oxidative stress in two rice (Oryza sativa L.) varieties. Biomed Res Int 2014(1):757219

Hayashi T, Makino T, Sato N, Deguchi K (2015) Barrenness and changes in tassel development and flowering habit of hybrid maize associated with low air temperatures. Plant Prod Sci 18(1):93–98

Hill CB, Li C (2016) Genetic architecture of flowering phenology in cereals and opportunities for crop improvement. Front Plant Sci 7:1906

Hong J, Wang C, Wagner DC, Gardea-Torresdey JL, He F, Rico CM (2021) Foliar application of nanoparticles: mechanisms of absorption, transfer, and multiple impacts. Environ Sci Nano 8(5):1196–1210

Hou Z, Li Y, Cheng Y, Li W, Li T, Du H, Kong F, Dong L, Zheng D, Feng N, Liu B (2022) Genome-wide analysis of DREB genes identifies a novel salt tolerance gene in wild soybean (Glycine soja). Front Plant Sci 13:821647

Huang W (2024) The current situation and future of using GWAS strategies to accelerate the improvement of crop stress resistance traits. Mol Plant Breed 15:52

Huang S, Jin S (2025) Enhancing drought tolerance in horticultural plants through plant hormones: a strategic coping mechanism. Front Plant Sci 15:1502438

Hussain I, Afzal S, Ashraf MA, Rasheed R, Saleem MH, Alatawi A, Ameen F, Fahad S (2023) Effect of metals or trace elements on wheat growth and its remediation in contaminated soil. J Plant Growth Regul 42(4):2258–2282

Hyles J, Bloomfield MT, Hunt JR, Trethowan RM, Trevaskis B (2020) Phenology and related traits for wheat adaptation. Heredity 125(6):417–430

Isayenkov S, Hilo A, Rizzo P, Tandron Moya YA, Rolletschek H, Borisjuk L, Radchuk V (2020) Adaptation strategies of halophytic barley Hordeum marinum ssp. marinum to high salinity and osmotic stress. Int J Mol Sci 21(23):9019

Jeddi K, Siddique KH, Hessini K (2025) Impact of salinity on plant growth, photosynthesis, Cell Wall elasticity and osmotic adjustment in damask rose. Russ J Plant Physiol 72(5):171

Jha MK, Dev M (2024) Impacts of climate change. In: Smart Internet of things for environment and healthcare. Springer Nature Switzerland, Cham, pp 139–159

Joshi S, Nath J, Singh AK, Pareek A, Joshi R (2022) Ion transporters and their regulatory signal transduction mechanisms for salinity tolerance in plants. Physiol Plant 174(3):e13702

Khan A, Kanwal F, Ullah S, Fahad M, Tariq L, Altaf MT, Riaz A, Zhang G (2025) Plant secondary metabolites—central regulators against abiotic and biotic stresses. Metabolites 15(4):276

Kharshiing EV, Mawphlang OI, Lama V, Bhattacharjee R, Sahoo L (2022) Manipulation of light environment for optimising photoreceptor activity towards enhancing plant traits of agronomic and horticultural importance in crops. J Hortic Sci Biotechnol 97(5):535–551

Kothari M, Singh T, Shankhdhar SC, Guru SK (2024) Effect of heat stress during reproductive stage on plant growth, Total biomass production and their correlation with grain yield in Rice. J Exp Agric Int 46(8):762–773

Kuanar SR, Sarkar RK, Panigrahi R, Mohapatra PK (2023) Introgression of SUB1 aggravates the susceptibility of the popular rice cultivars Swarna and Savitri to stagnant flooding. Sci Rep 13(1):9032

Kumar G (2024) Unraveling the dynamics of antioxidant defense in plants under drought conditions. In: Plant secondary metabolites and abiotic Stress, pp 35–79

Kumar P, Paul D, Jhajhriya S, Kumar R, Dutta S, Siwach P, Das S (2024) Understanding heat-shock proteins' abundance and pivotal function under multiple abiotic stresses. J Plant Biochem Biotechnol 33(4):492–513

Kumar M, Kumar R, Sarvade S, Gautam K, Yadav G, Banyal R, Mann A, Dobhal S (2025) Underlying survival mechanisms in model trees for enhanced abiotic stress tolerance. In: Cutting edge technologies for developing future crop plants. Springer Nature Singapore, Singapore, pp 251–279

Kumari VV, Banerjee P, Verma VC, Sukumaran S, Chandran MA, Gopinath KA, Venkatesh G, Yadav SK, Singh VK, Awasthi NK (2022) Plant nutrition: an effective way to alleviate abiotic stress in agricultural crops. Int J Mol Sci 23(15):8519

Laad P, Patel P, Guruprasad KN (2023) UVR8 signaling, mechanism, and integration with other pathways. In: UV-B radiation and crop growth. Springer Nature Singapore, Singapore, pp 193–221

Leakey AD, Ferguson JN, Pignon CP, Wu A, Jin Z, Hammer GL, Lobell DB (2019) Water use efficiency as a constraint and target for improving the resilience and productivity of C3 and C4 crops. Annu Rev Plant Biol 70(1):781–808

Leegood RC (2017) I. The production of triose phosphate. In: Photoassimilate distribution plants and crops source-sink relationships, vol 21. Routledge

Li Y, Li H, Li Y, Zhang S (2017) Improving water-use efficiency by decreasing stomatal conductance and transpiration rate to maintain higher ear photosynthetic rate in drought-resistant wheat. Crop J 5(3):231–239

Li X, Yang R, Li L, Liu K, Harrison MT, Fahad S, Wei M, Yin L, Zhou M, Wang X (2023) Physiological and molecular responses of wheat to low light intensity. Agronomy 13(1):272

Liu H, Song S, Zhang H, Li Y, Niu L, Zhang J, Wang W (2022a) Signaling transduction of ABA, ROS, and Ca2+ in plant stomatal closure in response to drought. Int J Mol Sci 23(23):14824

Liu S, Liu X, Zhang X, Chang S, Ma C, Qin F (2022b) Co-expression of ZmVPP1 with ZmNAC111 confers robust drought resistance in maize. Genes 14(1):8

Liu X, Zhong X, Liao J, Ji P, Yang J, Cao Z, Duan X, Xiong J, Wang Y, Xu C, Yang H (2023) Exogenous abscisic acid improves grain filling capacity under heat stress by enhancing anti-oxidative defense capability in rice. BMC Plant Biol 23(1):619

Liu L, Gai S, Liu C, Zeng Z, Tan X, Li J, Zhou Z (2025a) The effect of far-red light on the growth of tobacco leaves. Plants 14(16):2520

Liu Y, Qiao Y, Liao W (2025b) Calmodulin-binding transcription factors: roles in plant response to abiotic stresses. Plants 14(4):532

Lou D, Wang H, Yu D (2018) The sucrose non-fermenting-1-related protein kinases SAPK1 and SAPK2 function collaboratively as positive regulators of salt stress tolerance in rice. BMC Plant Biol 18(1):203

Maity K, Heumann JM, McGrath AP, Kopcho NJ, Hsu PK, Lee CW, Mapes JH, Garza D, Krishnan S, Morgan GP, Hendargo KJ (2019) Cryo-EM structure of OSCA1. 2 from Oryza sativa elucidates the mechanical basis of potential membrane hyperosmolality gating. Proc Natl Acad Sci 116(28):14309–14318

Mao H, Jiang C, Tang C, Nie X, Du L, Liu Y, Cheng P, Wu Y, Liu H, Kang Z, Wang X (2023) Wheat adaptation to environmental stresses under climate change: molecular basis and genetic improvement. Mol Plant 16(10):1564–1589

Matilla AJ (2021) The orthodox dry seeds are alive: a clear example of desiccation tolerance. Plants 11(1):20

Mishra G, Mohapatra SK, Rout GR (2024a) Plant membrane transporters function under abiotic stresses: a review. Planta 260(6):125

Mishra S, Sharma A, Srivastava AK (2024b) Ascorbic acid: a metabolite switch for designing stress-smart crops. Crit Rev Biotechnol 44(7):1350–1366

Mohanty S, Hembram P (2025) An overview of LEA genes and their importance in combating abiotic stress in Rice. Plant Mol Biol Rep 43(2):337–351

Monostori I, Heilmann M, Kocsy G, Rakszegi M, Ahres M, Altenbach SB, Szalai G, Pál M, Toldi D, Simon-Sarkadi L, Harnos N (2018) LED lighting–modification of growth, metabolism, yield and flour composition in wheat by spectral quality and intensity. Front Plant Sci 9:605

Mu Z, Xu M, Manda T, Yang L, Hwarari D, Zhu FY (2024) Genomic survey and evolution analysis of calcium-dependent protein kinases in plants and their stress-responsive patterns in populus. BMC Genomics 25(1):1108

Muitire C, Kamutando C, Moyo M (2021) Building stress resilience of cereals under future climatic scenarios:'the case of maize, wheat, rice and sorghum'. In: Cereal grains-volume 1. IntechOpen

Nguyen HC, Lin KH, Ho SL, Chiang CM, Yang CM (2018) Enhancing the abiotic stress tolerance of plants: from chemical treatment to biotechnological approaches. Physiol Plant 164(4):452–466

Noor F, Rehman A, Ashfaq UA, Saleem MH, Okla MK, Al-Hashimi A, AbdElgawad H, Aslam S (2022) Integrating network pharmacology and molecular docking approaches to decipher the multi-target pharmacological mechanism of Abrus precatorius L. acting on diabetes. Pharmaceuticals 15(4):414

Ofori KF, Antoniello S, English MM, Aryee AN (2022) Improving nutrition through biofortification–a systematic review. Front Nutr 9:1043655

Oubohssaine M, Hnini M, Rabeh K (2025) Phospholipid signaling in plant growth and development: insights, biotechnological implications and future directions. J Plant Physiol 307:154454

Ozfidan-Konakci C, Yildiztugay E, Elbasan F, Kucukoduk M, Turkan I (2020) Hydrogen sulfide (H2S) and nitric oxide (NO) alleviate cobalt toxicity in wheat (Triticum aestivum L.) by modulating photosynthesis, chloroplastic redox and antioxidant capacity. J Hazard Mater 388:122061

Quinn LD, Straker KC, Guo J, Kim S, Thapa S, Kling G, Lee DK, Voigt TB (2015) Stress-tolerant feedstocks for sustainable bioenergy production on marginal land. Bioenergy Res 8(3):1081–1100

Rademacher W (2015) Plant growth regulators: backgrounds and uses in plant production. J Plant Growth Regul 34(4):845–872

Rahman MM, Keya SS, Bulle M, Ahsan SM, Rahman MA, Roni MS, Al Noor MM, Hasan M (2025) Past trauma, better future: how stress memory shapes plant adaptation to drought. Funct Plant Biol 52(6):FP24355

Rai N, Sarma BK, Rai SP (2024) Transcriptional regulation of biotic and abiotic stress responses: challenges and potential mechanism for stress tolerance and chickpea improvement. Trop Plant Biol 17(2):83–107

Rao MJ, Zheng B (2025) The role of polyphenols in abiotic stress tolerance and their antioxidant properties to scavenge reactive oxygen species and free radicals. Antioxidants 14(1):74

Rao MJ, Duan M, Zhou C, Jiao J, Cheng P, Yang L, Wei W, Shen Q, Ji P, Yang Y, Conteh O (2025) Antioxidant defense system in plants: reactive oxygen species production, signaling, and scavenging during abiotic stress-induced oxidative damage. Horticulturae 11(5):477

Rasheed A, Khan AA, Nawaz M, Mahmood A, Arif U, Hassan MU, Iqbal J, Saleem MH, Ali B, Fahad S (2023) Development of aluminium (Al)-tolerant soybean using molecular tools: limitations and future directions. J Plant Growth Regul 42(12):7403–7417

Ravi B, Foyer CH, Pandey GK (2023) The integration of reactive oxygen species (ROS) and calcium signalling in abiotic stress responses. Plant Cell Environ 46(7):1985–2006

Raza A, Ashraf F, Zou X, Zhang X, Tosif H (2020) Plant adaptation and tolerance to environmental stresses: mechanisms and perspectives. In: Plant ecophysiology and adaptation under climate change: mechanisms and perspectives I: general consequences and plant responses. Springer Singapore, Singapore, pp 117–145

Ren H, Zhang Y, Zhong M, Hussian J, Tang Y, Liu S, Qi G (2023) Calcium signaling-mediated transcriptional reprogramming during abiotic stress response in plants. Theor Appl Genet 136(10):210

Sahil KR, Patra A, Mehta S, Abdelmotelb KF, Lavale SA, Chaudhary M, Aggarwal SK, Chattopadhyay A (2021) Expression and regulation of stress-responsive genes in plants under harsh environmental conditions. In: Harsh environment and plant resilience: molecular and functional aspects. Springer International Publishing, Cham, pp 25–44

Saleem MH, Mfarrej MF, Alatawi A, Mumtaz S, Imran M, Ashraf MA, Rizwan M, Usman K, Ahmad P, Ali S (2023) Silicon enhances morpho–physio–biochemical responses in arsenic stressed spinach (Spinacia oleracea L.) by minimizing its uptake. J Plant Growth Regul 42(3):2053–2072

Saleem MH, Mfarrej MF, Khan KA, Ercisli S, Elsharkawy MM, Fahad S (2024) Advances in physiochemical and molecular mechanisms of abiotic stress tolerance in plants. J Crop Health 76(4):753–767

Saleem MH, Noreen S, Ishaq I, Saleem A, Khan KA, Ercisli S, Anas M, Khalid A, Ahmed T, Hassan A, Fahad S (2025) Omics technologies: unraveling abiotic stress tolerance mechanisms for sustainable crop improvement. J Plant Growth Regul 44:1–23

Sarraf M, Vishwakarma K, Kumar V, Arif N, Das S, Johnson R, Janeeshma E, Puthur JT, Aliniaeifard S, Chauhan DK, Fujita M (2022) Metal/metalloid-based nanomaterials for plant abiotic stress tolerance: an overview of the mechanisms. Plants 11(3):316

Shah SH, Islam S, Mohammad F, Siddiqui MH (2023) Gibberellic acid: a versatile regulator of plant growth, development and stress responses. J Plant Growth Regul 42(12):7352–7373

Shaw RG, Etterson JR (2012) Rapid climate change and the rate of adaptation: insight from experimental quantitative genetics. New Phytol 195(4):752–765

Shirdelmoghanloo H, Chen K, Paynter BH, Angessa TT, Westcott S, Khan HA, Hill CB, Li C (2022) Grain-filling rate improves physical grain quality in barley under heat stress conditions during the grain-filling period. Front Plant Sci 13:858652

Shrestha S, Mahat J, Shrestha J, Madhav KC, Paudel K (2022) Influence of high-temperature stress on rice growth and development. A review. Heliyon 8(12):e12651

Sine B, Ndiaye A, Kanfany G, Grondin A, Ndiaye A, Faye A, Sy O, Diack O, Kane NA, Laplaze L (2024) Physiological and molecular bases of drought and heat tolerance in pearl millet. In: Pearl millet in the 21st century: food-nutrition-climate resilience-improved livelihoods. Springer Nature Singapore, Singapore, pp 247–278

Singh V (2020) Environmental plant physiology: botanical strategies for a climate smart planet. CRC Press

Singh M, Kumar J, Singh S, Singh VP, Prasad SM (2015) Roles of osmoprotectants in improving salinity and drought tolerance in plants: a review. Rev Environ Sci Biotechnol 14(3):407–426

Singh AK, Pal P, Sahoo UK, Sharma L, Pandey B, Prakash A, Sarangi PK, Prus P, Paşcalău R, Imbrea F (2024a) Enhancing crop resilience: the role of plant genetics, transcription factors, and next-generation sequencing in addressing salt stress. Int J Mol Sci 25(23):12537

Singh H, Kumar N, Kumar A (2024b) Enhancing resource use efficiency in crops through plant functional traits. In: Plant functional traits for improving productivity. Springer Nature Singapore, Singapore, pp 97–117

Singhal RK, Saha D, Skalicky M, Mishra UN, Chauhan J, Behera LP, Lenka D, Chand S, Kumar V, Dey P, Indu (2021) Crucial cell signaling compounds crosstalk and integrative multi-omics techniques for salinity stress tolerance in plants. Front Plant Sci 12:670369

Szyller J, Bil-Lula I (2021) Heat shock proteins in oxidative stress and ischemia/reperfusion injury and benefits from physical exercises: a review to the current knowledge. Oxidative Med Cell Longev 2021(1):6678457

Tadele Z (2016) Drought adaptation in millets. In: Abiotic and biotic stress in plants-recent advances and future perspectives. IntechOpen

Usman B, Derakhshani B, Jung KH (2023) Recent molecular aspects and integrated omics strategies for understanding the abiotic stress tolerance of rice. Plants 12(10):2019

Varadharajan V, Rajendran R, Muthuramalingam P, Runthala A, Madhesh V, Swaminathan G, Murugan P, Srinivasan H, Park Y, Shin H, Ramesh M (2025) Multi-omics approaches against abiotic and biotic stress—a review. Plants 14(6):865

Venkataraman G, Shabala S, Véry AA, Hariharan GN, Somasundaram S, Pulipati S, Sellamuthu G, Harikrishnan M, Kumari K, Shabala L, Zhou M (2021) To exclude or to accumulate? Revealing the role of the sodium HKT1; 5 transporter in plant adaptive responses to varying soil salinity. Plant Physiol Biochem 169:333–342

Verma V, Kumar A, Partap M, Thakur M, Bhargava B (2023) CRISPR-Cas: a robust technology for enhancing consumer-preferred commercial traits in crops. Front Plant Sci 14:1122940

Wahab A, Munir A, Saleem MH, AbdulRaheem MI, Aziz H, Mfarrej MF, Abdi G (2023) Interactions of metal-based engineered nanoparticles with plants: an overview of the state of current knowledge, research progress, and prospects. J Plant Growth Regul 42(9):5396–5416

Wang Y, Sheng D, Zhang P, Dong X, Yan Y, Hou X, Wang P, Huang S (2021) High temperature sensitivity of kernel formation in different short periods around silking in maize. Environ Exp Bot 183:104343

Wang X, Liu X, Su Y, Shen H (2025) Rice responses to abiotic stress: key proteins and molecular mechanisms. Int J Mol Sci 26(3):896

Waseem M, Hasan MM, Hazzazi Y, Alharbi BM, Ghani MU, Ahmad P, Carriquí M (2025) Potential mechanisms for the rapid post-drought reversal of ABA-induced stomatal closure by melatonin, 5-aminolevulinic acid, and brassinosteroids. Photosynthetica 63(2):104

Wei H, Wang X, Wang K, Tang X, Zhang N, Si H (2024) Transcription factors as molecular switches regulating plant responses to drought stress. Physiol Plant 176(3):e14366

Wood KA, Stillman RA, Clarke RT, Daunt F, O'Hare MT (2012) Understanding plant community responses to combinations of biotic and abiotic factors in different phases of the plant growth cycle. PLoS One 7(11):e49824

Xie Q, Zhou Y, Jiang X (2022) Structure, function, and regulation of the plasma membrane Na+/H+ antiporter salt overly sensitive 1 in plants. Front Plant Sci 13:866265

Yadav S, Modi P, Dave A, Vijapura A, Patel D, Patel M (2020) Effect of abiotic stress on crops. Sustainable crop production 3(17):5–16

Zahedi SM, Karimi M, Venditti A, Zahra N, Siddique KH, Farooq M (2025) Plant adaptation to drought stress: the role of anatomical and morphological characteristics in maintaining the water status. J Soil Sci Plant Nutr 25(1):409–427

Zhang J, Zhang H, Srivastava AK, Pan Y, Bai J, Fang J, Shi H, Zhu JK (2018) Knockdown of rice microRNA166 confers drought resistance by causing leaf rolling and altering stem xylem development. Plant Physiol 176(3):2082–2094

Zhang Y, Xu J, Li R, Ge Y, Li Y, Li R (2023) Plants' response to abiotic stress: Mechanisms and strategies. Int J Mol Sci 24(13):10915

Zheng Q, Xin J, Zhao C, Tian R (2024) Role of methylglyoxal and glyoxalase in the regulation of plant response to heavy metal stress. Plant Cell Rep 43(4):103

Zhou L, Ullah F, Zou J, Zeng X (2025) Molecular and physiological responses of plants that enhance cold tolerance. Int J Mol Sci 26(3):1157

Phytoremediation Mechanisms of Submerged Macrophytes (*Vallisneria spiralis* and *Ceratophyllum demersum*) in Remediating Inorganic Pollutant-Contaminated Surface Water Environments and Their Further Enhancements

5

Weihong Hu, Demus Matheus Huang, Yinqiu Wang, Widad Fadhullah, and Mohd Rafatullah

Abstract

Several previous research studies have proved the effective phytoremediation mechanisms demonstrated by several submerged macrophytes in absorbing water pollutants using their roots. In this book chapter, we aim to elucidate the phytoremediation mechanisms of two major submerged macrophytes (*Vallisneria spiralis* and *Ceratophyllum demersum*) in purifying surface water environments from several inorganic water pollutants based on the findings from previous research studies. Also, we further highlighted some recent advanced treatments that may enhance their phytoremediation capabilities, including the incorporation of maifanite, lanthanum-modified bentonite (LMB), biological treatment based on the symbiosis of different organisms, and monitoring systems and machine learning. Overall, these two submerged macrophytes have been identified to be highly capable of removing the inorganic water pollutants from surface water environments by implementing various mechanisms, which reflects their high-water purification or remediation efficiency. Moreover, the combination of the submerged macrophytes with the aforementioned recent advanced treatments

W. Hu · D. M. Huang · Y. Wang · W. Fadhullah (✉)
Environmental Technology Division, School of Industrial Technology, Universiti Sains Malaysia, Penang, Malaysia
e-mail: widad@usm.my

M. Rafatullah
Department of Mathematics and Natural Sciences, Prince Mohammad Bin Fahd University, Al-Khobar, Al-Khobar, Saudi Arabia

H. C. Joshi et al. (eds.), *Climate-Responsive Plant Biology*,
https://doi.org/10.1007/978-981-95-7354-7_5

may further enhance their phytoremediation efficiency in removing the inorganic pollutants from the surface water environments.

Keywords

Surface water · Submerged macrophytes · *Vallisneria spiralis* · *Ceratophyllum demersum* · Phytoremediation · Inorganic water pollutants

5.1 Introduction

In recent years, excessive inorganic pollutants generated by various pollutant sources (i.e., industrial wastewater, agricultural wastes, municipal wastes, etc.) have been continuously discharged into various aquatic environments, especially surface water environments (i.e., lakes, ponds, rivers, estuaries, coastal waters) (Mishra et al. 2023). The persistent introduction of these inorganic pollutants into surface water environments has periodically induced various environmental problems that pose detrimental risks to humans and other surrounding organisms within the affected environments.

Inorganic pollutants are non-biodegradable polluting chemical substances that do not possess any carbon–hydrogen bond in their chemical structure. Several types of inorganic pollutants that are widely identified to pollute surface water environments are heavy metals, alkaline substances, inorganic salts, acidic substances, fluorides, and radionuclides. Heavy metal pollutants (i.e., Fe, Pb, Hg, Zn, As) are predominantly derived from in-land anthropogenic heavy industrial and municipal activities, which may enter the aquatic environments via several pathways (i.e., run-off, atmospheric deposition) and may exist in either dissolved or particulate phase states (Ammar et al. 2024; Kumar et al. 2023; Tong et al. 2022; Xu et al. 2016). Long-term accumulation of heavy metals in the aquatic environments may further induce serious deterioration of aquatic environments, instability of aquatic ecosystems, and death of a vast number of aquatic creatures. On the other hand, the introduction of alkaline substances (i.e., NH_3) and inorganic salts (i.e., PO_4^{3-}, NO_3^-) is mainly caused by intensive and poorly managed agricultural activities (i.e., excessive usage of fertilizers). Under high concentration, these pollutants may induce abnormal nutrient enrichment in the water bodies, potentially leading to substantial algal proliferation that corresponds to eutrophication (Fadhullah et al. 2020; Isiuku and Enyoh 2020; Kumar et al. 2023; Li et al. 2020). Primary acidic pollutants such as SO_2, NO_x, and CO_2 mostly originate from combustion of fossil fuels and other related anthropogenic activities, which may end up being precipitated into surface waters through the acid rain process and become dissolved in water as H_2SO_4, HNO_3, and H_2CO_3, respectively. These secondary acidic pollutants are mainly responsible for causing water acidification that corresponds to declining water pH, which may disrupt the water-nutrient cycling process, suppression of microbial decomposition, and severely harm the living aquatic creatures in the long-term period (Huang et al. 2025; Xie et al. 2024). Lastly, radionuclide pollutants, which are potentially originated from both natural (i.e., ^{238}U and ^{232}Th from Earth's crust and mantle) and anthropogenic sources (i.e., ^{137}Cs and ^{90}Sr from nuclear and radiation accidents), may contaminate aquatic creatures and surface water environments via bioaccumulation and bioconcentration, which may end up being deposited in human body via biomagnification in food web system and direct consumption of the affected water (Shahroudi and Pourimani 2023; Tan et al. 2023). Prolonged exposure to radioactive elements with high doses may cause DNA and cellular damage or mutations in the body that subsequently lead to the occurrence of a number of serious diseases, especially cancers.

To reduce the effect of these inorganic pollutants in the aquatic environment, series of traditional methods (physical method, chemical method and biological method) have been periodic utilized. Physical and chemical method, it could effectively remove the inorganic pollutants in surface water. Biochar, which is synthesized by organic material calcinated in vacuum condition can sorb the inorganic salt in surface water. As one of the physical methods, surface aromatic groups of biochar can attract positively charged ions existing in the surface water, generating cation–π interactions with the C=C aromatic bonds to remove inorganic salt in the aquatic environment (Sizmur et al. 2017). On the other hand, Fenton, one of advanced oxidation methods, it can oxidize the inorganic contaminants adequately based on generation of OH ion throughout oxidation of Fe^{2+} (Liu et al. 2025). Overall, physical and chemical methods are identified as effective measurements to decline the concentration of inorganic pollution in the aquatic environment.

However, the physical method and chemical method are considered as containing series demerits. Particularly, synthesis of biochar may consume excessive energy in view of the fact that biochar is calcinated into 1000°, which consume vast amount of energy. Also, in terms of the Fenton reaction, byproduct of Fe^{2+} may contain high toxicity, bringing adverse effect of aquatic environment. Therefore, the biological method is identified as an eco-friendly measurement to inactivate inorganic contaminants compared with the aforementioned methods. Submerged macrophytes are macroscopic living aquatic plants with most of their body parts (i.e., stem, roots) are sunk below the water surface layer. These plants are well-adapted to inhabit various surface water environments (i.e., lakes, swamps, ponds, estuaries, coastal waters) and possess multi-functional roles that contribute to the improvement of water quality and the stability of aquatic ecosystems (Khanna et al. 2024; Li et al. 2024). Some examples of submerged macrophyte along with their habitat distribution are given in Fig. 5.1.

Vallisneria spiralis (tape grass) and *Ceratophyllum demersum* (hornwort), two prominent species of submerged macrophytes, possess great number of advantages in remediating surface water environments that are contaminated by inorganic pollutants. *V. spiralis* may effectively function in purifying various inorganic pollutants in surface waters even under high-eutrophic condition (Morini et al. 2024; Tang et al. 2021). Whereas *C. demersum* possesses an excellent ability to remove heavy metals and inorganic salts from surface water environments (Gałczyńska et al. 2019; Qadri et al. 2022). Due to their capabilities, these submerged macrophytes have utilized in surface water treatment. In this paper, the phytoremediation mechanisms

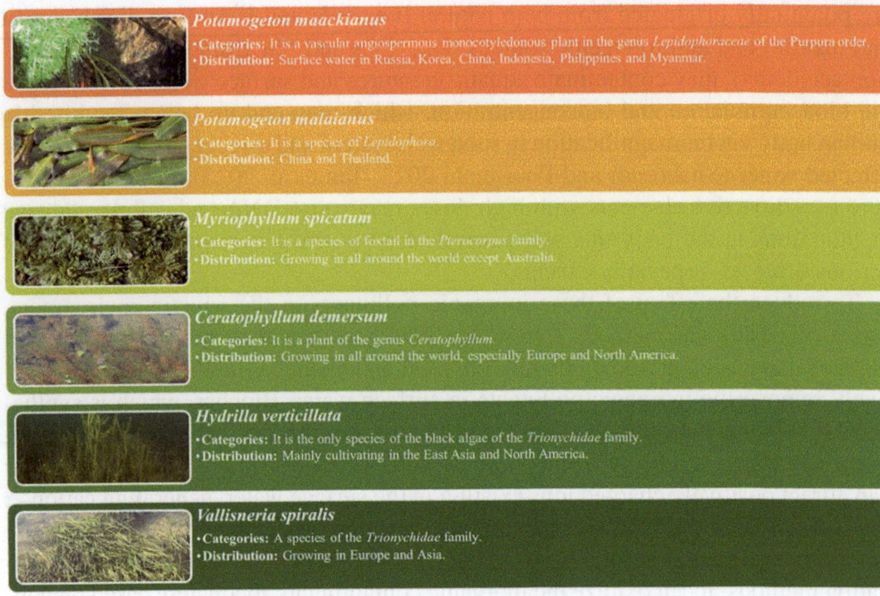

Fig. 5.1 Overview of some different species of submerged macrophyte and their habitat distribution. (Source: iNaturalist (2024) (https://www.inaturalist.org/))

of both *V. spiralis* and *C. demersum* in purifying the inorganic pollutant-contaminated surface waters will be explored in detail through reviewing the previous research studies which were published within the latest 14-year timespan (2010 to 2024), which will be sourced from both Google Scholar and China National Knowledge Infrastructure (CNKI) databases. Furthermore, several recent advanced treatments that may further enhance the phytoremediation efficiency of these submerged macrophytes will be presented.

5.2 Recent Research on Phytoremediation Mechanisms of Two Submerged Macrophytes in Treating Inorganic Pollutants in Surface Water Environments

5.2.1 *Vallisneria spiralis*

Several previous research studies had identified the high-potential phytoremediation capabilities *Vallisneria spiralis* in resolving the eutrophication and other types of pollution caused by inorganic pollutants that commonly happen in surface water environments. This submerged macrophyte generally implements radial oxygen loss (ROL) regulation (Magri et al. 2023; Yan et al. 2022) and direct uptake or absorption mechanisms (Tang et al. 2021; Yan et al. 2022) to remove the inorganic pollutants from the surrounding environments, which could be illustrated as in Fig. 5.2.

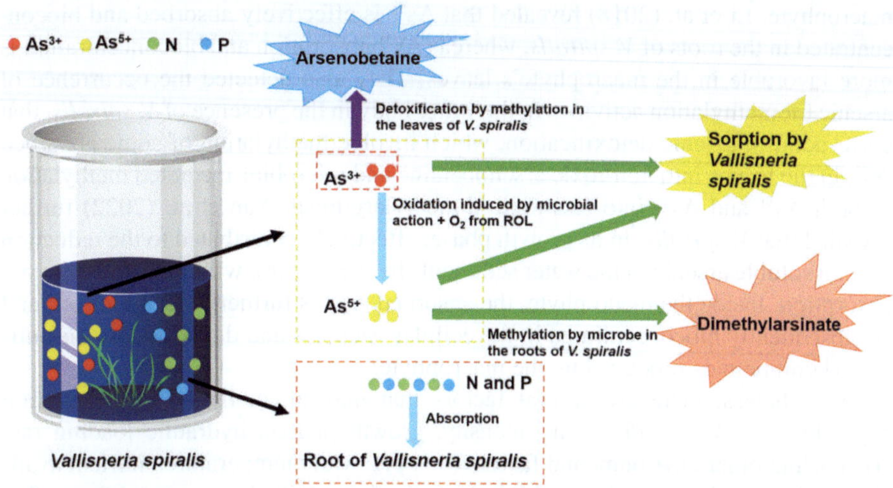

Fig. 5.2 Phytoremediation mechanisms of Vallisneria spiralis in removing inorganic pollutants such as inorganic nutrients (i.e., nitrogen N and phosphorus P) and soluble arsenic (As) forms. (Adapted from Li et al. (2018), Magri et al. (2023), and Yan et al. (2022))

ROL regulation mechanism involves the modulated release of dissolved oxygen from the roots of *V. spiralis* to oxidize heavy metal pollutants in the water sediment environment, which lead to detoxification and higher precipitatibility, immobilization, and sorption of those pollutants. Magri et al. (2023) confirmed that *V. spiralis* could release considerable amount of oxygen from its roots to the water sediments via seasonal regulation of radial oxygen loss, which further induces the occurrence of microoxic environment surrounding its roots that are favorable for metal-oxidizing microbial activities. This allows the oxidation of the existing heavy metal ions (i.e., Fe^{2+} and Mn^{2+}) in the sediments and their subsequent precipitation and co-precipitation with dissolved inorganic phosphorus (DIP) on the root surface; thus, reducing the concentration of those inorganic pollutants and improving the water sediment environment. Similarly, Yan et al. (2022) also found that *V. spiralis* is capable in implementing ROL regulation mechanism, which allows oxidation of As^{3+} to As^{5+} in the water sediments to minimize its toxicity and to promote its precipitatibility and immobilization.

Direct uptake or absorption mechanism allows the inorganic pollutants to be absorbed into *V. spiralis* mainly via roots, although other parts of the macrophyte (i.e., leaves, stems) also possess absorption functionality. Tang et al. (2021), based on their mesocosm study, validated that *V. spiralis* could effectively absorb nutrients (i.e., nitrogen, phosphorus) from both water sediments and water column using its roots, leaves, and stems during the rapid growth, stable growth, and withering and death phases. At the end, total nitrogen TN in the water sediment samples treated by the macrophyte is decreased by approximately 40.7%. Also, they identified significant depletion of NH_4^+-N concentration in the water sediments treated by *V. spiralis*, which could be mainly due to both ROL and absorption mechanisms of the macrophyte. Li et al. (2018) revealed that As^{5+} is effectively absorbed and bioconcentrated in the roots of *V. spiralis*, whereas As^{3+} absorption and bioconcentration is more favorable in the macrophyte's leaves. They also detected the occurrence of arsenic biomethylation activities in the water body in the presence of *V. spiralis*, that correspond to arsenic detoxification, which involve methylation of some absorbed As^{3+} in the leaves into non-toxic arsenobetaine and microbial-mediated methylation of both As^{3+} and As^{5+} into less-toxic dimethylarginines. Yan et al. (2022) further revealed that *V. spiralis*, in its growth phase, effectively contributed to the reduction of total soluble arsenic in the water sediments by nearly 50% within 66 days via root absorption. Inside the macrophyte, the absorbed As^{5+} is further reduced to As^{3+} and is subsequently bioconcentrated in the cellular vacuoles and detoxified by the sulfhydryl compounds produced by the macrophyte.

Nevertheless, there are several factors that may affect the phytoremediation capabilities of *V. spiralis*: light intensity, growth phases, hydraulic loading rate (HLR), and other environmental factors (i.e., pH, water temperature, nutrient availability, dissolved oxygen level). Light intensity is identified as a crucial factor that influences the macrophyte's phytoremediation capabilities in purifying inorganic pollutants in surface waters. Du et al. (2020) revealed that *V. spiralis* could provide maximum removal rate of total nitrogen (TN) (60.59%) when irradiated by white light with 5600 lx for 8 hours, whereas maximum removal rate of PO_4^{3-} (100%) was achieved when subjected to two conditions: white light with 5600 lx for 8 hours and blue light with 2700 lx for 16 hours. Sufficiently high amount of light irradiation under surface water environment with dynamic water flow is vital to initiate and promote the photosynthesis of *V. spiralis*; thus, allowing better growth and productivities along with optimal phytoremediation functionalities of the macrophyte (Du et al. 2020; Wei et al. 2018). Furthermore, Tang et al. (2021) suggested the water purification efficiency of *V. spiralis* may depend on its growth phases. They elaborated that different growth phases of *V. spiralis* have dissimilar efficiency to purify the inorganic nutrients in surface water environments, with effective removal of those pollutants is generally significant during rapid growth, stable growth, and withering and decay phases. Moreover, HLR, the measurement of volume of water or wastewater for treatment within a specific period of time by the surface area of treatment facility, may also influence the nutrients (i.e., total nitrogen TN, total phosphorus TP, NH_4^+-N, NO_3^--N) removal performance of *V. spiralis*. Cai et al. (2024) highlighted significant negative correlation between HLR and the nutrients removal efficiency of *V. spiralis* in vegetative drainage ditches, which indicates decreasing nutrients absorption efficiency of the individual macrophyte following the increase in HLR that corresponds to shorter residence time of those nutrients. Nonetheless, they revealed significant positive correlation between HLR and the areal nutrients removal rate by *V. spiralis*, which implies that elevated HLR may increase the possibility for higher amount of the nutrients to be removed by the group of *V. spiralis* in those ditch areas. Lastly, several environmental factors such as pH, water temperature, nutrient availability and dissolved oxygen level may influence the physiological functions of *V. spiralis* as the macrophyte prefers environmental condition with pH of 7–8 (slightly alkaline), warm water, and considerable levels of nutrients and dissolved oxygen (Rimac et al. 2020). Table 5.1 had concluded the removal efficiency of inorganic contaminant in surface water in the latest 14-year timespan (2010 to 2024).

5.2.2 *Ceratophyllum demersum*

Unlike *Vallisneria spiralis*, *Ceratophyllum demersum* is a submerged macrophyte that does not possess true roots, but it forms rhizomes. Nonetheless, *C. demersum* also possesses effective phytoremediation capabilities in purifying inorganic pollutant-contaminated surface waters. The phytoremediation mechanisms of *C. demersum* include absorption-secretion-competition (Jin et al. 2022; Wang and Zhou 2022) and bioconcentration mechanisms (Qadri et al. 2022; Ran and Chen 2020; Sun et al. 2020).

Absorption-secretion-competition mechanism of *C. demersum* (Fig. 5.3) involves accelerated sorption (absorption and adsorption) of nitrogen and phosphorus forms by *C. demersum* with the mediation of microbial degradation (Wang and Zhou 2022). Specifically, Wang and Zhou (2022) highlight that this mechanism incorporates the accelerated sorption of water inorganic pollutants by the macrophyte's rhizomes, stems, and leaves, which is mediated by an adhesive layer that is created from the macrophyte's secreted gel. This layer may promote adsorption or enrichment of those inorganic nutrient pollutants on the body parts of *C. demersum*, which will be further degraded by the surrounding microorganisms and directly absorbed by the macrophyte. This mechanism may thus prevent the release of excessive nitrogen and phosphorus forms from the water sediments into the surface water body, further promoting aquatic habitat improvement (Jin et al. 2022; Wang and Zhou 2022).

Moreover, *C. demersum* can effectively purify heavy metals in the aquatic environment via bioconcentration mechanism. Sun et al. (2020) selected six different species of submerged macrophytes to investigate their purification efficiency in purifying the water body contaminated with Hg. They later revealed that *C. demersum* could effectively purify Hg in the water body with maximum 91.51% removal rate. Meanwhile, Ran and Chen (2020) revealed strong absorption and enrichment capacity of *C. demersum* in remediating soluble As in the aquatic environment. They found that the buds of *C. demersum* are the macrophyte's part with the greatest As bioconcentration intensity, with the bioconcentration coefficient and enrichment coefficient are calculated to be 20,000 times and higher than 1, respectively. Furthermore, Qadri et al. (2022) confirmed the high phytoremediation efficiency of *C. demersum* in removing several heavy metals from eutrophic lake water samples of Dal Lake via sorption mechanism. When subjected to surface water environment with 6 ppm heavy metal pollutants, they evaluated that *C. demersum* could effectively remove Co^{2+}, Cd^{2+} and Mn^{2+} by approximately 45%, 40%, and 37.6%, respectively.

Despite of its high phytoremediation capabilities, the productivities and phytoremediation efficiency of *C. demersum* may be affected by several factors such as

Table 5.1 Reduction of water parameters utilizing *V. spiralis* in previous study (from 2010 to 2024)

Author	Database	Type of method	Description	Research target	Group	Reduction of water parameter (%)					
						TN	NH_4^+	NO_3^-	NO_2^-	TP	PO_4^{3-}
Huo et al. (2010)	CNKI	Hybrid treatment	*V. spiralis* combine with three species of submerged macrophytes (*Hydrilla verticillata, Potamogenton lucens, Potamogenton cripus*)	Di Shui Lake	Combination group	48.2	72.2	90.3	77.3	80.7	76.9
Liu et al. (2011)	CNKI	Single treatment	Five species of submerged plants are placed into different experimental groups	Lake	*Vallisneria spiralis*	68.75				48	
Wang et al. (2012)	CNKI	Hybrid treatment	Three major experimental groups are set up in the invesigation(Vallisneria spiralis, Elodea canadensis and Vallisneria spiralis+ Elodea canadensis)	Dong jing River	*Vallisneria spiralis*	83.4	78.8	70.2		62.5	
					Vallisneria spiralis + Elodea canadensis	86.2	88.8	78.8		83.0	
Lin et al. (2012)	CNKI	Single treatment	Four species of submerged plants are placed into different experimental groups	Surface water in Jilin Agricultural University reservoir	*Vallisneria spiralis*	66.1					76.1

Cao et al. (2017)	CNKI	Single treatment	Four species of submerged plants are placed into different experimental groups	Tributaries of the Jaru River	*Vallisneria spiralis*		92.7				
Zhou et al. (2018)	CNKI, Dimension AI	Single treatment	*V. spiralis* was subjected to treatment with high nutrient concentrations for 25 days [TN = 12 mg/L; NO3- = 8 mg/L; NH4+ = 4 mg/L; TP = 1 mg/L]	Laboratory mesocosm with sediments from Dongtai Lake	*Vallisneria spiralis*	95				95	
Lin et al. (2019)	CNKI	Hybrid treatment	Adjustable capping bed for eutrophication pollutant sediment is created by Vallisneria spiralis, Hydrilla verticillata and activated carbon	Bai Lu lake	*Combination group*		98.8				
Fu et al. (2021)	CNKI	Single treatment	Ecological module consisting 4 different species emergent aquatic plant, submerged plant and floating plant is utilized to absorb the nutrients in water body.	Bao Dai River	*Vallisneria spiralis*	86	82.5			92	

(continued)

Table 5.1 (continued)

Author	Database	Type of method	Description	Research target	Group	Reduction of water parameter (%)					
						TN	NH_4^+	NO_3^-	NO_2^-	TP	PO_4^{3-}
Lv et al. (2024)	DimensionAI	Single treatment	In-situ enclosure experiments with duration of 120 days involving different treatments	Lake Mochou, Jiangsu	*V. spiralis*	57.8	27.1	35.5		32.2	
		Hybrid treatment			*V. spiralis + LMC*	51.3	38.1	29		77.4	72.6
Zhang et al. (2024)	CNKI	Single treatment	Five species of submerged plants are placed into different experimental groups	Tong Long River	*Experimental group (Y2) the main species is Vallisneria spiralis (60%), 10% of the others)*	15					42

Single Treatment: solely one species of macrophyte (*V. spiralis*) is utilized to degrade inorganic pollutant in surface water
Hybrid Treatment: various species of macrophytes are mentioned to inactivate the inorganic contaminant in previous investigation

Fig. 5.3 Purification of nitrogen pollutants by absorption-secretion-competition mechanism of *Ceratophyllum demersum*. (Adapted from Wang and Zhou (2022))

nutrient availability and other environmental factors (i.e., light availability, pH, temperature). Although nutrients are vital for the growth and productivities of *C. demersum* and other aquatic organisms, their excessive concentration in surface waters may cause several problems. Excessive enrichment of nutrients and other suspended solids may cause intensive eutrophication and turbidity, which further lead to attenuated light irradiance into the water body that may reduce photosynthesis efficiency of *C. demersum* and other autotrophs, oversaturation of dissolved oxygen in the water caused by algal blooms, and impairment of physiological functions of the habituating organisms that may lead to their death (Guo et al. 2020; Huang et al. 2025; Tian et al. 2022; Wang and Zhou 2022; Wen et al. 2023). Based on their review study, Wang and Zhou (2022) highlighted that *C. demersum* might experience significant stress, impaired physiological functions and phytoremediation efficiency for prolonged period when the concentrations of NH_4^+-N and TP in the surface water environment exceed 3 mg/L and 0.4 mg/L, respectively. Ranawakage and Asaeda (2020) further revealed that *C. demersum* possess optimal growth and photosynthesis efficiency under static open water environment with significant amount of light irradiance and negligible water flow. Moreover, *C. demersum* may achieve favorable growth and productivities in the warm surface water environment with slightly alkaline pH between 7.0 to 8.5 (Širić et al. 2023; Syed et al. 2018).

Table 5.2 shows the purification efficiency of different inorganic pollutants in previous investigation. Additionally, six major water indicators are measured to assess the purification capability of *C. demersum* in terms of the reduction of these water parameters.

Table 5.2 Reduction of water parameters utilizing *Ceratophyllum demersum* in previous study (from 2010 to 2024)

Author	Database	Type of method	Description	Research target	Group	Reduction of water parameter (%)					
						TN	NH_4^+	NO_3^-	NO_2^-	TP	PO_4^{3-}
Fang et al. (2010)	CNKI	Single treatment	Five species of submerged macrophytes are utilized in the experiment	River close to Shanghai university	*Ceratophyllum demersum*	83.2	81.0	95.8		95.2	99.5
Ren et al. (2011)	CNKI	Single treatment	Four species of submerged macrophytes are utilized in the experiment	Baiyang Lake	*Ceratophyllum demersum*	88.6	79.0			37.4	
Lin et al. (2012)	CNKI	Single treatment	Four species of submerged macrophytes are utilized in the experiment respectively	Surface water in Jilin Agricultural University reservoir	*Ceratophyllum demersum*	69.4				78.0	
Chang et al. (2016)	CNKI	Hybrid treatment	Three major species of submerged plants are combined in the investigation	Reservoir water in Tianjin Binhai New Area	*Ceratophyllum demersum* + *Myriophyllum*	18.8	45.5			52.4	

Chang et al. (2016)	CNKI	Hybrid treatment	Three major species of submerged plants are combined in the investigation	Reservoir water in Tianjin Binhai New Area	*Ceratophyllum demersum + Myriophyllum + Potamogeton*	21.3	54.8			62.3	
Sha et al. (2016)	CNKI	Hybrid treatment	Three major species of submerged plants are combined in the investigation	River in Ningbo	*Ceratophyllum demersum*	72				84	
					Ceratophyllum demersum + Hydrocotyle vulgari + Dracaena sanderiana	75.5				41.2	
					Ceratophyllum demersum + Dracaena sanderiana	78.2				68.2	
Xiang et al. (2016)	CNKI	Hybrid treatment	Three major species of aquatic plants are combined in the investigation	Qiantang River	*Ceratophyllum demersum + Pistia stratiotes*	59.9				82.5	
					Ceratophyllum demersum + Pistia stratiotes + Myriophyllum aquaticum	66.8				84.4	
Cao et al. (2017)	CNKI	Single treatment	Four species of submerged macrophytes are utilized in the experiment respectively	Jialu River	*Ceratophyllum demersum*		89.5				

(continued)

Table 5.2 (continued)

Author	Database	Type of method	Description	Research target	Group	Reduction of water parameter (%)					
						TN	NH_4^+	NO_3^-	NO_2^-	TP	PO_4^{3-}
Fu et al. (2021)	CNKI	Single treatment	Ecological module consisting 4 different species emergent aquatic plant, submerged plant and floating plant is utilized to absorb the nutrients in water body.	Bao Dai River	*Ceratophyllum demersum*	89	87			90.1	
Zhang et al. (2023)	CNKI	Hybrid treatment	Three major species of aquatic plants are combined in the investigation	Cui Liu Lake	*Ceratophyllum demersum*	25.9	59.7			23	
					Ceratophyllum demersum + *Hydrilla verticillata*	30.9	50			33.5	
					Ceratophyllum demersum + *Myriophyllum verticillatum*	27.2	63.6			39.5	
					Ceratophyllum demersum + *Hydrilla verticillata* + *Myriophyllum aquaticum*	24.2	67			42	

Qadri et al. (2022)	SCOPUS, DimensionAI	Single treatment	Acclimatized C. demersum were grown in treated enclosures subjected with 6 different heavy metals with 6 ppm concentration.	Dal Lake in Kashmir valley	*Ceratophyllum demersum*	37.6	22.3	45	6.1	40	24.6

Single Treatment: solely one species of macrophyte (*V. spiralis*) is utilized to degrade inorganic pollutant in surface water
Hybrid Treatment: various species of macrophytes are mentioned to inactivate the inorganic contaminant in previous investigation

5.3 Recent Advanced Treatments for Enhancing Phytoremediation Efficiency of Submerged Macrophytes in Removing Inorganic Pollutants from Surface Water Environments

5.3.1 Maifanite

Maifanite is a silicate mineral with gullies and pores on the surface that comprises quartz (SiO_2), margarite ($CaAl_2(Si_2Al_2)\ O_{10}(OH)_2$), and other trace metal elements such as Mg, Zn, Ti, and Mn (Han et al. 2020; Liu et al. 2018). This mineral has been previously researched to possess effective remediation capability in removing inorganic pollutants from surface waters via adsorption mechanism. Di et al. (2019) revealed that the removal rates of SO_4^{2-}, Cr^{3+}, and Cr^{6+} in the surface water by utilizing maifanite were 91.37%, 64.1%, and 70.1%, respectively. Similarly, Jin et al. (2021) further confirmed that maifanite could purify various heavy metals in the surface water such as Cr^{6+}, Pb^{2+}, Cd^{2+}, Zn^{2+}, Ni^{2+}, Cu^{2+}, and As^{5+} with maximum adsorption mass fractions of 54.45, 63.84, 56.22, 68.41, 56.87, 54.78, and 39.57 mg/kg, respectively. Moreover, maifanite has been proved to provide beneficial growth-promoting effect on submerged macrophytes, especially *V. spiralis*. Han et al. (2020) revealed that the amount of biomass, height, number of leaves, and leave width of *V. spiralis* treated with maifanite were considerably increased throughout the treatment period. Also, they emphasized that *V. spiralis* experienced optimal growth process after being treated with raw and modified maifanite with 1 cm thickness. These findings further imply the potential combination of maifanite and *V. spiralis* in remediating inorganic pollutant-contaminated surface water environments.

5.3.2 Lanthanum-Modified Bentonite (LMB)

Lanthanum-modified bentonite (LMB) is regarded a passivation agent that primarily consists of 95% of bentonite and 5% rare earth lanthanum La. The surface of LMB is characterized by its considerable roughness and porosity, which not solely possess large cation exchange capacity but also numerous adsorption sites that improve its adsorption capacity on several inorganic pollutants such as phosphorus forms and heavy metals; thus, preventing their release flux to wider environments (Chen et al. 2024; Yan et al. 2023; Yao et al. 2023).

Some previous research studies have successfully demonstrated the efficiency of LMB when combined with *V. spiralis*. Chen et al. (2024) revealed the effectiveness of the combined LMB-*V. spiralis* treatment in controlling the release of P (i.e., DIP), soluble tungsten W, dissolved organic matter (DOM), and other heavy metals in eutrophic lakes. They found that the aforementioned combined treatment resulted in effective removal of DIP, soluble W, and DOM by 74.49%, 78.58%, and 54.78%, respectively. Similarly, Yan et al. (2023) investigated three treatment methods to reduce soluble As forms in water sediments: treatment with sole LMB, treatment with sole *V. spiralis*, and combined treatment of LMB-*V.spiralis*. After 66 days of treatment period, they obtained highest As removal efficiency with LMB-*V. spiralis* (59.73%) followed by sole *V. spiralis* treatment (48.81%) and sole LMB treatment (12.71%). Yan et al. (2023) further elaborated that the combination of both LMB and *V. spiralis* may not only effectively provide enhanced sorption and detoxification of As, but may further stimulate the sediment microbial growth and activities in oxidizing and immobilizing As.

5.3.3 Symbiotic Treatment

The symbiotic treatment mainly relies on the symbiotic relationship between the submerged macrophytes and other different organisms to further enhance the phytoremediation of inorganic pollutant-contaminated surface water environments (Saravanan et al. 2021). This method is evaluated to significantly result in higher inorganic pollutant removal rate than that of sole macrophyte (Zhu et al. 2024). Previous research studies have revealed the efficiency of symbiotic treatment in removing the inorganic pollutants from the surface waters. For instance, Yang et al. (2019) implemented the combination of both *C. demersum* and Chinese daffodil (*Narcissus tazetta* subsp. *chinesis*) to remove nitrogen and phosphorus from eutrophic surface water. This treatment resulted in the removal of both nitrogen and phosphorus by approximately 42% and 80%, respectively. Furthermore, Zhu et al. (2024) integrated *V. spiralis* and other three species of macrophytes (waterthymes *Hydrilla verticillata*, Louisiana iris hybrids and sweet flag *Acorus calamus* L.) in removal of TN and TP from eutrophic surface water. They revealed that the surface water TN was removed by 91.2%, 87.2%, and 81.1% after symbiotic treatments with *V. spiralis*-Lousiana iris hybrids, *V. spiralis-A. calamus* L. and *V. spiralis-H. verticillata* respectively, whereas the TP was removed by 88.9%, 86.1% and 73.3% following the symbiotic treatments with *V. spiralis-A. calamus* L., *V. spiralis*-Lousiana iris hybrids, and *V. spiralis-H. verticillata* respectively. Separately, Han et al. (2024) compared the nitrogen removal efficiency between symbiotic treatment with combined three organisms (*V. spiralis*, silver carp *Hypophthalmichthys molitrix* and Chinese freshwater snail *Bellamya aeruginosa*), symbiotic treatments with combined two organisms (*V. spiralis-H. molitrix* and *V. spiralis-B. aeruginosa*) and sole *V. spiralis* treatment on eutrophic lake water and sediment samples collected from Taihu Lake. They identified that the symbiotic treatment with those three organisms yielded highest nitrogen removal efficiency (17.65%), followed by the treatments with *V. spiralis-H. molitrix* (14.57%), sole *V. spiralis* (11.96%) and *V. spiralis-B. aeruginosa* (9.05%). In the end, the symbiotic treatment is evaluated to be highly prospective green technology and is feasible for phytoremediation of surface waters polluted with inorganic pollutants.

5.3.4 Monitoring Systems and Machine Learning

Monitoring systems and machine learning have been considered as potential tools to monitor and optimize the water phytoremediation efficiency of submerged macrophytes. These advanced monitoring systems can integrate Internet of Things (IoT) technologies (Kumar et al. 2019), sensor networks and remote sensing technologies (Adilakshmi and Venkatesan 2024) in monitoring several complex key parameters that affect surface water in real-time, such as pollutant concentrations, dissolved oxygen, pH and macrophyte growth status (Mahmoud et al. 2023). Humans can use data acquisition and analysis while using machine learning algorithms to build relevant predictive models that can predict phytoremediation efficiency of the submerged macrophytes (Maurya et al. 2024). For instance, Aasim et al. (2023) has successfully incorporated machine learning modelling that involves the artificial intelligence-based predictive models such as random forest (RF), extreme gradient boosting (XGBoost), and multilayer perceptron (MLP) to accurately predict the cadmium (Cd) removal efficiency of in-vitro regenerated *C. demersum* under different conditions. The implementation of decision support system may also provide scientific advice to the users or management team, may assist in develop efficient maintenance and management strategies, and may produce more scientific judgments in order to achieve optimal submerged macrophytes' physiological developments and phytoremediation performance. Conjecturably, as the concentrations of inorganic pollutants in a surface water environment elevate, the system can automatically suggest and provide adjustment on the planting density of macrophytes or increase their aeration intensity (Oruganti et al. 2023).

5.4 Conclusion

Overall, two submerged macrophytes *V. spiralis* and *C. demersum* have been researched to possess effective phytoremediation mechanisms in removing various inorganic pollutants from polluted surface water environments to a certain degree. The phytoremediation mechanisms of *V. spiralis* mainly involve ROL regulation that promotes higher sorption of the inorganic pollutants by the macrophyte and direct uptake or absorption mechanisms, whereas *C. demersum* incorporates absorption-secretion-competition and bioconcentration mechanisms. Nonetheless, the macrophytes' phytoremediation efficiency on inorganic pollutants is dependent on various factors, mainly light intensity, nutrient availability, and other environmental factors such as pH and water temperature.

Furthermore, the recent advanced treatments with maifanite, LMB, symbiotic treatment, and monitoring systems and machine learning have been shown to be highly effective in enhancing the phytoremediation efficiency of submerged macrophytes in removing the inorganic pollutants from surface waters as well as their physiological developments. These treatments are not only proved to be beneficial in enhancing the phytoremediation, but are also more sustainable and environmental-friendly. The innovation and implementation of these advanced treatments may further reform the current paradigm of phytoremediation research and development;

thus, encouraging more researchers and decision-makers to further investigate, develop and refine any current or novel technologies that may promote high phytoremediation efficiency of submerged plants in remediating polluted surface water environments in the future.

5.4.1 Limitations and Recommendations for Future Studies

Although some progress has been made by the previous studies in researching the phytoremediation efficiency of both *V. spiralis* and *C. demersum* in removing the inorganic pollutants from various surface water environments, there are still some limitations. Firstly, most of the aforementioned research studies were conducted in the lab. However, in realistic natural surface water environments, the phytoremediation efficiency of the submerged macrophytes is mainly affected by complex correlations between different environmental factors, which tend to be onerous to be simulated by laboratory-based microcosm studies. This further encourages widespread implementation of mesocosm study as this type of study may effectively address the aforementioned limitation. Furthermore, the research studies about the application of maifanite, LMB, machine learning and other novel technologies or treatments that may further improve the phytoremediation efficiency of the submerged macrophytes, especially other than *V. spiralis*, are still relatively sparse. Future studies might want to consider and overcome the aforementioned limitations to further expand the current knowledges.

References

Aasim M, Ali SA, Aydin S, Bakhsh A, Sogukpinar C, Karatas M, Khawar KM, Aydin ME (2023) Artificial intelligence–based approaches to evaluate and optimize phytoremediation potential of in vitro regenerated aquatic macrophyte *Ceratophyllum demersum* L. Environ Sci Pollut Res 30:40206–40217. https://doi.org/10.1007/s11356-022-25081-3

Adilakshmi A, Venkatesan V (2024) Effective monitoring of Noyyal River surface water quality using remote sensing and machine learning and GIS techniques. Desalin Water Treat 320:e100630. https://doi.org/10.1016/j.dwt.2024.100630

Ammar M, Yousef E, Ashraf S, Baltrusaitis J (2024) Removal of inorganic pollutants and recovery of nutrients from wastewater using electrocoagulation: a review. Separations *11*(11):e320. https://doi.org/10.3390/separations11110320

Cai M, Cui N, Zhang X, Chen G, Zhou L, Zou G (2024) Butong shuili fuhe dui liang zhong shengtai gouqu nei chen shui zhiwu ku cao jinghua xiaoguo de yingxiang yanjiu [Influence of different hydraulic load rates on the purification performance of submerged macrophyte *Vallisneria natans* in two vegetated drainage ditch types]. Nongye huanjing kexue xuebao 43(7):1621–1631. https://doi.org/10.11654/jaes.2023-1112

Cao QQ, Yu LJ, Lü CM, Fan Z, Li TM (2017) Sizhong chenshui zhiwu dui chengshi wuran shuiti de jinghua xiaoguo yanjiu. Purification effects of four submerged macrophytes on urban polluted water bodies. People's Yellow River, (5):76–80. https://doi.org/10.3969/j.issn.1000-1379.2017.05.019

Chang SY, Wu T, Zhao JJ (2016) Butong chenshui zhiwu zupei dui Beidagang shuiku shuiti jinghua xiaoguo de yingxiang. Effects of different submerged macrophyte assemblages on water purification in Beidagang Reservoir. Chinese J Environ Eng, (1):439–444

Chen X, Liu L, Wang Y, Zhou L, Xiao J, Yan W, Li M, Li Q, He X, Zhang L, You X, Zhu D, Yan J, Wang B, Hang X (2024) The combined effects of lanthanum-modified bentonite and *Vallisneria spiralis* on phosphorus, dissolved organic matter, and heavy metal(loid)s. Sci Total Environ *917*:e170502. https://doi.org/10.1016/j.scitotenv.2024.170502

Di J, Xu H, Zhao W, Jiang G, Guo J, Liu J, Lin X (2019) Shengwu mai fan shi ke li shengwu huoxing fenxi ji chuli han ge feishui shiyan [Bioactivity analysis of biological medical stone particles and its treatment of chromium-containing wastewater]. Gongye shui chuli 39(12):33–36. https://doi.org/10.11894/iwt.2018-0973

Du J, Xue R, Gu L (2020) Butong guangzhao huanjing xia ku cao de shengzhang texing jiqi dui shuizhi de jinghua zuoyong [Growth characteristics of *Vallisneria* under different light environments and its purification effect on water quality]. Nongye yu jishu 40(11):18–22. https://doi.org/10.19754/j.nyyjs.20200615006

Fadhullah W, Yaccob NS, Syakir MI, Muhammad SA, Yue F-J, Li S-L (2020) Nitrate sources and processes in the surface water of a tropical reservoir by stable isotopes and mixing model. Sci Total Environ 700:e134517. https://doi.org/10.1016/j.scitotenv.2019.134517

Fang YX, He CQ, Liang X, Jin H, Ma ZF, Zhan YW, et al (2010) Shuisheng zhiwu dui wuran shuiti dan lin de jinghua xiaoguo yanjiu. Purification effects of aquatic plants on nitrogen and phosphorus in polluted water bodies. J Hydroecology, (6):36–40. https://doi.org/10.15928/j.1674-3075.2010.06.007

Fu GX, Cao WZ, Tao J (2021) Shier zhong shuisheng zhiwu dui fuyingyanghua shuiti de jinghua xiaoguo. Purification effects of twelve aquatic plants on eutrophic water bodies. Environ Sci Technol, (S2):308–315. https://doi.org/10.19672/j.cnki.1003-6504.0982.21.338

Gałczyńska M, Mańkowska N, Milke J, Buśko M (2019) Possibilities and limitations of using *Lemna minor*, *Hydrocharis morsus-ranae* and *Ceratophyllum demersum* in removing metals with contaminated water. J Water Land Dev 40(I–III):161–173. https://doi.org/10.2478/jwld-2019-0018

Guo Y, Xue J, Wu Y, Zhou Z, Hua J (2020) Chen shui zhiwu dui fu yingyang hua shuiti de jinghua zuoyong ji xiufu jishu yanjiu jinzhan [Research progress on purification effects and restoration technologies of submerged macrophytes on eutrophic water]. Zhiwu ziyuan yu huanjing xuebao 29(3):58–68. https://doi.org/10.3969/j.issn.1674-7895.2020.03.08

Han F, Zhang Y, Liu Z, Wang C, Luo J, Liu B, Qiu D, He F, Wu Z (2020) Effects of maifanite on growth, physiological and phytochemical process of submerged macrophytes *Vallisneria spiralis*. Ecotoxicol Environ Saf *189*:e109941. https://doi.org/10.1016/j.ecoenv.2019.109941

Han X, Wu W, Huang T, Xu X, Zhuang J (2024) Ku cao-lian yu-tong xiu huan leng luo zuhe xitong dui hubo dan su de quchu xiaoguo he tujing yanjiu [The study on the nitrogen removal efficiency and pathways of the combination system of *Vallisneria natans-Hypophthalmichthys molitrix-Bellamya aeruginosa* in lakes]. Huanjing Kexue Xuebao 44(12):37–48. https://doi.org/10.13671/j.hjkxxb.2024.0225

Huang DM, Sigid MF, Yusup Y, Fadhlullah W, Salleh S (2025) An assessment of live hard coral cover distribution and its physicochemical factors in the Strait of Malacca from 1995 to 2016. Mar Environ Res 203:e106833. https://doi.org/10.1016/j.marenvres.2024.106833

Huo Y, He W, Luo K, Wang Y, Zhang Y, Tian Q, He P (2010) Da xing sao yin dao de chen shui zhi bei sheng tai xiu fu dui Dishui Hu shui zhi de jing hua xiao guo Water purification effects of submerged macrophyte ecological restoration guided by *Daphnia magna* in Dishui Lake. Chin J Appl Ecol, 21(2):495–499. https://doi.org/10.13287/j.1001-9332.2010.0006

Isiuku BO, Enyoh CE (2020) Pollution and health risks assessment of nitrate and phosphate concentrations in water bodies in South Eastern, Nigeria. Environ Adv 2:e100018. https://doi.org/10.1016/j.envadv.2020.100018

Jin X, He J, Huang Y, Lu Y, Liang D (2021) Mai fan shi zai xiufu shuiti he turang zhong zhongjinshu wuran fangmian de yanjiu [The study on immobilization of heavy metal contaminated water and soils by manfan stone]. Huanjing keji 34(4):23–28. https://doi.org/10.19824/j.cnki.cn32-1786/x.2021.0052

Jin G, Zhang Z, Yuan H, Tang H, Chen H, Zhang S (2022) Heliu qianliu dai dibiao-dixiashui guocheng dui dianxing shuisheng shengwu huodong de xiangying yanjiu jinzhan [Advances in the response of surface-subsurface water exchange to activities of typical aquatic organisms in the hyporheic zone of rivers]. Shui Kexue Jinzhan 33(5). https://doi.org/10.14042/j.cnki.32.1309.2022.05.013

Khanna S, Hestir EL, Bellvert J, Boyer JD, Shapiro KD, Ustin SL (2024) A new index for detection of submerged aquatic plants under variable quality water: an extension of the soil-line concept. GIScience Remote Sens *61*(1):e2399386. https://doi.org/10.1080/15481603.2024.2399386

Kumar S, Tiwari P, Zymbler M (2019) Internet of Things is a revolutionary approach for future technology enhancement: a review. J Big Data *6*(1):e111. https://doi.org/10.1186/s40537-019-0268-2

Kumar V, Sharma M, Sondhi S, Kaur K, Sharma D, Sharma S, Utreja D (2023) Removal of inorganic pollutants from wastewater: innovative technologies and toxicity assessment. Sustainability *15*(23):e16376. https://doi.org/10.3390/su152316376

Li B, Gu B, Yang Z, Zhang T (2018) The role of submerged macrophytes in phytoremediation of arsenic from contaminated water: a case study on *Vallisneria natans* (Lour.) Hara. Ecotoxicol Environ Saf 165:224–231. https://doi.org/10.1016/j.ecoenv.2018.09.023

Li J, Li X, Li W, Zhao S (2020) Jiyu AQUATOX moxing de wuliangsuhai fu yingyang hua moni ji kongzhi yanjiu [Simulation and control of eutrophication in Wuliangsuhai Lake using AQUATOX model]. Shengtai Huanjing Xuebao *29*(6):1215–1224. https://doi.org/10.16258/j.cnki.1674-5906.2020.06.018

Li Z, Wu T, Chi W, Pam H (2024) Xuzhou jiuli hu chen shui zhiwu dui fu yingyang hua hubo de shengtai gaishan zuoyong [Ecological improvement of submerged macrophytes on eutrophication of Jiuli Lake in Xuzhou]. Huanjing jiankong yu yujing 16(1):80–86. https://doi.org/10.3969/j.issn.1674-6732.2024.01.013

Lin CF, Cao GJ, Wu P, Zhang Y, Wang CY (2012) Sizhong chenshui zhiwu dui fuyingyanghua shuiti de jinghua xiaoguo yanjiu. Purification effects of four submergedmacrophytes on eutrophic water bodies.J Anhui Agric Sci, (10):6083–6085. https://doi.org/10.13989/j.cnki.0517-6611.2012.10.015

Lin CY, Li T, Teng T, Lin JH, Pang ZJ, Zhou ZM, Yuan BL (2019) Kediaoshi fugai chuang dui fuyingyanghua shuiti de jinghua xiaoguo. Purification effects of an adjustable cover bed on eutrophic water bodies. J Huaqiao Univ (Natural Science Edition), (4):522–526. https://doi.org/10.11830/ISSN.1000-5013.201811072

Liu H, Zhu DT, Li MH, Qiao NN (2011) Wuzhong Chenshui Zhiwu Dui Fuyingyanghua Shuiti de Jinghua Xiaoguo Purification effects of five submerged macrophytes on eutrophic water bodies. Ecol Sci, (6):596–601. https://doi.org/10.3969/j.issn.1008-8873.2011.06.006

Liu Z, Zhang Y, Han F, Yan P, Liu B, Zhou Q, Wu Z (2018) Investigation on the adsorption of phosphorus in all fractions from sediment by modified maifanite. Sci Rep *8*:e15619. https://doi.org/10.1038/s41598-018-34144-w

Liu H, Guo Y, Wang Y, Wang Y, Li P, Chen L et al (2025) Preparation of heterogeneous catalysts from honeycomb-structured activated carbon for the treatment of dye wastewater in chemical and electrical Fenton advanced oxidation. J Water Process Eng 76:108308. https://doi.org/10.1016/j.jwpe.2025.108308

Lv Y, Wang W, Yin H (2024) Efficacy of P-sorbent material combined with aquatic plants in controlling nutrient release from urban lake sediment: field investigation. Environ Res, 263:120233. https://doi.org/10.1016/j.envres.2024.120233

Magri M, Benelli S, Bartoli M (2023) *Vallisneria spiralis* promotes P and Fe retention via radial oxygen loss in contaminated sediments. Water *15*(24):e4222. https://doi.org/10.3390/w15244222

Mahmoud MMM, Darwish R, Bassiuny AM (2023) Development of an economic smart aquaponic system based on IoT. J Eng Res 12(4):886–894. https://doi.org/10.1016/j.jer.2023.08.024

Maurya BM, Yadav N, T A, J S, A S, V P, Iyer M, Yadav MK, Vellingiri B (2024) Artificial intelligence and machine learning algorithms in the detection of heavy metals in water and wastewater: methodological and ethical challenges. Chemosphere *353*:e141474. https://doi.org/10.1016/j.chemosphere.2024.141474

Mishra RK, Mentha SS, Misra Y, Dwivedi N (2023) Emerging pollutants of severe environmental concern in water and wastewater: a comprehensive review on current developments and future research. Water-Energy Nexus 6:74–95. https://doi.org/10.1016/j.wen.2023.08.002

Morini L, Ferrari C, Bartoli M, Zilius M, Broman E, Visioli G (2024) *Vallisneria spiralis* L. adaptive capacity improves pore water chemistry and increases potential nitrification in organic polluted sediments. Ecol Process *13*:e26. https://doi.org/10.1186/s13717-024-00506-8

Oruganti RK, Biji AP, Lanuyanger T, Show PL, Sriariyanun ML, Upadhyayula VKK, Gadhamshetty V, Bhattacharyya D (2023) Artificial intelligence and machine learning tools for high-performance microalgal wastewater treatment and algal biorefinery: a critical review. Sci Total Environ *876*:e162797. https://doi.org/10.1016/j.scitotenv.2023.162797

Qadri H, Uqab B, Javeed O, Dar GH, Bhat RA (2022) *Ceratophyllum demersum*-an accretion biotool for heavy metal remediation. Sci Total Environ 806:e150548. https://doi.org/10.1016/j.scitotenv.2021.150548

Ran Y, Chen G (2020) Ji zhong changjian chen shui zhiwu dui shen fu ji de yanjiu jinzhan [Arsenic accumulation by submerged plants: a review]. Shengwu gongcheng xuebao 36(3):407–415. https://doi.org/10.13345/j.cjb.190237

Ranawakage VP, Asaeda T (2020) Evaluation of the physiological alterations in *Ceratophyllym demersum* L along a diurnally changing solar irradiance gradient. J Plant Interact 15(1):8–16. https://doi.org/10.1080/17429145.2020.1719223

Ren WJ, Tian, ZF, Ning GH, Liu X, Hu XB, Xie JZ (2011) Sizhǒng chenshui zhiwu dui Baiyangdian fuyingyanghua shuiti jinghua xiaoguo de yanjiu. Purification effects of our submerged macrophytes on eutrophic water bodies in Baiyangdian Lake. Ecol Environ Sci, (2):345–352. https://doi.org/10.16258/j.cnki.1674-5906.2011.02.024

Rimac A, Alegro A, Šegota V, Koletić N, Stanković I, Bogdanović S, Vuković N (2020) Distribution and habitat characteristics of *Vallisneria spiralis* L. Croatia Hacquetia 19(1):1–12. hacq-2020-0014

Saravanan A, Kumar PS, Varjani S, Jeevanantham S, Yaashikaa PR, Thamarai P, Abirami B, George CS (2021) A review on algal-bacterial symbiotic system for effective treatment of wastewater. Chemosphere *271*:e129540. https://doi.org/10.1016/j.chemosphere.2021.129540

Sha HL, Shou JC, Cai LX, Xie GJ (2016) Sanzhong shuisheng zhiwu dui heichou heshui de jinghua xiaoguo yanjiu. Purification effects of three aquatic plants on blackodorous river water. Sichuan Environ, (4):17–21. https://doi.org/10.14034/j.cnki.schj.2016.04.005

Shahroudi SMM, Pourimani R (2023) Pollution distribution caused by natural and artificial radionuclide in water and sediments of the International Miankaleh Lagoon in north of Iran. Environ Adv 13:e100413. https://doi.org/10.1016/j.envadv.2023.100413

Širić I, Taher MA, Kumar P, Fayssal SA, Bachheti RK, Mioč B, Andabaka Ž, Singh J, Eid EM (2023) Comparative assessment of treatment of mushroom farm wastewater using plant (*Ceratophyllum demersum* L.) and algae (*Chlorella vulgaris*): experimental and kinetic studies. Horticulturae 9(10):e1081. https://doi.org/10.3390/horticulturae9101081

Sizmur T, Fresno T, Akgül G, Frost H, Moreno-Jiménez E (2017) Biochar modification to enhance sorption of inorganics from water. Bioresour Technol 246:34–47. https://doi.org/10.1016/j.biortech.2017.07.082

Sun X, Hu Y, Zhang Z, Zheng Y, Chen Y, Yu T, Wu L, Chen F, Xiao G (2020) Liu zhong shuisheng zhiwu dui Hg wuran shuiti de jinghua yanjiu [Study on the purification of Hg polluted water body by six aquatic plants]. 40(5):42–48. https://doi.org/10.3969/j.issn.1672-1098.2020.05.007

Syed I, Fatima H, Mohammed A, Siddiqui MA (2018) *Ceratophyllum demersum* a free-floating aquatic plant: a review. Indian J Pharm Biol Res 6(2):10–17. https://doi.org/10.30750/ijpbr.6.2.3

Tan K, Cai X, Tan K, Kwan KY (2023) A review of natural and anthropogenic radionuclide pollution in marine bivalves. Sci Total Environ *896*:e165030. https://doi.org/10.1016/j.scitotenv.2023.165030

Tang H, Dai Y, Fan Y, Song X, Wang F, Liang W (2021) Effect of *Vallisneria spiralis* on water quality and sediment nitrogen at different growth stages in eutrophic shallow lake mesocosms. Pol J Environ Stud 30(3):2341–2351. https://doi.org/10.15244/pjoes/127418

Tian P, Song L, Ji D, Li Y, Yang F, Guo Y (2022) Sanxia ku qu shennong xi butong shiqi rongjie yang yu yelüsu a chui xiang fenbu tezheng [Vertical distribution of dissolved oxygen and chlorophyll-a in Shennong River above Three Gorges Reservoir at low and high water levels]. Shui shengtai xue zazhi 43(2):1–8. https://doi.org/10.15928/j.1674-3075.202006120171

Tong Y, Huang Z, Janssen ABG, Wishart M, He W, Wang X, Zhao Y (2022) Influence of social and environmental drivers on nutrient concentrations and ratios in lakes: a comparison between China and Europe. Water Res 227:e119347. https://doi.org/10.1016/j.watres.2022.119347

Wang YY, Wu HL, Huo YZ, Qu XC, Chen QF, He PM (2012) Dongji kushui yu yilezao dui Gonghu shuiyuandi shuizhi jinghua xiaoguo yanjiu. Study of the depuration effect of *Vallisneria spiralis* and *Elodea canadensis* in water sources area of Gong Hu in winter. J Shanghai Ocean Univ, (1):73–77

Wang Z, Zhou M (2022) Jinyu zao dui shuiti wuran shengwu xiufu de yanjiu jinzhan [Research progress on bioremediation of water pollution by *Ceratophyllum demersum*]. Zhongguo huanjing kexue xuehui 2022 nian kexue jishu nianhui—huanjing gongcheng jishu chuangxin yu yinyong fen huichang lunwen ji (si). 254–260. https://doi.org/10.26914/arthurc.nkihy.2022.042720

Wei H, He F, Xu D, Zhou Q, Xiao E, Zhang L, Wu Z (2018) A comparison of the growth and photosynthetic response of *Vallisneria natans* (Lour.) Hara to a long-term water depth gradient under flowing and static water. J Freshw Ecol 33(1):223–237. https://doi.org/10.1080/02705060.2018.1432509

Wen G, Wang R, Wen C, Bai Y, Cao R, Si F, Li K, Huang T (2023) Danshui jia zao shui hua xingcheng yuanyin ji yuan wei kongzhi yanjiu jinzhan [A review on the formation mechanism and *in-situ* control of freshwater dinoflagellate bloom]. Zhongguo huanjing kexue 43(5):2239–2253. https://doi.org/10.19674/j.cnki.issn1000-6923.20230131.006

Xiang C, Lu JB, Liu F, Li WD (2016) Shuisheng zhiwu dui fuyingyanghua shuiti jinghua xiaoguo de chubu yanjiu. A preliminary study on the purification effects of aquatic plants on eutrophic water bodies]. Environ Sci Guid, (5):32–37. https://doi.org/10.13623/j.cnki.hkdk.2016.05.010

Xie D, Zhao B, Kang R, Ma X, Larssen T, Jin Z, Duan L (2024) Delayed recovery of surface water chemistry from acidification in subtropical forest region in China. Sci Total Environ *912*:e169126. https://doi.org/10.1016/j.scitotenv.2023.169126

Xu G, Liu J, Pei S, Gao M, Kong X (2016) Transport pathway and depocenter of anthropogenic heavy metals off the Shandong Peninsula, China. Estuar Coast Shelf Sci 180:168–178. https://doi.org/10.1016/j.ecss.2016.07.003

Yan W, He X, Wu T, Chen M, Lin J, Chen X, Li Q, Li M, Yan Y, Yao Q (2022) A combined study on *Vallisneria spiralis* and lanthanum modified bentonite to immobilize arsenic in sediments. Environ Res, 216:114689. https://doi.org/10.1016/j.envres.2022.114689

Yan W, He X, Wu T, Chen M, Lin J, Chen X, Li Q, Li M, Yan Y, Yao Q (2023) A combined study on *Vallisneria spiralis* and lanthanum modified bentonite to immobilize arsenic in sediments. Environ Res *216*:e114689. https://doi.org/10.1016/j.envres.2022.114689

Yang C, Zhang R, Feng X (2019) Shuixian jinyu zao zuhe zhiwu qiudong xiufu jingguan shuiti de kexing xing yanjiu [Feasibility study on the restoration of landscape water bodies by plant combination of *Narcissus tazetta-Ceratophyllum demersum* in autumn and winter seasons]. Shandong huagong 48(16):244–246. https://doi.org/10.19319/j.cnki.issn.1008-021x.2019.16.107

Yao J, Tan Q, Shi S, Han L, Zhi Y (2023) Shengwu raodong dui lan xi suo lin ji kong lin xiaoneng de yingxiang [Study on the effect of bioturbation on the efficacy of lanthanum modified bentonite for phosphorus control]. Zhongguo huanjing kexue 43(7):3696–3703. https://doi.org/10.19674/j.cnki.issn1000-6923.20230220.018

Zhang J, Li JJ, Ouyang XS, He YF, Liu L (2023) Chenshui zhiwu ji zuhe dui Cuiliu Hu shuiti de jinghua xiaoguo yanjiu. Purification effects of submerged macrophytes and their combinations on water quality in Cuiliu Lake. China Resour Compr Util, (6):192–197. https://doi.org/10.3969/j.issn.1008-9500.2023.06.025

Zhang C, Chen XY, Bai GL, You LX (2024) Chenshui zhiwu qunluo dapei dui heliu shuiti jinghua xiaoguo yanjiu. Effects of submerged macrophyte community assemblages on river water purification. Waterw Harb, (2):271–276

Zhou YW, Xu XG, Han RM, Zhou XH, Feng DY, Li ZC, Wang GX (2018) Shuiti dan lin yingyang fuhe dui kushui jinghua nengli he guanghe yingguang texing de yingxiang. Effects of nitrogen and phosphorus nutrient loads on the purification capacity and photosynthetic fluorescence characteristics of *Vallisneria natans*. Environ Sci, (3):1180–1187. https://doi.org/10.13227/j.hjkx.201705111
Zhu J, Li J, Mu C, He J, Zhang Y, Gao C (2024) Butong shuisheng zhiwu zuhe dui fu yingyang hua shuiti de jinghua yanjiu [Study on purification of eutrophic water by different aquatic plant combinations]. Huanjing baohu yu xunhuan jingji 44(7):48–52. https://doi.org/10.3969/j.issn.1674-1021.2024.07.011

Conservation Strategies for Plant Diversity

6

Megha Sharma, Samiksha Joshi, Anroop Kaur,
Sanjay Gupta, Ganesh Pandey, and Saurabh Gangola

Abstract

Plant diversity encompasses a wide range of plant species, from genes to ecosystems, and is primarily found in hotspots and crop diversity centres. However, different anthropogenic activities severely threatened the plant biodiversity and have a negative impact on human life as it depends on the plants up to a varying extent. The loss of biodiversity on a global, national, and regional scale is a complex problem that impacts social, economic, organizational, political, scientific, and communicational communities. Hence, there is an urgent need to preserve the plant biodiversity from extinction, thereby ensuring that the present and future generations to use plant material sustainably. The growing extinction and vulnerability of various plant species in the twenty-first century have made conservation areas such as sanctuary, parks, etc., as an essential part of the global biodiversity protection plan. To guarantee comprehensive worldwide biodiversity conservation strategies, the International Union for Conservation of Nature (IUCN) emphasizes the necessity of utilizing all available resources and methods, including collaborative and indigenous forest conservation techniques. To effectively focus on *both* in situ and ex situ conservation, the completion of a

M. Sharma (✉) · S. Joshi · S. Gupta
School of Biosciences, Swami Rama Himalayan University, Jollygrant, Dehradun, Uttarakhand, India

A. Kaur
School of Agricultural Biotechnology, Punjab Agricultural University, Ludhiana, Punjab, India

G. Pandey
Faculty of Agricultural Sciences, Shree Guru Gobind Singh Tricentenary University, Chandu-Budhera, Gurugram, Haryana, India

S. Gangola
Department of Microbiology, Graphic Era (Deemed to be University), Dehradun, India

H. C. Joshi et al. (eds.), *Climate-Responsive Plant Biology*,
https://doi.org/10.1007/978-981-95-7354-7_6

global botanical inventory and an evaluation of the conservation status of 94% of plant species that have not yet been assessed are the most urgent demands. This chapter includes several in situ and ex situ strategies, as well as the growing trend of integrated conservation (which connects in situ and ex situ) activities, in an attempt to prevent the ongoing loss of plant diversity worldwide.

Keywords

Biodiversity · In situ · Ex situ · Conservation · Plant species · Endanger

6.1 Introduction

Biodiversity includes a variety of plants, animals, and microbes in different ecosystems. Their deprivation has come out as one of the most important challenges of the environment in twenty-first century. Diverse species of plants constitute an essential component to life on earth and serves multiple functions to ecosystems such as food, timber, medicines, shelter oxygen production, sequestration of CO_2 and regulation of climate change (Isbell et al. 2011). The services offered by plant biodiversity are usually ignored and not valued by people in terms of the long-term sustainability and various benefits (Cardinale et al. 2012). Deterioration of agriculture land and rapid decrease in forest area (i.e., 129 million ha to 4 billion ha) has resulted into inbreeding, isolation, competition, parasitism, predation, and disease for plants (Arshad et al. 2020). However, the loss of plant diversity in the world has been reported at alarming rate further negatively affecting human population and planets biodiversity (Primack 2020). Several factors, such as pollution, overharvesting of certain species, increasing population, invasive species, habitat loss, fragmentation, and climate change, are responsible for declining plant diversity and increased extinction leading to the decline in ecosystem functioning (Biggs et al. 2008). In addition, several plant species are also expected to be at risk of extinction due to rapid urbanization, industrialization, anthropogenic activities, over-exploitation, genetic erosion, deforestation, and thus are considered endangered plant species (Wilson 2016). Introducing non-indigenous or non-native species of plants into new habitats or environmental conditions by human activities often surpasses indigenous plants in competition and causes significant damage, leading to their extinction (Raven and Wagner 2021). Change in climate conditions like precipitation patterns, growing season, and temperature conditions aggravate such threats by modifying the optimum conditions required for plant survival (Bradshaw et al. 2009). According to Goettsch et al. (2015), strategies designed for conservation of plants has received less attention as compared to animals as plants lack the widespread appeal demonstrated by many animal groups. Therefore, conservation of plant diversity lacks resources in comparison to conservation of animal diversity (Havens et al. 2014). Plants with almost 500,000 species provide a huge diversity of essential products and services to human beings, livestock and is a basis for all terrestrial ecosystems. Moreover, many wild species of plants contribute to nutrition and food security, and are found effective in traditional medicine.

Uneven distribution of biodiversity throughout the geographical areas has resulted in the distribution of 36 hotspots of biodiversity. This involves 1500 endemic species and has lost more than 70% species of their habitat. GSPC (Global Strategy for Plant Conservation) targets to achieve 'an online flora of all known plants' (www.cbd.int/gspc/targets.shtml) and their conservation status by 2020. Data according to several reports worldwide demonstrated a total of near 500,000 species of land plants which covers 1000 species of gymnosperms (Christenhusz et al. 2011), 450,000 species of angiosperm out of which few are still not known (Pimm and Joppa 2015), 10,000 species of pteridophytes (Ranker and Sundue 2015), 1300 species of lycophytes, 9000 species of mosses, 250 species of hornworts (Villarreal et al. 2010), and 7500 species of liverworts (Von Konrat et al. 2010). Based on IUCN Red List criteria less than 20,000 plant species have been evaluated globally and those that are threatened are still not known clearly. Some studies showed that one third of total flowering plants are at risk of the extinction, whereas among 7000 plant species chosen randomly (bryophytes, ferns, gymnosperms, and angiosperms) 22% were found threatened and 30% near threatened (Pimm and Joppa 2015; Brummitt et al. 2015). On comparison with other groups like amphibians, birds, mammals, etc., plants are found to be more threatened. Tropical Rainforest is considered to be the major habitat with the highest number of threatened species. Fall in biodiversity is observed to be at faster rate than that found in the fossil record. According to the International union for conservation of nature more then 120,000 species are reported in the IUCN red list with almost 32,000 species under the category of threatened with extinction (Butchart 2008).

The rapid decline of vulnerable plant species can cause severe risk to human well-being and environmental services. Despite the fact that there is an urgent need for conservation, efforts are being overshadowed by problems like lack of public awareness, insufficient funding, and lack of political support (Balding and Williams 2016). Different approaches for the conservation of plants can be distinguished into in situ and ex situ. In situ strategies focus on maintenance and preservation of genetic variability of plants in their natural habitats or at their location (habitat restoration and protected areas), and ex situ strategies involve maintenance and preservation of genetic variability of plants away from their location or natural habitats (e.g., botanical gardens and seed banks) (Rodrigues et al. 2006).

Although both strategies have their own limitations but are considered essential for long-term survival of plants and protection of endangered species (Young et al. 2014). The severe impacts of different threats on the vegetation demands for a critical analysis of the conservation status of flora, including different rare, indicator, and endemic species. This chapter emphasizes the biodiversity along with major threats to plant species and assesses the types as well as the effectiveness of different conservation methods (Fig. 6.1).

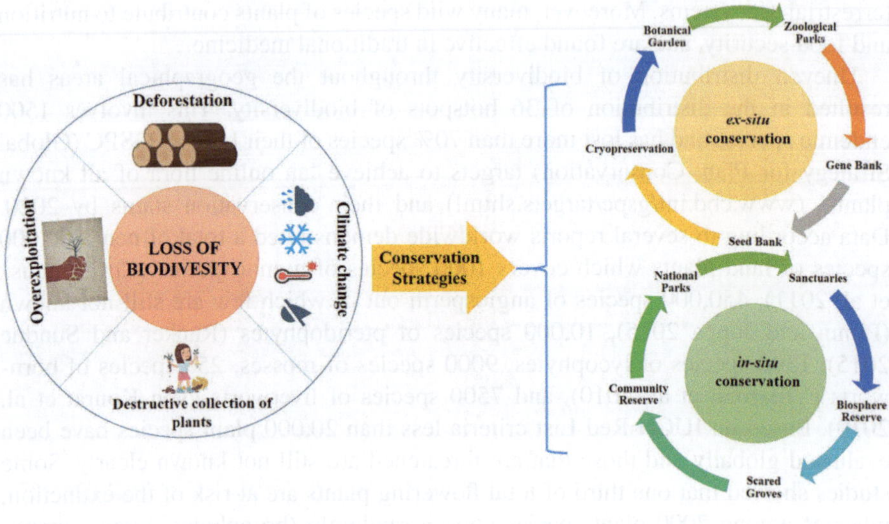

Fig. 6.1 An overview of conservation strategies

6.2 Major Threats to Plants

6.2.1 Human Population

The extensive expansion of the world's population has resulted in increased per capita demand for timber, wood fuels, food, decorations, fodder, shelter, and clothes. This explosion may lead to the conversion of nearly 1 billion hectares of natural habitat into agricultural land. Also, it has shown negative side effects on plant biodiversity and pressure on limited land resulting in increased usage of fertilisers and pesticides further leading to soil erosion (Scherra and McNeely 2008). Moreover, forest fires are found to be harmful for native plant species and provide optimum conditions for invasive species to invade. Some species considered as fire indicator and found tolerant to withstand fire include *Carrya arborea, Tectona grandis, Acacia catechu, Dalbergia latifolia, Terminalia spp, Dillenia pentagyna* and *Xylia xylocarpa* (Daniels 2001). Few other species like *Sarracenia* and *Drosera spp* are reported to disappear due to fire suppression in the southeastern United States (Jennings and Rohr 2011). Different activities of humans, such as building dams and modifications with nature has resulted into decline of tree diversity for example, *Cranther acarnalica, Pinanga dicksonii, Myristica fatua* and *Semacarpus* sp. disappeared along with swamps (Daniels 2001). Monocultures practices followed of *Eucalyptus obliqua, Anacardium excelsum, H. brasiliensis Acacia pycnantha, Cinchona officinalis, C. odorata* and *Syzygium aromaticum* have reduced natural forests throughout the globe (Daniels and Vencatesan 2008). List of some Indian threatened species on the IUCN red list has been presented in Table 6.1. Anthropogenic activities are also shown to severely impact aquatic ecosystems by modifying streams into drought conditions leading to groundwater loss and

Table 6.1 Enlist of some Indian threatened species on the IUCN red list

Common Name	Scientific Name	Stage of extinction	Current status	IUCN Red Listed Year
Agarwood	*Aquilaria malaccensis*	Critically Endangered	Decreasing	2018
Andaman Fingerroot	*Boesenbergia siphonantha*	Vulnerable	Stable	2018
Ativisha	*Aconitum heterophyllum*	Endangered	Decreasing	2014
Atlas Daisy	*Anacyclus pyrethrum*	Vulnerable	Decreasing	2014
Cinnamomum wightii	*Cinnamomum wightii*	Endangered	Decreasing	2014
Congested Diplocentrum	*Diplocentrum congestum*	Vulnerable	Decreasing	2024
Cylindrolobus lohitensis	*Cylindrolobus lohitensis*	Critically Endangered	Decreasing	2021
Diplomeris josephi	*Diplomeris josephi*	Critically Endangered	Decreasing	2021
Fan-Shaped Belly-Lip Orchid	*Gastrochilus flabelliformis*	Endangered	Decreasing	2024
Gastrodia arunachalensis	*Gastrodia arunachalensis*	Critically Endangered	Decreasing	2021
Gurmar	*Gymnema khandalense*	Endangered	Decreasing	2014
Himalayan Gentian	*Gentiana kurroo*	Critically Endangered	Decreasing	2014
Himalayan Mayapple	*Podophyllum hexandrum*	Endangered	Decreasing	2023
Himalayan soap pod tree	*Gymnocladus assamicus*	Critically Endangered	Decreasing	2014
Indian Nard	*Nardostachys jatamansi*	Critically Endangered	Decreasing	2020
Kaitha	*Pandanus palakkadensis*	Critically Endangered	Decreasing	2023
Kattukurumulagu	*Piper barberi*	Endangered	Decreasing	2014
Khasia plum	*Syzygium khasianum*	Critically Endangered	Decreasing	2023
Kurunjaval	*Syzygium densiflorum*	Vulnerable	Decreasing	1998
Kulavu	*Prioria pinnata*	Vulnerable	Stable	2022
Kuth	*Saussurea costus*	Critically Endangered	Decreasing	2014
Lata Sanchi	*Aquilaria khasiana*	Critically Endangered	Decreasing	2018
Leschenault's Valerian	*Valeriana leschenaultii*	Critically Endangered	Decreasing	2014
Mahalikizhangu	*Utleria salicifolia*	Critically Endangered	Decreasing	2014

(continued)

Table 6.1 (continued)

Common Name	Scientific Name	Stage of extinction	Current status	IUCN Red Listed Year
Malampunna	*Calophyllum polyanthum*	Near Threatened	Decreasing	2023
Merbau	*Intsia bijuga*	Near Threatened	Decreasing	2020
Neelakurinji	*Strobilanthes kunthiana*	Vulnerable	Decreasing	2024
Pand	*Pinda concanensis*	Near Threatened	Decreasing	2021
Panicled ebony/ Tembruni	*Diospyros paniculata*	Vulnerable	Decreasing	2014
Robust Bead Tree	*Ormosia robusta*	Critically Endangered	Decreasing	2023
Vatsanabha	*Aconitum chasmanthum*	Critically Endangered	Decreasing	2014
White Cedar	*Dysoxylum malabaricum*	Endangered	Decreasing	2014
Wight's Vanda	*Vanda wightii*	Vulnerable	Decreasing	2024

affecting survival of plant in a negative way (Bond et al. 2008). Modifications or alterations done to nature for residential and commercial developments by humans showed direct correlation with the threatened or extinction of species (Zhao et al. 2015).

6.2.2 Habitat Loss, Fragmentation, and Degradation

In tropical forests habitat loss and associated fragmentation are key driver of forest loss and threat to plant diversity converting them to monocultures of commercial crops and pastures (Ter Steege et al. 2015). According to Kettle and Koh (2014) few species adapted to forest can survive complete deforestation or even when a small fraction of the forest cover exists, fragmentation results into decline in plant diversity. Forest remains after fragmentation are mostly degraded by fire, fuelwood harvesting, and logging of trees (Specht et al. 2015). Similarly, habitats such as savannas, grasslands, and deserts are also threatened by fragmentation and degradation in agriculture. Habitat loss due to natural, manmade activities, changes in land use pattern is linked to fragmentation and is considered as one of the major reasons for biodiversity loss worldwide (Loewe and Hill 2010). Fragmentation in long term can cause permanent loss of forest ecosystem as well as can result into species extinction (Haddad et al. 2015). More than 50% of natural vegetation has been altered because of exploitation of pastures, croplands, plantations, and urban areas by humans (Foley et al. 2005). The need of world forest tree covers to maintain ecological stability has fallen down and resulted into elimination of endemic species (Caro and Sherman 2011). Habitat fragmentation is responsible for disturbing ecosystems via modifications in climatic conditions such as wind exposure, higher penetration of light, and increased rate of evaporation. Moreover, many studies have also reported changes in dispersal patterns and down fall in pollinators on comparison to non-fragmented habitats (Giannini et al. 2015). Therefore, it is essential to prevent habitat from fragmentation for the survival of living organisms so that they can serve a variety of ecological functions essential for a sustainable environment.

6.2.3 Overexploitation

Overexploitation is considered as one of the major threats to plants, more or less specific to particular species except for a few that can be grouped together for specific usage such as dipterocarps showing same type of wood like timber or medicinal plants with the same traits. Exploitation by people of plants species valued because of rarity or their limited range for sale or home use can bring them to point of extinction. Similarly, plant collections done for private and horticulture sectors can become a big threat to orchids, cacti, cycads, and many other ornamental plant species (Goettsch et al. 2015). Another activity affecting most of the tropical forests is logging. Logging may also lead to extinction of plants, less likely to happen if harvesting is done only of the largest individuals (Sharrock and Kiehn 2020). Overexploitation of animals may also interfere with the process of pollination and seed dispersal, further creating a threat to plant species (Harrison et al. 2013).

6.2.4 Invasive Species

Another potential threat to native plants includes invasive alien species/non-native species/non-indigenous, which needs to be addressed on an urgent basis to cease the downfall in biodiversity. These species cause irreversible destruction to biodiversity of plant leading to negative impact on entire ecosystem. A recent report showed that anthropogenic activities have resulted into most of the vascular plant flora to naturalise outside their native range (Van Kleunen et al. 2015). Such cases are seen more in Tropical than temperate regions (Corlett and Primack 2011). According to several investigations, such alien species have massive negative effects on native diversity, regional diversity, and global diversity of plants, as well as cause alterations in biogeochemical cycling and fire regimes (Pysek et al. 2012). Very few evidences are available on continental areas threatened or facing extinction of native plants because of alien species (Thomas and Palmer 2015). Gilbert and Levine (2013) suggested that the exclusion of native plant species via competition is a time-consuming process with minimum risk of extinction in comparison to more threatening invasive animals like herbivores (Chynoweth et al. 2013). Some well-known non-indigenous species includes *Eupatorium adenophorum, Caulerpa taxifolia, Galinsoga parviflora Ageratum conyzoides*, *Lantana camara*, *Mikania micrantha*, *Parthenium hysterophorus* and *Eupatorium odoratum* (Rai and Gaur 2004). Such species are causal drivers of biodiversity loss, propagates fast and are highly competent causing biological pollution as well as development of non-invasive species in different regions According to the reports by (CBD 1992) a mediterranean mussel *Mytilus galloprovincialis* in 1970s was introduced to the south coast as invasive species outcompeting local mussels (Hanekom 2008). This issue was called upon by governments to prevent the introduction of invasive species (CBD 1992). Therefore, political, administrative, and scientific coordination is necessarily required for effective management of invasive species.

6.2.5 Air Pollution and Nitrogen Deposition

Global environmental pollution is another key driver of biodiversity loss, leading to species extinction and destruction of their habitat (Lotze et al. 2006). Overpopulation, rapid urbanisation, and industrialisation are increasing pollution at a large level. Amongst different types of pollution air pollution has the most adverse effects on biodiversity, water quality, and soil, further reducing crop yield (Manisalidi et al. 2020). Plants are exposed to different concentrations of greenhouse gases and airborne pollutants with a negative impact on their diversity (Corlett 2014). Various primary and secondary pollutants causing air pollution, such as heavy metals (HMs), sulphur dioxide (SO_2), inorganic nutrients, ozone (O_3), pesticides, herbicides, and N compounds are competent enough of affecting ecological processes in air, soil and aquatic system. Nitrogen compounds deposited in dry and wet forms can alter the cycling of nutrients and can also result in soil acidification affecting diversity of plants in a negative way (Zhu et al. 2015). Indeed, there are limited studies demonstrating effect of different pollutants on the biodiversity of tropical forests. Several plant species, such as *Campanula rotundifolia, Corchorus olitorius*, *Hylocomium splendens, Calluna vulgaris, Medicago sativa, Impatiens* spp., *Lotus corniculatus, Nicotiana tabacum,* and *Phaseolus vulgaris* have been affected by high levels of ozone and nitrogenous compounds (Bhatnagar and Kumari 2013). Lichens act as biomonitors or indicators of environmental pollution and get reduced in their population with increase in concentration of nitrogen oxides and sulphur (Loppi et al. 2002). High concentration of majorly nitrogen and few other nutrients magnify algal growth and cause eutrophication leading to disbalance in the aquatic ecosystem. Pollution has also showed severe decline in population of *Cystoseira* sp. found in Mediterranean regions (Sales et al. 2011). Few reports showed the effects of inorganic nutrients and heavy metals on structural changes of species composition and littoral communities. Other consequences related to HMs contamination may include their accumulation by algae and transfer to higher trophic levels and ultimately to humans (Baumann et al. 2009).

6.2.6 Climate Change

Climate change as one of the key contributors for causing biodiversity loss is demonstrated by Intergovernmental Panel on Climate Change. Climate change can alter sea-level, precipitation, and rainfall patterns, leading to pollution, land degradation, over-exploitation and invasive species. In addition, it has also influenced other factors like fragmentation and fire in an adverse manner with rise in CO_2 levels (Corlett 2014). Issues faced to climate change like wildfires, cyclones and floods will be more frequent in the coming future further causing disturbance in the functioning and balance of ecosystem (Lindenmayer et al. 2010). Numerous biological processes involving interactions among two or more different species and entire flora as well as fauna are thought to be the major cause for species extinctions (Hangerman et al. 2010). Furthermore, climate change has impacted flowering and leaf fall timings of temperate zone plants in autumn and spring as well as have extended the range of few species towards higher altitude (Ge et al. 2015; Hijioka et al. 2014). Temperature fluctuation has also affected growth rate of plants. Several reports have demonstrated that no global extinctions of plants, but local extinctions have been reported at climatic margins of species ranges (Buse et al. 2015). Mostly, it is observed that fluctuation in climatic conditions when exceed the limit of natural variation plants usually survive by adjusting physiologically, making evolutionary changes, moving to some other suitable conditions, or die (Corlett 2015). Elevation in temperature range due to emission of GHGs via industrial and man-made activities has resulted into droughts conditions for long time, mainly in tropical and subtropical regions leading to population decline and extinction of plant species (Halofsky et al. 2020). Similarly rise in temperature of oceans affects herbivores and their foraging activities, spreading of pathogens with enhanced virulence and decreased resilience of host organisms. Wernerg and coworkers (2011), observed a massive decline of *kelps Ecklonia radiata* and *Laminaria religiosa* in New Zealand and Japan due to the outbreaks of disease (Wernberg et al. 2011). Thomas et al. (2004) suggested that in coming years climate change can become a major reason for the extinction of more than a million of plant species in terrestrial ecosystems. Some other reports demonstrated rise in sea level due to greenhouse gases causing storm, acidification, physical disturbance, salinization and coastal erosion. Such conditions have adverse impacts on marshy habitats that are considered as major areas of vulnerable and ecologically valuable biodiversity (Wernberg and Connell 2008).

6.3 Conservation Strategies for Plant Diversity: Integrating Traditional and Biotechnology Approaches

Plant diversity forms the basis of ecosystem stability, agricultural productivity, and cultural traditions. However, the latest *Global Tree Assessment* reports that nearly 38% of tree species are at risk of extinction (IUCN 2024). Similar challenges are evident in agriculture, where genetic erosion, particularly the loss of landraces and crop wild relatives, threatens resilience to pests, diseases, and climate variability. Responding to these concerns, the *Kunming–Montreal Global Biodiversity Framework (GBF)* has committed signatory nations to conserving at least 30% of terrestrial and marine areas by 2030, restoring degraded ecosystems, and safeguarding genetic diversity. National initiatives such as India's *Biological Diversity (Amendment) Act, 2023,* and the *National Biodiversity Strategy and Action Plan (2024–2030)* complement these international goals.

At COP16 (2025), countries also reached a milestone agreement on *Digital Sequence Information (DSI),* which refers to genetic data stored in digital repositories. A new international 'Cali Fund' was established to ensure fair distribution of benefits arising from the use of such data, extending the principles of the *Nagoya Protocol* beyond physical samples (such as seeds or tissues) to cover digital genetic resources as well. For example, if genetic sequences from a rare medicinal plant collected in India are used to develop pharmaceuticals, part of the benefits monetary or otherwise must return to India and the local community that conserved the species. For researchers and conservation practitioners, this means that DNA sequencing for purposes such as building pan-genomes, conducting eDNA surveys, or developing seed bank databases must now also comply with benefit-sharing requirements.

6.3.1 Traditional Conservation Frameworks

6.3.1.1 In Situ Conservation: Protecting Species in Natural Habitats

In situ conservation is the process of protecting and maintaining species within their natural habitats, allowing them to evolve and interact naturally with their ecosystems while safeguarding ecological processes, genetic diversity, and cultural values (CBD 2024). Unlike ex situ conservation methods such as gene banks or botanical gardens, it ensures that organisms remain in their adaptive environments (Table 6.2). This approach is vital as it preserves ecological interactions among pollinators, prey, and predators, maintains evolutionary processes by enabling species to adapt to environmental changes in real time, and protects entire habitats and ecosystems rather than just individual species. Additionally, it supports local communities by providing essential ecosystem services like water security, carbon sequestration, and livelihoods, while also respecting cultural and Indigenous knowledge, as many sacred landscapes simultaneously serve as biodiversity reservoirs.

6.3.1.1.1 Protected Areas (PAs)

PAs are legally recognized spaces such as national parks, wildlife sanctuaries, biosphere reserves, and nature reserves that are managed by governments or co-managed with communities to conserve biodiversity and ecosystems. For example, Corcovado National Park in Costa Rica, expanded under integrated restoration programs, combines protected areas with payment-for-ecosystem-services and reforestation initiatives, resulting in the recovery of jaguars, tapirs, and macaws along with significant forest cover increase (Forest Landscape Restoration 2022). Similarly, Kaziranga National Park in India, a UNESCO World Heritage Site, safeguards the world's largest population of the one-horned rhinoceros (*Rhinoceros unicornis*) through active patrolling, anti-poaching measures, and wetland restoration, making it a global model of species recovery.

Table 6.2 List of different botanical garden, Genebanks/Seed banks, National Parks in India

Conservation Strategies	Name of Center	Location	Foundation	No. of Taxa	Herbarium	Ex-situ Collection
Botanical Gardens	Botanical Garden	Agra, Uttar Pradesh	1930	300	No	Trees, shrubs, Ephedra, ferns, flowering plants, Leptadaema, Salvadora
	The Retreat Botanical Garden of Sarabhai Foundation	Shahibag, Ahmedabad, Gujarat	1978	160	Yes (Approx. 100 Herbarium specimens)	Palms, cactus and other succulents, roses, orchids
	Gujarat University USSC Botanic Garden	Ahmedabad, Gujarat	1964	+150	Yes (Approx. 1212 Herbarium specimens)	Trees, shrubs, ferns, aquatics
	Aligarh Muslim University Botanic Garden	Aligarh, Uttar Pradesh	1962	210	Yes (Approx. 100 Herbarium specimens)	812 trees accessions, 3882 shrubs accessions, foliage and flowering plants
	Experimental Garden	Allahabad, Uttar Pradesh	1962	500	Yes (Approx. 50,000 Herbarium specimens)	*Crotons*, *Dioscorea,* Hibiscus, roses, orchids, medicinal plants, Economic plants of India
	Botanical Garden Guru Nanak Dev University	Amritsar, Punjab	1975	5000	Yes (Approx. 4000 Herbarium specimens)	Trees & shrubs (5000 accessions) such as *Acacia* spp., *Butea monosperma Dalbergia sisso, Tectona grandis* and *Terminalia bellerica*, bamboos, cacti, lilies and other succulents (+100 spp.), medicinal plants (*Emblica officinalis & Aegle marmelos*)
	Lalbagh Botanical Garden	Bangalore	1856	1800	Yes (Approx. 3000 Herbarium specimens)	Cassia, Dioscorea, Indian flora, Ornamental flowering trees, Orchids,, Shrubs, Climbers, Medicinal plants, annuals
	Botanical Garden Gandhi Krishi Vignana Kendra	Bangalore	1973	1600	Yes (Approx. 3000 Herbarium specimens)	Medicinal plants (+600 species), weeds, Native and exotic trees and shrubs
	Regional Plant Resources Centre Bhubaneswar	Bhubaneswar	1985	1200	Yes (Approx. 5000 Herbarium specimens)	Bamboos, Dioscorea, Orchids
	Indian Botanic Garden	Calcutta	1787	5000	Yes (Approx. 1500,000 Herbarium specimens)	Bambuseae (26 spp.), Bougainvillea (148 cvs.), Jasmines (25 spp.), water lilies (30 cvs of 4 spp.), Palmae (109 spp.), Citrus, succulents (100 spp.), Orchidaceae (80 spp.), Economic plants, medicinal plants (450 spp.)
	Eden Gardens	Calcutta	1836	103	No	Climbers, shrubs, ferns, trees, orchids, shade plant
	National Botanical Research Institute	Lucknow	1953	6000	Yes (Approx. 90,000 Herbarium specimens)	*Bentinckia nicobarica, Cycas beddomei, C. pectinata, Commiphora wightii, Erythrina resupinata, Frerea indica, Hoya pendula, H. wightii, Phoenix rubicola, Rauvolfia serpentina, Tecomella undullata Sophora mollis* and *Vanilla planifolia*

(continued)

Table 6.2 (continued)

Conservation Strategies	Name of Center	Location	Foundation	No. of Taxa	Herbarium	Ex-situ Collection
Genebanks/ Seed banks	ICAR-National Bureau of Plant Genetic Resources (ICAR-NBPGR)	Delhi	1996	More than 511,825 accessions	–	Millets, Legumes, Forages, Cereals, Pseudocereals, Vegetables, Oilseeds, Fruits & Nuts etc.
	International Crops Research Institute for the Semi-Arid Tropics (ICRISAT)	Patancheru, Hyderabad, Telangana	1972	132,087 germplasm accessions	–	Sorghum, chickpea, groundnut, barnyard millet, finger millet, foxtail millet, little millet, kodo millet, proso millet and pearl millet
	CSIR-Central Institute for Medicinal and Aromatic Plants (CIMAP)	Lucknow	–	3334 Pan-Indian accessions	Yes	Medicinal and aromatic plants (MAP)
	Tropical Botanical Garden and Research Institute (TBGRI)	Thiruvan anthapuram	1979	50,000 accessions	–	Medicinal and aromatic plants (MAP)
	Central Rice Research Institute (CRRI)	Cuttack, Odisha	1946	109,000 accessions	–	Rice germplasm
	Central Institute for Cotton Research (CICR)	Nagpur, Maharashtra	1976	12,336 accessions	–	Cotton germplasm
National Parks	Great Himalayan National Park	Himachal Pradesh	1984	805 Vascular plant species	–	25 distinct forest types and associated with rich diversity of fauna species
	Nanda Devi National Park	Uttarakhand	1982	Over 1000 plant species	–	Alpine vegetation consist of scrub communities (*Rhododendron campanulatum, R. anthopogon* and *Salix denticulate*) as well as herbaceous species

	Silent Valley National Park	Kerala	1984	1000 species of flowering plants, 107 species of orchids, 100 ferns and fern allies		West-coast and Southern tropical and subtropical evergreen and broad leaved hill forest, Southern high altitude wet temperate forest-grassland ecosystem
	Desert National Park	Rajasthan	1980	168 plant species belonging to 48 families	–	Includes *Acacia* spp., *Ammannie desertorum, Commiphora wightii, Dipcadi erythraem, Ephedra foliata, Glossonema varians, Helitropium rariflorum, Limeum indicum, Moringa concanensis, Tecomella undulata brachystachyus, Tephrosia falciformis, Tribulus rajasthanensis Rhynchosia schimpari, Seddera latifolia, Sesuvium sesuvioides* and *Ziziphus truncate* etc.
	Sundarbans National Park	West Bengal	1984	78 species of mangroves,		Highly diverse fauna species (including threatened with extinction)

6.3.1.1.2 Key Biodiversity Areas (KBAs)

KBAs are sites that contribute significantly to the global persistence of biodiversity, identified using scientific criteria such as endemism, presence of threatened species, and ecological integrity. For instance, the Western Ghats in India, designated as a KBA hotspot, host more than 325 globally threatened species, with conservation efforts focused on endemic amphibians, freshwater fishes, and plants while addressing agricultural pressures. Similarly, BirdLife KBAs in Africa, including Lake Tana in Ethiopia and key sites in Kenya, support endemic and migratory bird species, where community-based water management has sustained habitats while also providing local livelihoods.

6.3.1.1.3 Other Effective Area-Based Conservation Measures (OECMs)

OECMs are areas not officially designated as protected but still delivering significant conservation outcomes, such as community forests, restored lands, private reserves, or biodiversity parks. A notable example is the Aravalli Biodiversity Park in India, a 392-acre former mining site restored with native species through citizen-led initiatives and recognized in 2022 as the country's first urban OECM; today, it supports over 200 native plants, pollinators, and mammals while functioning as a carbon sink. Likewise, Indonesia's Marine OECMs, comprising over 390 sites managed under customary and community systems, safeguard fisheries, mangroves, and coral reefs while simultaneously sustaining local livelihoods.

6.3.1.1.4 Community Conserved Areas (CCAs) and Sacred Groves

CCAs and Sacred Grooves are regions protected by local communities through traditional laws, spiritual beliefs, or cultural norms, ensuring the preservation of biodiversity alongside cultural heritage. The Mawphlang Sacred Grove in Meghalaya, India, conserved under Khasi customary law, safeguards orchids, rare medicinal plants, and endemic tree species, with strict prohibitions on cutting or hunting (Environment & Society Portal 2020). In Nagaland, over 400 CCAs covering approximately 1700 sq km are managed by communities that enforce bans on hunting and logging, thereby conserving hornbills, orchids, and endemic mammals while sustaining customary practices. Similarly, in Uttarakhand, the Forest Department documented 161 sacred natural sites in 2025, including alpine meadows and high-altitude lakes, which protect endangered species and integrate traditional conservation practices into state-level strategies.

6.3.1.1.5 Habitat Restoration in Natural Settings

These settings involve reviving degraded ecosystems within the species' natural range by using native species and engaging local communities. An example is Parashurama Vana in Karnataka, India, established in 2025 near the Dharmasthala temple, where culturally significant and native trees were planted on degraded land to restore biodiversity while reinforcing cultural traditions. Similarly, Costa Rica's Forest Landscape Restoration program has implemented large-scale reforestation with native species, supported by payment-for-ecosystem-services schemes, which has improved ecological connectivity and facilitated the return of key fauna such as peccaries and jaguars (Forest Landscape Restoration 2022).

In situ conservation is the backbone of biodiversity protection, ensuring species survive within their ecosystems while maintaining natural processes. Protected areas safeguard large habitats, KBAs highlight global priorities, OECMs and CCAs integrate human–nature relationships, and restoration projects revive degraded ecosystems. Successful examples from Costa Rica, India, and Indonesia show that a combination of legal protection, community stewardship, and cultural traditions is essential. With global goals such as the *30 × 30 Target* under the Kunming–Montreal Global Biodiversity Framework, in situ conservation remains central to halting biodiversity loss and achieving climate resilience.

6.3.1.2 Ex Situ Conservation: Safeguarding Diversity Beyond Natural Habitats

Ex situ conservation refers to the preservation of plant genetic material outside its natural habitat, acting as a safety net against extinction risks by maintaining living collections or genetic resources under controlled conditions where they can be safeguarded, studied, and made available for future use. Its functions are multifaceted: it serves as a genetic backup to ensure the long-term survival of species outside their natural ecosystems; provides material for research and breeding programs supporting crop improvement, biotechnology, and ecological studies; supplies resources for restoration and reintroduction of species into degraded habitats; and strengthens food and livelihood security by conserving landraces and crop wild relatives that are crucial for agriculture and local adaptation.

6.3.1.2.1 Seed Banks

Seed banks are the most cost-effective and widely used method for conserving orthodox seeds that can tolerate drying and freezing. The Svalbard Global Seed Vault functions as a backup repository for national and international genebanks, ensuring global crop security. Similarly, the Millennium Seed Bank at Kew partners with organizations worldwide to conserve and study wild plant seeds, while also supplying material for ecological restoration projects.

6.3.1.2.2 Field Genebanks

For crops propagated vegetatively (e.g., banana, cassava, yam) and species with recalcitrant seeds (which cannot be dried or frozen), field genebanks maintain live collections under cultivation. These serve as readily accessible genetic resources for farmers and researchers.

6.3.1.2.3 In Vitro Tissue Collections

Plant tissues are maintained in controlled laboratory conditions on nutrient media. This technique is particularly useful for clonal crops and short- to medium-term storage, ensuring the conservation of genetic material without requiring field space (Nagel et al. 2024).

6.3.1.2.4 Cryopreservation

Advances in cryobiotechnology have enabled long-term conservation of plant tissues in liquid nitrogen at −196 °C. This technique has expanded options for conserving species with recalcitrant seeds, orchids, tree species, and crop wild relatives (Ballesteros et al. 2021; Nagel et al. 2024).

6.3.1.2.5 Community Seedbanks

At the grassroots level, community seed banks play a vital role in preserving landraces and ensuring farmer access to local varieties. In Nepal, such banks have strengthened food security and built resilience against natural disasters by providing locally adapted planting material.

Ex situ conservation plays a critical role in safeguarding biodiversity beyond natural habitats by providing a secure backup system and a resource base for future use. Through diverse techniques ranging from global-scale seed banks like the Svalbard Global Seed Vault and the Millennium Seed Bank, to field genebanks, in vitro collections, and advanced cryopreservation methods, it ensures the survival of species that cannot always be conserved in situ. Equally important are community seed banks, which not only preserve local landraces but also empower farming communities, strengthen food security, and build resilience against natural disasters. Together, these approaches create a complementary system that both protects genetic resources from extinction risks and makes them available for research, restoration, and sustainable development, highlighting ex situ conservation as an indispensable pillar of global biodiversity conservation (Ballesteros et al. 2021).

6.4 Biotechnology in Plant Conservation

Biotechnological tools expand the scope of conservation beyond traditional practices. Genomic research, including pangenome analyses, has identified structural variants and rare alleles that inform breeding and genetic rescue. For instance, studies in alfalfa revealed adaptive variants overlooked by single-reference genomes, offering new avenues for crop conservation (Kaur et al. 2024a, b). Environmental DNA (eDNA) metabarcoding has revolutionized biodiversity monitoring. A national airborne eDNA survey in France proved its effectiveness in detecting diverse plant taxa across landscapes, supporting large-scale GBF-aligned assessments (Tournayre et al. 2025). Portable nanopore sequencing devices now facilitate field-based DNA barcoding, phytosanitary checks, and community-led biodiversity monitoring (Jiménez et al. 2025). Cryobiotechnology complements these advances by providing reliable long-term conservation of species with complex seed biology. Standardized cryopreservation protocols for embryos, shoot tips, and pollen are now being integrated into major seed banks worldwide (Nagel et al. 2024). In parallel, in vitro propagation techniques such as shoot culture, somatic embryogenesis, and temporary immersion systems provide rapid clonal multiplication of endangered species and support ex situ conservation. Recent protocols developed for the endangered orchid *Coelogyne mossiae* in the Western Ghats of India, for two rare

Dianthus species in Europe, and for *Peumus boldus* using temporary immersion culture in Chile highlight the role of tissue culture in restoring wild populations while reducing harvesting pressure. Indian studies on *Withania somnifera* (Ashwagandha) have also demonstrated successful micropropagation from nodal explants with high survival and genetic fidelity, showing how in vitro approaches can reduce pressure on overharvested medicinal plants. Emerging research also emphasizes the importance of seed microbiota and mycorrhizal fungi for germination and restoration, suggesting that microbiome-aware approaches should be incorporated into conservation strategies (Lethielleux-Juge et al. 2025). Case studies demonstrate biotechnology's impact: pangenomes aiding stress-tolerance breeding (Kaur et al. 2024a, b), genomic tools accelerating American chestnut restoration (Sandercock et al. 2024; Speed Breeding Consortium 2025), cryopreservation expanding coverage for recalcitrant species (Nagel et al. 2024), and in vitro propagation ensuring large-scale production and reintroduction of threatened taxa.

6.5 Toward Integrated Conservation of Plant Diversity

Future strategies for conserving plant diversity require a holistic integration of in situ, ex situ, and biotechnology-based approaches. Protected areas serve as ecological baselines, while seed banks and cryo-collections ensure long-term genetic redundancy (CBD 2024). Seed banks act as 'genetic libraries', preserving seeds of crops and their wild relatives under controlled conditions to safeguard them from environmental catastrophes, pests, or diseases. Cryo-collections complement these efforts by storing seeds, pollen, embryos, or tissue cultures at ultra-low temperatures, thereby maintaining viability for centuries, including for recalcitrant species that cannot be dried and stored conventionally (Nagel et al. 2024). A striking example of this system in action is the Svalbard Global Seed Vault in Norway, which currently safeguards more than 1.2 million seed samples. Its value was demonstrated during the Syrian civil war, when researchers from the International Center for Agricultural Research in the Dry Areas (ICARDA) were able to restore lost collections using seeds withdrawn from Svalbard. Similarly, the Millennium Seed Bank (MSB) at Kew has conserved seeds of more than 40,000 species, including rare and endemic plants. At the national level, India's National Bureau of Plant Genetic Resources (NBPGR) maintains over 450,000 accessions and has developed cryo-preservation facilities for long-term conservation of recalcitrant seeds and tissue cultures, thereby securing the genetic diversity of crops such as rice, wheat, and pulses. At community scales, seed banks in regions such as Nepal and Ethiopia empower farmers to conserve and exchange locally adapted varieties, thus strengthening resilience against climate variability.

While ex situ approaches secure the genetic material of plants, the conservation of species in their natural habitats increasingly relies on emerging technologies such as remote sensing and artificial intelligence (AI). Remote sensing enables large-scale, repeatable, and non-invasive monitoring of vegetation, using satellites, drones, and airborne sensors to capture patterns of habitat fragmentation, phenological changes, and stress responses to climate change. In Madagascar, high-resolution satellite data from Sentinel-2 and Landsat, complemented by drone surveys, were used to map forest patches that harbor endemic orchids. These remotely sensed data not only revealed orchid-rich microhabitats such as moist montane forests but also tracked deforestation over two decades, providing an urgent basis for targeted conservation. In India, similar methods have been used in the Himalaya and Western Ghats biodiversity hotspots. Chitale et al. (2019) integrated vegetation indices derived from satellite data with climate and topographic variables to predict plant species richness through machine learning, identifying biodiversity-rich regions that are most at risk from deforestation and climate change.

Artificial intelligence complements these efforts by processing the vast datasets generated by remote sensing and herbarium digitization. Through machine learning, deep learning, and computer vision, AI can classify plant species, extract morphological traits, predict suitable habitats, and detect ecological threats such as invasive species or illegal harvesting. In Madagascar, AI models trained on herbarium specimens and field photographs enabled automatic identification of orchid species and generated predictive habitat suitability maps that guided the discovery of new populations. Computer vision further extracted floral traits from digitized herbarium sheets, aiding taxonomic revision and prioritization for conservation (Cornwell et al. 2023). In India, AI has been applied to link alpine vegetation changes in the western Himalaya to climate shifts, providing early warning signals of grassland degradation (Pandey et al. 2021). Similar approaches have also been used in East Africa, where AI models such as Random Forests classified areas invaded by *Prosopis juliflora* with more than 90% accuracy, enabling targeted eradication and habitat restoration (Paliwal et al. 2024).

Taken together, these approaches demonstrate that conservation cannot rely on a single strategy. Ex situ collections such as seed banks and cryo-preservation ensure that plant genetic resources are preserved for future generations, while remote sensing and AI enable large-scale, dynamic monitoring and predictive modeling in natural habitats. The integration of these approaches linking global facilities such as the Svalbard Vault with national repositories like India's NBPGR and harnessing cutting-edge technologies for habitat monitoring offers a resilient and adaptive framework for safeguarding plant diversity in the face of climate change, habitat loss, and socio-political instability.

6.6 Conclusions and Future Prospects

The worldwide growing population demands for rapid urbanisation and development, leading to widespread deforestation, loss of natural habitats and increased pollution which together contributes for significant decline in plant diversity. Subsequently, the overexploitation and destructive collection of important plants are also major threats in biodiversity loss. Different strategies, such as, in situ or ex situ, have been adopted to conserve the plant diversity, thereby preventing the extinction of various important species. Conserving plant diversity in the twenty-first century demands an integrated framework that bridges in situ protection, ex situ repositories, and biotechnology. Protected areas and Key Biodiversity Areas (KBAs) provide ecological baselines, while seed banks and cryo-collections ensure long-term redundancy. Linking global initiatives such as the *Millennium Seed Bank* and *Svalbard Global Seed Vault* with national and community-level repositories creates resilience against biodiversity loss. Furthermore, different biotechnology approaches pave the way towards precise conservation strategies such as, pangenomics identifies adaptive alleles, eDNA enables real-time monitoring, nanopore sequencing allows local diagnostics, and microbiome research supports ecosystem restoration. In addition, recent technologies such as, remote sensing and artificial intelligence also have been applied to herbarium specimens allow large-scale mapping of plant distributions and traits. Embedding these scientific approaches within equitable policy frameworks such as the GBF, ABS rules, extension of the Nagoya Protocol and DSI regulations to ensure the effective conservation, thereby preventing biodiversity loss. Besides this, sustainable development instead of unregulated development to fulfil the necessity of expanding population needs to be addressed in order to preserve and protect the diversity in their natural habitat.

References

Arshad Z, Robaina M, Shahbaz M, Veloso AB (2020) The effects of deforestation and urbanization on sustainable growth in Asian countries. Environ Sci Pollut Res 27(9):10065–10086

Balding M, Williams KJ (2016) Plant blindness and the implications for plant conservation. Conserv Biol 30(6):1192–1199

Ballesteros D, Pence VC, Walters C (2021) Cryopreservation of plant germplasm: recent advances and applications. Plant Cell Rep 40(5):879–894

Baumann HA, Morrison L, Stengel DB (2009) Metal accumulation and toxicity measured by PAM—chlorophyll fluorescence in seven species of marine macroalgae. Ecotoxicol Environ Saf 72(4):1063–1075

Bhatnagar S, Kumari R (2013) Bioremediation: a sustainable tool for environmental management–a review. Ann Rev Res Biol 3(4):974–993

Biggs R, Simons H, Bakkenes M, Scholes RJ, Eickhout B, van Vuuren D, Alkemade R (2008) Scenarios of biodiversity loss in southern Africa in the 21st century. Glob Environ Chang 18(2):296–309

Bond NR, Lake PS, Arthington AH (2008) The impacts of drought on freshwater ecosystems: an Australian perspective. Hydrobiologia 600(1):3–16

Bradshaw CJ, Sodhi NS, Brook BW (2009) Tropical turmoil: a biodiversity tragedy in progress. Front Ecol Environ 7(2):79–87

Brummitt NA, Bachman SP, Griffiths-Lee J, Lutz M, Moat JF, Farjon A, Donaldson JS, Hilton-Taylor C, Meagher TR, Albuquerque S, Aletrari E (2015) Green plants in the red: a baseline global assessment for the IUCN sampled Red List Index for plants. PLoS One 10(8):e0135152

Buse J, Boch S, Hilgers J, Griebeler EM (2015) Conservation of threatened habitat types under future climate change–lessons from plant-distribution models and current extinction trends in southern Germany. J Nat Conserv 27:18–25

Butchart SH (2008) Red list indices to measure the sustainability of species use and impacts of invasive alien species. Bird Conserv Int 18(S1):S245–S262

Cardinale BJ, Duffy JE, Gonzalez A, Hooper DU, Perrings C, Venail P, Narwani A, Mace GM, Tilman D, Wardle DA, Kinzig AP (2012) Biodiversity loss and its impact on humanity. Nature 486(7401):59–67

Caro T, Sherman PW (2011) Endangered species and a threatened discipline: behavioural ecology. Trends Ecol Evol 26(3):111–118

CBD (Convention of Biological Diversity) (1992) United Nations Environment Programme Nairobi Kenya

Chitale VS, Behera MD, Roy PS (2019) Deciphering plant richness using satellite remote sensing: a study from three biodiversity hotspots. Biodivers Conserv 28:2183–2196

Christenhusz MJ, Reveal JL, Farjon A, Gardner MF, Mill RR, Chase MW (2011) A new classification and linear sequence of extant gymnosperms. Phytotaxa 19:55–70

Chynoweth MW, Litton CM, Lepczyk CA, Hess SC, Cordell S (2013) Biology and impacts of Pacific Island invasive species. 9. Capra hircus, the feral goat (*Mammalia: Bovidae*). Pac Sci 67(2):141–156

Convention on Biological Diversity (CBD) (2024) In situ and ex situ conservation frameworks. Retrieved from https://www.cbd.int/

Corlett RT (2014) Forest fragmentation and climate change. In: Global forest fragmentation. CABI, Wallingford, pp 69–78

Corlett RT (2015) Plant movements in response to rapid climate change. In: Routledge handbook of forest ecology. Routledge, pp 517–526

Corlett RT, Primack RB (2011) Tropical rain forests: an ecological and biogeographical comparison. Wiley

Cornwell WK, Bragg JG, Wilde B et al (2023) Studying plant–climate relationships using machine learning: leaf size variation via herbarium specimens. Am J Bot

Daniels RJR (2001) A report on the national biodiversity strategy and action plan-the Western Ghats ecoregion. Rep. to the Ministry of Environment and Forest, India

Daniels RR, Vencatesan J (2008) Western Ghats: biodiversity, people, conservation. Rupa & Company

Environment & Society Portal (2020) Sacred groves and biodiversity conservation in India. Rachel Carson Center for Environment and Society

Foley JA, DeFries R, Asner GP, Barford C, Bonan G, Carpenter SR, Chapin FS, Coe MT, Daily GC, Gibbs HK, Helkowski JH (2005) Global consequences of land use. Sci 309(5734):570–574

Forest Landscape Restoration (2022) Costa Rica's restoration success stories. International Union for Conservation of Nature (IUCN)

Ge Q, Wang H, Rutishauser T, Dai J (2015) Phenological response to climate change in China: a meta-analysis. Glob Chang Biol 21(1):265–274

Giannini TC, Boff S, Cordeiro GD, Cartolano EA Jr, Veiga AK, Imperatriz-Fonseca VL, Saraiva AM (2015) Crop pollinators in Brazil: a review of reported interactions. Apidologie 46(2):209–223

Gilbert B, Levine JM (2013) Plant invasions and extinction debts. Proc Natl Acad Sci 110(5):1744–1749

Goettsch B, Hilton-Taylor C, Cruz-Piñón G, Duffy JP, Frances A, Hernández HM, Inger R, Pollock C, Schipper J, Superina M, Taylor NP (2015) High proportion of cactus species threatened with extinction. Nat Plants 1(10):1–7

Haddad NM, Brudvig LA, Clobert J, Davies KF, Gonzalez A, Holt RD, Lovejoy TE, Sexton JO, Austin MP, Collins CD, Cook WM (2015) Habitat fragmentation and its lasting impact on earth's ecosystems. Sci Adv 1(2):e1500052

Hagerman S, Dowlatabadi H, Satterfield T, McDaniels T (2010) Expert views on biodiversity conservation in an era of climate change. Glob Environ Chang 20(1):192–207

Halofsky JE, Peterson DL, Harvey BJ (2020) Changing wildfire, changing forests: the effects of climate change on fire regimes and vegetation in the Pacific Northwest, USA. Fire Ecol 16(1):1–26

Hanekom N (2008) Invasion of an indigenous Perna perna mussel bed on the south coast of South Africa by an alien mussel Mytilus galloprovincialis and its effect on the associated fauna. Biol Invasions 10(2):233–244

Harrison RD, Tan S, Plotkin JB, Slik F, Detto M, Brenes T, Itoh A, Davies SJ (2013) Consequences of defaunation for a tropical tree community. Ecol Lett 16(5):687–694

Havens K, Kramer AT, Guerrant EO Jr (2014) Getting plant conservation right (or not): the case of the United States. Int J Plant Sci 175(1):3–10

Hijioka Y, Lin E, Pereira JJ, Corlett RT, Cui X, Insarov G, Lasco R, Lindgren E, Surjan A, Aizen EM, Aizen VB (2014) IPCC fifth assessment report: chapter 24 Asia. In: Climate change 2014-impacts, adaptation and vulnerability: Part B: Regional Aspects Working Group II Contribution to the IPCC Fifth Assessment Report, pp 1327–1370

Isbell F, Calcagno V, Hector A, Connolly J, Harpole WS, Reich PB, Scherer-Lorenzen M, Schmid B, Tilman D, Van Ruijven J, Weigelt A (2011) High plant diversity is needed to maintain ecosystem services. Nature 477(7363):199–202

IUCN (2024) Global tree assessment: state of the world's trees. International Union for Conservation of Nature

Jennings DE, Rohr JR (2011) A review of the conservation threats to carnivorous plants. Biol Conserv 144(5):1356–1363

Jiménez A, González R, Torres L (2025) Portable nanopore sequencing for community-led biodiversity monitoring. Mol Ecol Resour 25(3):451–463

Kaur H, Shannon LM, Samac DA (2024a) A stepwise guide for pangenome development in crop plants: an alfalfa (Medicago sativa) case study. BMC Genomics 25(1):1022

Kaur P, Sharma R, Gill K (2024b) Pangenome analyses reveal adaptive variants in alfalfa overlooked by reference genomes. Plant Biotechnol J 22(5):890–905

Kettle CJ, Koh LP (2014) Global forest fragmentation: introduction. In: Global forest fragmentation. CABI, Wallingford, pp 1–4

Lethielleux-Juge M, Moreau A, Fontaine C (2025) Microbiome-aware strategies for plant germination and restoration. New Phytol 235(4):1567–1579

Lindenmayer DB, Steffen W, Burbidge AA, Hughes L, Kitching RL, Musgrave W, Smith MS, Werner PA (2010) Conservation strategies in response to rapid climate change: Australia as a case study. Biol Conserv 143(7):1587–1593

Loewe L, Hill WG (2010) The population genetics of mutations: good, bad and indifferent. Philos Trans R Soc B: Biol Sci 365(1544):1153–1167

Loppi S, Ivanov D, Boccardi R (2002) Biodiversity of epiphytic lichens and air pollution in the town of Siena (Central Italy). Environ Pollut 116(1):123–128

Lotze HK, Lenihan HS, Bourque BJ, Bradbury RH, Cooke RG, Kay MC, Kidwell SM, Kirby MX, Peterson CH, Jackson JB (2006) Depletion, degradation, and recovery potential of estuaries and coastal seas. Science 312(5781):1806–1809

Manisalidis I, Stavropoulou E, Stavropoulos A, Bezirtzoglou E (2020) Environmental and health impacts of air pollution: a review. Front Public Health 8:14

Nagel M, Ellis D, Dulloo M (2024) Cryopreservation protocols for global seed banks: expanding coverage for recalcitrant species. Plant Genetic Resour 22(2):101–115

Paliwal A, Mhelezi M, Galgallo D, Banerjee R, Malicha W, Whitbread A (2024) Utilizing artificial intelligence and remote sensing to detect *Prosopis juliflora* invasion: environmental drivers and community insights in rangelands of Kenya. Plants 13(13):1868

Pandey A, Singh G, Palni S, Chandra N, Rawat JS, Singh AP (2021) Application of remote sensing in alpine grasslands cover mapping of western Himalaya, Uttarakhand, India. Environ Monit Assess 193(4):166

Pimm SL, Joppa LN (2015) How many plant species are there, where are they, and at what rate are they going extinct? Ann Mo Bot Gard 100(3):170–176

Primack RB (2020) Essentials of conservation biology, 7th edn. Sinauer Associates

Pyšek P, Jarošík V, Hulme PE, Pergl J, Hejda M, Schaffner U, Vilà M (2012) A global assessment of invasive plant impacts on resident species, communities and ecosystems: the interaction of impact measures, invading species' traits and environment. Glob Chang Biol 18(5):1725–1737
Rai LC, Gaur JP (2004) Invasive alien species and biodiversity in India. In: Proceedings of the workshop held at the Department of Botany, Banaras Hindu University, Varanasi, during August, pp 18–20
Ranker TA, Sundue MA (2015) Why are there so few species of ferns? Trends Plant Sci 20(7):402–403
Raven PH, Wagner DL (2021) Agricultural intensification and climate change are rapidly decreasing insect biodiversity. Proc Natl Acad Sci U S A 118(2):e2002548117
Rodrigues AS, Pilgrim JD, Lamoreux JF, Hoffmann M, Brooks TM (2006) The value of the IUCN red list for conservation. Trends Ecol Evol 21(2):71–76
Sales M, Cebrian E, Tomas F, Ballesteros E (2011) Pollution impacts and recovery potential in three species of the genus Cystoseira (Fucales, Heterokontophyta). Estuar Coast Shelf Sci 92(3):347–357
Sandercock AM, Westbrook JW, Zhang Q, Holliday JA (2024) A genome-guided strategy for climate resilience in American chestnut restoration populations. Proc Natl Acad Sci U S A 121(30):e2403505121
Scherr SJ, McNeely JA (2008) Biodiversity conservation and agricultural sustainability: towards a new paradigm of 'ecoagriculture' landscapes. Philos Trans R Soc B: Biol Sci 363(1491):477–494
Sharrock S, Kiehn M (2020) Plant conservation report 2020: a review of progress in implementation of the global strategy for plant conservation 2011–2020: a contribution to the fifth edition of the global biodiversity outlook (GBO-5). CBD Technical Series
Specht MJ, Pinto SRR, Albuquerque UP, Tabarelli M, Melo FP (2015) Burning biodiversity: fuelwood harvesting causes forest degradation in human-dominated tropical landscapes. Glob Ecol Conserv 3:200–209
Speed Breeding Consortium (2025) Accelerating crop adaptation through genomic selection and speed breeding. Nat Plants 11(1):56–65
Ter Steege H, Pitman NC, Killeen TJ, Laurance WF, Peres CA, Guevara JE, Salomão RP, Castilho CV, Amaral IL, de Almeida Matos FD, de Souza Coelho L (2015) Estimating the global conservation status of more than 15,000 Amazonian tree species. Sci Adv 1(10):e1500936
Thomas CD, Palmer G (2015) Non-native plants add to the British flora without negative consequences for native diversity. Proc Natl Acad Sci 112(14):4387–4392
Thomas CD, Cameron A, Green RE, Bakkenes M, Beaumont LJ, Collingham YC, Erasmus BF, De Siqueira MF, Grainger A, Hannah L, Hughes L (2004) Extinction risk from climate change. Nature 427(6970):145–148
Tournayre O, Littlefair JE, Garrett NR, Allerton JJ, Brown AS, Cristescu ME, Clare EL (2025) First national survey of terrestrial biodiversity using airborne eDNA. Sci Rep 15(1):19247
Van Kleunen M, Dawson W, Essl F, Pergl J, Winter M, Weber E, Kreft H, Weigelt P, Kartesz J, Nishino M, Antonova LA (2015) Global exchange and accumulation of non-native plants. Nature 525(7567):100–103
Villarreal JC, Cargill DC, Hagborg A, Soderstrom L, Renzaglia KS (2010) A synthesis of hornwort diversity: patterns, causes and future work. Phytotaxa 9:150–166
Von Konrat M, Soderstrom L, Renner MA, Hagborg A, Briscoe L, Engel JJ (2010) Early land plants today (ELPT): how many liverwort species are there? Phytotaxa 9(1):22–40
Wernberg T, Connell SD (2008) Physical disturbance and subtidal habitat structure on open rocky coasts: effects of wave exposure, extent and intensity. J Sea Res 59(4):237–248
Wernberg T, Russell BD, Moore PJ, Ling SD, Smale DA, Campbell A, Coleman MA, Steinberg PD, Kendrick GA, Connell SD (2011) Impacts of climate change in a global hotspot for temperate marine biodiversity and ocean warming. J Exp Mar Biol Ecol 400(1–2):7–16
Wilson EO (2016) Half-earth: our planet's fight for life. WW Norton & Company

Young RP, Hudson MA, Terry AMR, Jones CG, Lewis RE, Tatayah V, Zuël N, Butchart SHM (2014) Accounting for conservation: using the IUCN red list index to evaluate the impact of a conservation organization. Biol Conserv 180:84–96

Zhao J, Zhang C, Deng L, Ren Y, Yan J, Luo Y, Zuo S, Zhang K, Wang H (2015) Impact of human activities on plant species composition and vegetation coverage in the wetlands of Napahai, Shangri-La County, Yunnan Province, China. Int J Sustain Dev World Ecol 22(2):127–134

Zhu J, He N, Wang Q, Yuan G, Wen D, Yu G, Jia Y (2015) The composition, spatial patterns, and influencing factors of atmospheric wet nitrogen deposition in Chinese terrestrial ecosystems. Sci Total Environ 511:777–785

Agriculture, Sustainability, and Climate Change: Interconnected Futures

7

Aashu Rajput, Pardeep Kumar, Gaurav Ayodhya Singh, and Shambhavi Tiwari

Abstract

One of the biggest threats to agriculture is climate change, which has an impact on rural livelihoods, food security, and the sustainability of natural resources worldwide. Traditional farming methods are being disrupted by rising temperatures, changed precipitation patterns, and an increase in the frequency of extreme weather events. The idea and application of sustainable agriculture as a calculated reaction to climate change adaptation and mitigation are examined in this chapter. In order to create farming systems that are resilient, productive, and environmentally sound, sustainable agriculture combines ecological, economic, and social concepts. Important strategies are covered in detail, including crop diversification, integrated nutrient and pest management, agroforestry, organic farming, conservation agriculture, and climate-smart agriculture. These methods improve soil health and resource efficiency while also lowering greenhouse gas emissions and increasing carbon sequestration. The chapter also highlights how the shift to climate-resilient farming systems can be aided by farmer-led innovations, indigenous knowledge, and supporting governmental frameworks. Case studies are provided to illustrate real-world uses and difficulties in climate-vulnerable areas, especially in the Global South. The chapter concludes by dis-

A. Rajput (✉)
School of Agriculture, Uttaranchal University, Dehradun, Uttarakhand, India

P. Kumar
Department of Agriculture, Motherhood University, Roorkee, Uttarakhand, India

G. A. Singh
EF Polymer Pvt. Ltd, Udaipur, Rajasthan, India

S. Tiwari
Department of Plant Pathology, SVPUAT, Meerut, Uttar Pradesh, India

Department of Agriculture, Tulas Institute, Dehradun, India

H. C. Joshi et al. (eds.), *Climate-Responsive Plant Biology*,
https://doi.org/10.1007/978-981-95-7354-7_7

cussing the necessity of interdisciplinary research, institutional assistance, and capacity building to mainstream sustainable agriculture into national and international climate strategies. Sustainable agriculture is a key strategy for attaining food security in the face of changing climate circumstances by coordinating agricultural development with environmental sustainability and climate resilience.

Keywords

Sustainable agriculture · Climate change · Soil health · Climate-smart agriculture · Carbon sequestration · Agroecology

7.1 Introduction

The foundation of human society is still agriculture, which provides a variety of ecological services that support life, including food, fibre, fodder, and fuel. Farming systems have changed over millennia in response to regional socioeconomic, ecological, and climatic factors. However, the growing effects of climate change provide an unprecedented challenge for agriculture in the twenty-first century. Global agricultural productivity and food security are in danger due to rising global temperatures, changed precipitation patterns, and an increase in the frequency of extreme weather events, including heatwaves, droughts, and floods. Millions of smallholder farmers depend on agriculture for their livelihoods in areas like South Asia, Sub-Saharan Africa, and parts of Latin America, where these climate disruptions are especially concerning.

The demand for food, water, and other natural resources will rise dramatically as the world's population is expected to approach 10 billion people by 2050 (FAO 2017). It needs innovative agricultural practices that can guarantee productivity while preserving environmental sustainability to meet this demand in the face of climate stress. Despite their success in increasing yields during the Green Revolution, conventional high-input farming methods have led to the depletion of water resources, biodiversity loss, soil degradation, and greenhouse gas (GHG) emissions. Therefore, it is imperative that agricultural development be reoriented toward sustainability while tackling the dual challenges of mitigating and adapting to climate change.

A workable approach for addressing these interconnected issues is sustainable agriculture. Sustainable agriculture, which is based on the ideas of ecology, equity, and economic viability, seeks to increase the resilience of farming systems, optimize resource usage, and improve ecosystem services. Sustainable techniques take environmental stewardship and long-term resilience into account, in contrast to traditional models that emphasize short-term productivity. In addition to sequestering carbon in soils and vegetation, practices like crop diversification, organic farming, integrated nutrient and pest management, agroforestry, and conservation agriculture have the potential to improve soil fertility, conserve water, lower chemical inputs, and foster biodiversity (Smith et al. 2014).

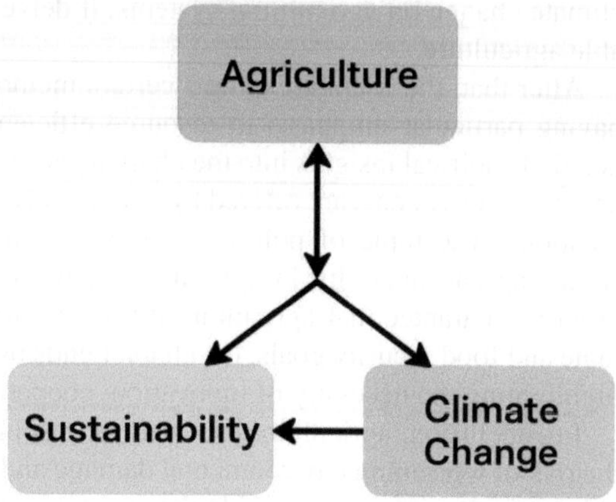

Fig. 7.1 Interconnections between agriculture, climate change, and sustainability

Climate change, sustainability, and agriculture are intricately and multidimensionally related (Fig. 7.1). On the one hand, the effects of climate change pose a serious danger to agricultural productivity, food supply, and farmer earnings. However, when combined with forestry and other land use (AFOLU), the sector accounts for around 23% of global greenhouse gas emissions, making it a major contributor to climate change (IPCC 2019). Agriculture is therefore both a cause and a sufferer of climate change. Understanding this dual role is crucial for creating strategies and policies that lessen agriculture's ecological imprint while simultaneously adapting to changing climatic circumstances.

Furthermore, vulnerable groups are disproportionately impacted by climate change. Climate-related hazards, including crop failures and animal losses, disproportionately affect smallholder farmers, who frequently lack access to infrastructure, technology, and credit. Vulnerability is further increased by gender inequality since, although being essential to agricultural productivity, women farmers typically have restricted access to land, inputs, and decision-making procedures. This emphasizes how crucial socially inclusive tactics are for empowering underrepresented populations and advancing equity in frameworks for sustainable agriculture.

The necessity of adapting agriculture for climate resilience and sustainability has been emphasized by international initiatives like the Paris Agreement, the Sustainable Development Goals (SDGs), and the UNFCCC's Koronivia Joint Work on Agriculture (UNFCCC 2018). Climate-smart agriculture methods are also becoming more and more integrated into national governments' development plans. However, obstacles including low funding, shoddy institutional frameworks, and sociopolitical impediments make it difficult to scale up sustainable solutions (Thornton et al. 2018). This calls for a multidisciplinary, multi-level strategy that incorporates conventional ecological knowledge, policy innovation, farmer participation, and scientific research.

Examining how agriculture, sustainability, and climate change are intertwined is the aim of this chapter. After examining the regional and worldwide effects of climate change on agricultural systems, it delves into the fundamentals of sustainable agriculture.

After that, the chapter explores certain methods that improve climate resilience, paying particular emphasis to resource efficiency, carbon sequestration, and soil health. Empirical insights into the challenges and achievements of adopting sustainable farming systems are offered by real-world case studies from climate-vulnerable locations. The topic of policy, governance, and institutional assistance in mainstreaming sustainability in agriculture is further explored in the conversation. In order to guarantee that agriculture makes a positive contribution to the global climate and food security goals, the chapter ends by describing future perspectives and highlighting the necessity of innovation, cooperation, and capacity-building.

In conclusion, agriculture is at a turning point. While conventional methods run the risk of worsening environmental damage and increasing food shortages, sustainable agriculture models provide avenues for ecological balance, justice, and resilience. It is not only desirable but also necessary for agriculture to align with sustainability principles as climate change picks up speed. This chapter makes the case that societies may create a more resilient and equitable world by promoting linked futures in which agriculture supports livelihoods, protects ecosystems, and slows down climate change.

7.2 Agriculture and Climate Change: An Overview

There is a complex and reciprocal interaction between agriculture and climate change. On the one hand, food production is disrupted, and rural livelihoods are threatened by climate change, which has become one of the most urgent dangers to agricultural systems globally. Conversely, agriculture plays a major role in land degradation, biodiversity loss, and global greenhouse gas (GHG) emissions, all of which exacerbate climate change (IPCC 2019). Designing solutions that allow agriculture to both contribute to mitigation efforts and adapt to climate shifts requires an understanding of this dichotomy. This section examines how agriculture is a source of greenhouse gas emissions, regional vulnerabilities with an emphasis on the worldwide South, and the worldwide effects of climate change on agriculture.

7.2.1 Global Impacts of Climate Change on Agriculture

Climate has a fundamental impact on agriculture. Crop yields, animal health, fisheries production, and forestry outputs are all directly impacted by changes in temperature, precipitation, and extreme weather events. Research shows that in some areas, the yields of staple crops, including maize, rice, and wheat, have already started to decline due to global warming. With an example, it is estimated that the average output of rice will decrease by 3.2% and the yield of wheat by 6% with every degree Celsius increase in the world mean temperature (Zhao et al. 2017). These decreases are especially worrisome in light of the growing demand for food worldwide.

Crop physiology is impacted by rising temperatures in a number of ways. Higher temperatures shorten growing seasons, speed up crop maturation, and shorten the time it takes for grains to fill, all of which result in lower yields. Additionally, heat stress raises evapotranspiration and hinders pollination, which increases the need for irrigation (Challinor et al. 2014). Heat stress increases susceptibility to illnesses and decreases feed intake, fertility, and milk production in livestock systems. Similar to this, fish stocks and coastal livelihoods are being disrupted by ocean warming, acidity, and changing migration patterns (FAO 2020).

Precipitation patterns are also altered by climate change. While some areas are dealing with extended droughts, others are seeing heavier rains and flooding. These modifications decrease groundwater recharge, increase soil erosion, and interfere with planting dates. For example, South Asia regularly suffers unpredictable monsoon rains that impact the production of rice and wheat, while the Horn of Africa has witnessed severe droughts that destroy pastoral populations. Furthermore, severe weather phenomena like hurricanes, cyclones, and wildfires exacerbate productivity losses by causing further harm to agricultural ecosystems and infrastructure.

7.2.2 Regional Vulnerabilities: Focus on the Global South

Despite being a worldwide occurrence, the effects of climate change on agriculture are not evenly dispersed. Because of their strong reliance on agriculture, low potential for adaptation, and socioeconomic difficulties, developing regions—especially those in the Global South—are disproportionately vulnerable.

7.2.2.1 South Asia

One of the world's most climate-vulnerable regions is South Asia. Here, agriculture is mostly reliant on the increasingly unpredictable monsoonal rains. The production of the region's main staples, wheat and rice, is in danger due to rising temperatures. According to studies, if adaptation is not implemented, yields in South Asia might drop by as much as 50% by 2050 under high-emission scenarios (Knox et al. 2012). Due to high levels of poverty, dispersed landholdings, and restricted access to irrigation, smallholder farmers in nations like Bangladesh, Nepal, and India face increased hazards.

7.2.2.2 Sub-Saharan Africa

Climate change presents serious hazards in Sub-Saharan Africa, where more than 60% of the population makes their living from agriculture. Due to the prevalence of rainfed farming, agricultural output is extremely vulnerable to changes in rainfall. Maize, sorghum, and millet are among the staple crops that are at risk from drought and desertification. Water scarcity and decreasing pasture productivity are also putting strain on livestock-based systems (Thornton et al. 2014). Vulnerability is increased by the region's low adaptive capacity, which is typified by poor infrastructure, scarce financial resources, and political instability.

7.2.2.3 Latin America

Climate change has an impact on livestock systems and crop output in Latin America. Due to changing climatic zones, there are fewer appropriate sites for coffee cultivation in Central and South American coffee-growing regions (Bunn et al. 2015). Increased drought frequency is threatening maize and soybean production in Argentina and Brazil. Furthermore, deforestation and climate-related stress threaten agricultural sustainability and biodiversity protection in the Amazon rainforest, which is essential for controlling the global climate.

7.2.2.4 Small Island Developing States (SIDS)

Storm surges, salinization, and sea level rise provide difficulties for agricultural systems in SIDS. Food security is threatened by the loss of freshwater supplies and coastal cropland. For instance, taro and cassava yields have been observed to be lower in Pacific Island nations because of saltwater intrusion (Campbell 2014). Importing food makes one even more susceptible to disturbances in international trade brought on by climate change.

7.2.3 Greenhouse Gas Emissions from Agriculture

In addition to being negatively impacted by climate change, agriculture also plays a major role in its causes. According to Smith et al. (2014), the sector accounts for roughly 10–12% of direct world GHG emissions, and up to 23% when forestry and other land use (AFOLU) are considered. The following are the primary sources of emissions from agriculture:

- Methane (CH_4): Mainly released by rice paddy farming and enteric fermentation in ruminants. Over a 100-year period, methane's global warming potential (GWP) is 28–34 times higher than that of carbon dioxide (IPCC 2019).
- Nitric Oxide (N_2O): Expelled from soils as a result of agricultural residue breakdown, manure management, and synthetic fertilizer application. The GWP of N_2O is around 300 times that of CO_2.
- Carbon Dioxide (CO_2): Produced by deforestation, changing land use, and the use of fossil fuels for irrigation, farming equipment, and fertilizer production.

One of the biggest sources, making up 14.5% of all anthropogenic GHG emissions worldwide, is livestock production (Gerber et al. 2013). Another significant source of methane is rice fields, particularly in Asia, where flooded agriculture methods are common. Furthermore, overuse of nitrogen fertilizers has led to a dramatic rise in N_2O emissions, which has a major impact on ozone layer depletion and global warming.

Because land-use change reduces the capacity to sequester carbon, emissions are made worse, especially when deforestation occurs for agricultural growth. The need for soybean farming, palm oil plantations, and cow ranching is a major factor in tropical deforestation in the Amazon, Southeast Asia, and Africa. In addition to releasing stored carbon, these land-use shifts endanger ecosystem resilience and biodiversity.

A contradiction is highlighted by the relationship between agriculture and climate change: whereas agriculture is essential to maintaining human life, it is also one of the industry's most at risk from climate change and a significant source of greenhouse gas emissions. The Global South, which is most dependent on agriculture, is disproportionately affected by the disruptions caused by global warming to crop and animal systems. Unsustainable farming methods also cause emissions and changes in land usage, which hasten climate change. To direct policy and practice toward sustainable agricultural systems that improve resilience and lower emissions, it is imperative to acknowledge this two-way interaction.

7.3 Principles of Sustainable Agriculture

Sustainable agriculture is a comprehensive framework that incorporates social justice, economic viability, and ecological care rather than just a collection of farming methods. It is a method of producing food that satisfies current demands without endangering the capacity of future generations to satisfy their own. Following the Brundtland Commission Report (WCED 1987), which defined sustainable development as development that is socially just, environmentally sound, and economically viable, the idea attracted international attention. To apply these ideas to agriculture, methods that maximize the use of natural resources, sustain farmer livelihoods, and guarantee food security in the face of climate change must be developed (Table 7.1).

7.3.1 Ecological Sustainability

The focus of ecological sustainability is on preserving and improving the foundation of natural resources—soil, water, biodiversity, and ecosystems—that agriculture relies on. Soil deterioration, water pollution, and biodiversity loss are the results of conventional agricultural practices that have weakened ecological resilience, such as monocropping, excessive pesticide usage, and deforestation. Sustainable agriculture, on the other hand, seeks to balance ecological benefits with food production.

Table 7.1 Comparison of conventional vs. sustainable agriculture practices

Parameter	Conventional agriculture	Sustainable agriculture
Input use	High chemical inputs	Integrated/natural inputs
Soil fertility	Declines over time	Improves through organic matter
Water use efficiency	Low	High (mulching, micro-irrigation)
GHG emissions	High	Low/mitigation-focused
Resilience to climate	Low	High

7.3.1.1 Soil Health and Fertility

A major danger to sustainability is the deterioration of soils, which form the basis of agricultural systems. Fertility and resilience are compromised by practices like excessive tillage and reliance on synthetic fertilizers, which lower soil organic matter and microbial diversity. Crop rotation, conservation tillage, organic amendments, and integrated nutrient management are examples of sustainable practices that increase SOC, promote nutrient cycling, and fortify resilience to climate shocks. Furthermore, healthy soils are essential for carbon sinking, playing a pivotal role in climate change mitigation.

7.3.1.2 Water Conservation and Use Efficiency

Almost 70% of freshwater withdrawals worldwide are related to agriculture (FAO 2017). Especially in arid and semi-arid areas, unsustainable irrigation methods have caused salinization and groundwater depletion. To maximize water utilization, ecological sustainability encourages techniques including mulching, drip irrigation, rainwater collection, and deficit irrigation. In addition to preserving hydrological cycles, watershed management and wetland restoration lessen susceptibility to floods and droughts (Rockström et al. 2010).

7.3.1.3 Biodiversity and Ecosystem Services

Ecosystem services, including pollination, pest control, and genetic resources, are all supported by biodiversity. Agroecosystems are more vulnerable to pests and diseases as a result of intensive monocropping, which endangers biodiversity. Intercropping, agroforestry, and habitat corridor maintenance are examples of agroecological techniques that improve ecological resilience and biodiversity. To retain genetic variety that can survive climatic challenges, it is equally important to preserve historic animal breeds and agricultural varieties.

7.3.1.4 Climate Change Mitigation

By lowering greenhouse gas emissions and improving carbon sequestration, sustainable agriculture helps to mitigate the effects of climate change. The sector's carbon footprint is greatly decreased by methods, including agroforestry, cover crops, less fertilizer, and sustainable livestock management (Smith et al. 2014). Therefore, ecological sustainability ensures long-term productivity while being in line with more general climatic goals.

7.3.2 Economic Viability

For agriculture to be genuinely sustainable, producers must receive a steady and adequate income, and consumers must be able to buy it. Optimizing input costs, guaranteeing market access, and boosting value addition are all necessary for economic viability. Regardless of the ecological advantages of sustainable methods, farmers are unlikely to implement them if they are not profitable.

7.3.2.1 Input Efficiency and Cost Reduction

The focus of sustainable farming is on efficient use of resources. Farmers can reduce production costs by minimizing their reliance on expensive external inputs like synthetic herbicides and fertilizers. To maximize nutrition availability and minimize waste, integrated nutrient management (INM) blends organic and inorganic sources. Like this, integrated pest management (IPM) reduces the use of pesticides by using resistant types, crop rotation, and biological control, which lowers expenses and protects the environment.

7.3.2.2 Diversification and Risk Management

Because of price swings and pest outbreaks, monoculture farming sometimes exposes farmers to financial risks. Risk is distributed, and several revenue streams are produced by crop diversification, mixed farming, and the integration of livestock, fisheries, and agroforestry. Systems that are more diverse are also more resilient to climate shocks. As an example, adding legumes increases soil fertility and offers a further source of market revenue and protein (Snapp et al. 2010).

7.3.2.3 Market Access and Value Chains

The capacity of farmers to reach markets and obtain fair pricing for their produce is another factor that affects sustainability. Profitability is increased, and intermediary exploitation is decreased by creating inclusive value chains, bolstering cooperatives, and encouraging direct selling avenues like farmers' markets (Vorley et al. 2012). Market opportunities are further increased by certification programs for fair-trade and organic products, while smallholders still face accessibility issues.

7.3.2.4 Long-Term Productivity and Food Security

Sustainable agriculture places more emphasis on long-term production than intensive systems, which focus on short-term yield improvements. Future generations can continue farming if soil fertility, water supplies, and biodiversity are maintained. This is in line with the overarching objective of global food security, which is to guarantee food's availability, affordability, and accessibility over time.

7.3.3 Social Equity and Farmer Well-Being

Although it is sometimes disregarded, the social aspect of sustainability is essential to developing resilient and inclusive agricultural systems. The advantages of agricultural development are distributed equitably among many groups, such as women, smallholders, indigenous peoples, and marginalized communities, thanks to social equity.

7.3.3.1 Smallholder Farmers and Inclusivity

Although they produce a sizable amount of the world's food, smallholders sometimes do not have access to technology, credit, or extension services. Smallholders must be empowered by sustainable agriculture through supportive legislation, knowledge-sharing platforms, and participatory methods. Scaling sustainable methods requires farmer-led innovations, training, and capacity building.

7.3.3.2 Gender Equity in Agriculture

Despite making up about half of the agricultural workforce worldwide, especially in poor nations, women still confront structural obstacles in decision-making, land ownership, and input availability (FAO 2011). Promoting equity and enhancing household food security need gender-responsive policies and initiatives, such as protecting women's land rights, granting them access to microfinance, and encouraging their involvement in agricultural cooperatives (Meinzen-Dick et al. 2011).

7.3.3.3 Community Engagement and Knowledge Systems

Acknowledging and incorporating indigenous knowledge and customs is also essential to sustainability. Over the centuries, numerous communities have created resilient farming practices such as community-based irrigation management, shifting agriculture, and mixed crops. Trust is increased, and local ownership of sustainability projects is strengthened when these practices are respected and expanded upon.

7.3.3.4 Health, Nutrition, and Food Sovereignty

Sustainable agriculture must support nutrient-dense food systems to improve health outcomes. Incorporating fruits, vegetables, and pulses into agroecological diversification improves dietary diversity and combats malnutrition. Additionally, the idea of food sovereignty promotes cultural identity, autonomy, and social justice by highlighting farmers' rights to establish their own food systems.

The three pillars of sustainable agriculture—economic viability, social equality, and ecological sustainability—are interconnected and reinforce one another. Economic viability secures farmers' livelihoods and investment capacity; social equity enables inclusive participation and justice; and ecological sustainability offers the natural basis for productivity. When taken as a whole, these elements offer a comprehensive framework for changing agriculture to meet the concerns of global food security and climate change.

7.4 Sustainable Agricultural Practices for Climate Resilience

The concepts of ecological, economic, and social sustainability are translated into practical farming techniques through sustainable agriculture practices (Fig. 7.2). These methods are intended to increase resilience to climate shocks, preserve natural resources, and boost productivity. They are crucial tactics for mitigating and adapting to climate change since they provide chances to lower greenhouse gas emissions and store carbon (FAO 2017).

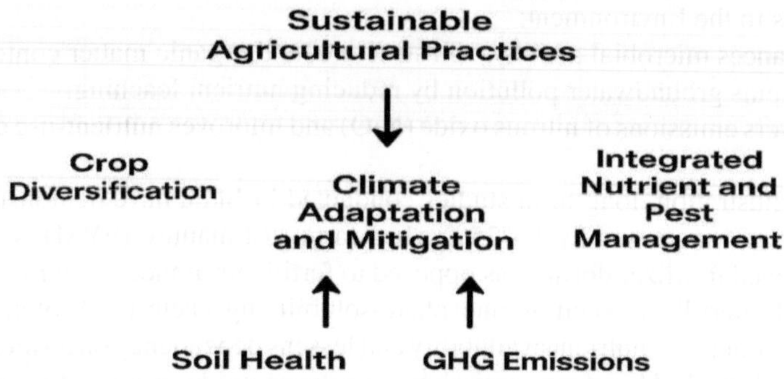

Fig. 7.2 Sustainable agricultural practices and role in climate adaptation and mitigation

7.4.1 Crop Diversification and Genetic Resources

Crop diversity is a tried-and-true method of risk management and resilience building in agriculture. Farmers distribute ecological and economic risks related to pests, illnesses, and climate unpredictability by growing a variety of crops rather than monocultures (Lin 2011).

Advantages for resilience:

- Breaks the cycles of disease and pests, reducing susceptibility to pest outbreaks.
- Crop rotation, especially with legumes that fix atmospheric nitrogen, improves soil fertility.
- Enhances household nutrition and nutritional diversity.
- Offers a variety of revenue sources, protecting farmers from fluctuations in the market.

Examples: Pulses like mungbean, which increase soil nitrogen content and lessen reliance on fertilizer, are being used to diversify rice–wheat cropping systems in India. Legumes such as groundnuts and cowpeas are interplanted with maize in sub-Saharan Africa to increase soil fertility and drought resistance (Snapp et al. 2010).

Climate link: By making farming systems less reliant on a single crop, crop diversification improves adaptive ability. Crop genetic variety, such as the use of heat- or drought-tolerant cultivars, increases resistance to climate extremes.

7.4.2 Integrated Nutrient Management (INM)

To sustain soil fertility and productivity, integrated nutrient management (INM) uses agricultural residues, organic manures, chemical fertilizers, and biofertilizers in balance. INM encourages a more sustainable nutrient cycle in contrast to traditional methods that mostly rely on synthetic fertilizers.

- Benefits to the Environment:
  - Enhances microbial activity, soil structure, and organic matter content.
  - Prevents groundwater pollution by reducing nutrient leaching.
  - Lowers emissions of nitrous oxide (N_2O) and improves nutrient-use efficiency.

As an illustration, long-term studies conducted in India have demonstrated that crop yields are increased by 10–25% when farmyard manure (FYM) is combined with approved fertilizer dosages as opposed to fertilizers alone. Likewise, the application of biofertilizers such as phosphate-solubilizing bacteria, Azotobacter, and Rhizobium improves nutrient availability and lessens dependency on artificial inputs.

Climate link: INM reduces emissions of nitrous oxide, a powerful greenhouse gas, by reducing an excessive reliance on synthetic nitrogen fertilizers. Additionally, organic inputs improve soil carbon sequestration, which aids in mitigation.

7.4.3 Integrated Pest Management (IPM)

To manage pests in an environmentally responsible way, integrated pest management, or IPM, is a comprehensive strategy that incorporates biological, cultural, physical, and chemical methods. The objective is to keep pest numbers below levels that are detrimental to the economy, not to eradicate them completely.

- *Strategies include*:
  - Using resistant crop types is one of the strategies.
  - Biological control by microbiological agents, parasitoids, and predators.
  - Crop rotation and habitat management can help minimize the accumulation of pests.
  - As a last option, apply pesticides sparingly and precisely.

Examples: Using IPM techniques like biological control and pheromone traps cut pesticide consumption in Indian cotton farming by 50% without sacrificing yield. In a similar vein, IPM for rice in Southeast Asia has effectively decreased reliance on pesticides while preserving yields.

Climate link: Using fewer chemical pesticides reduces emissions and the use of fossil fuels. Additionally, biological management makes ecosystems more resilient, which is important in the face of climatic variability.

7.4.4 Conservation Agriculture (CA)

According to Kassam et al. (2009), conservation agriculture places a strong emphasis on diversified crop rotations, permanent soil cover, and little soil disturbance. It has become a crucial technique for sustainable intensification, especially in areas where soil degradation is a problem.

- *Core principles*:
    1. *No-till or reduced tillage* to preserve soil structure.
    2. *Residue retention* for soil cover and organic matter enrichment.
    3. *Crop rotation* to enhance soil health and break pest cycles.
- *Benefits*:
    1. Improves soil organic matter and microbial diversity.
    2. Enhances water infiltration and reduces soil erosion.
    3. Reduces fuel use from tillage, cutting CO_2 emissions.

Examples: More than 100 million hectares of conservation agriculture have been implemented throughout Latin America, especially in Brazil and Argentina. According to Derpsch et al. (2010), farmers report increased drought resistance, reduced input costs, and higher yields.

Climate link: CA is a very climate-resilient technique because it helps with water conservation and carbon sequestration.

7.4.5 Agroforestry Systems

Agroforestry integrates trees with crops and/or livestock systems, creating multi-functional landscapes that provide food, fuelwood, fodder, and ecosystem services.

- *Benefits*:
    - Enhances soil fertility through leaf litter and nitrogen-fixing trees.
    - Provides shade and windbreaks, protecting crops and livestock.
    - Increases biodiversity and carbon sequestration.
    - Diversifies income for farmers.

Examples include the intercropping of *Faidherbia albida* with maize in sub-Saharan Africa, which increases yields without the need for extra fertilizer. Poplar and eucalyptus agroforestry in India has sustained agricultural output while giving farmers timber revenue.

Climate link: Systems of agroforestry are effective at absorbing carbon dioxide. According to the World Agroforestry Centre, if agroforestry were to become widely used, it could store between 1.1 and 2.2 Pg of carbon per year.

7.4.6 Organic and Regenerative Agriculture

Organic farming eliminates synthetic fertilizers and pesticides, relying on natural processes and inputs. Regenerative agriculture goes a step further, aiming to restore soil health, biodiversity, and ecosystem services (IFOAM 2018).

- Benefits:
  - Improves soil fertility through composting, green manures, and crop rotations.
  - Reduces pollution from agrochemicals.
  - Promotes biodiversity and ecological resilience.

Examples: With more than 14.6 million hectares managed organically, organic farming has greatly increased in Europe (Willer et al. 2021). Regenerative techniques like managed grazing and cover crops are becoming more popular in the United States as a result of carbon farming programs.

Climate link: Although yields may be lower, organic systems generally release fewer greenhouse gases per unit of land. Regenerative methods are essential for mitigating climate change because they place a special emphasis on carbon sequestration.

7.4.7 Climate–Smart Agriculture (CSA)

Climate–Smart Agriculture, promoted by the FAO, is an integrated approach that aims to:

1. Sustainably increase productivity.
2. Enhance resilience and adaptation.
3. Reduce GHG emissions where possible (FAO 2013).

Examples of CSA technologies:

- Drought-tolerant crop varieties.
- Improved livestock breeds and feed efficiency.
- Precision agriculture technologies for efficient input use.
- ICT-based weather advisory systems for farmers.

Case study: CSA measures like water harvesting, agroforestry, and improved maize varieties have been adopted in East Africa, increasing yields by 30–50% while decreasing drought susceptibility (Mwongera et al. 2017).

Climate link: CSA is a key component of future agricultural strategies since it is specifically created to solve the triple challenge of productivity, adaptation, and mitigation.

The methods offer various but complementary approaches to climate-resilient and sustainable agriculture. Broader systems like conservation agriculture, agroforestry, and regenerative farming address long-term ecological and climate problems, while crop diversification, INM, and IPM address urgent farm-level demands. These strategies are combined in climate-smart agriculture to create a coherent framework that harmonizes regional practices with international climate objectives.

7.5 Soil Health and Carbon Sequestration

The basis of terrestrial ecosystems and the vital component of agriculture is soil. In addition to supporting biodiversity, healthy soils control water flow, cycle nutrients, and act as a significant carbon sink. Resilience and soil health are closely related in the context of climate change: soils with high levels of organic matter improve nutrient-use efficiency, lower greenhouse gas emissions, and protect crops from drought and floods. On the other hand, soil degradation—due to erosion, nutrient depletion, or organic matter loss—endangers agricultural sustainability and increases the risks associated with climate change.

The function of soil organic carbon (SOC), microbial diversity, and soil and water conservation techniques are the three interconnected aspects of soil health and carbon sequestration that are examined in this section.

7.5.1 Role of Soil Organic Carbon (SOC) in Resilience

The most significant indication of soil health is soil organic carbon. It comes from microbial biomass, organic additives, plant waste, and root exudates. SOC affects almost every facet of soil function.

- *Functions of SOC*:
  - *Nutrient cycling*: SOC serves as a reservoir of nitrogen, phosphorus, and sulfur, which are slowly mineralized for plant uptake.
  - *Soil structure*: Organic matter improves aggregation, porosity, and water-holding capacity.
  - *Water regulation*: SOC-rich soils retain 20–40% more water, making crops more resilient to drought.
  - *Climate mitigation*: SOC sequestration offsets a portion of anthropogenic CO_2 emissions.

Global potential: The top meter of soils contains over 1500 Pg of carbon, which is more than twice as much as the atmosphere's carbon content. Food security might be improved, and climate change could be considerably reduced with even modest improvements in SOC of 0.4% annually, as encouraged by the "4 per 1000" campaign (Minasny et al. 2017).

As an illustration, long-term studies conducted in the Indo-Gangetic Plains of India reveal that methods such as residue retention and green manuring raise SOC by 20–30% in comparison to traditional plowing. In the Midwest of the United States, prairie restoration has restored SOC stocks that were depleted by intensive farming.

7.5.2 Soil Microbial Diversity and Ecosystem Services

The bacteria, fungi, archaea, and protozoa that make up soil are responsible for vital functions like disease prevention, nutrient cycling, and the breakdown of organic materials. One common characteristic of healthy soil is a varied and vibrant microbial population (Van der Heijden et al. 2008).

- *Functions of Microbial Communities*:
  - *Decomposition and nutrient release*: Microbes mineralize organic residues, releasing nutrients for plant uptake.
  - *Symbiosis with plants*: Mycorrhizal fungi enhance phosphorus uptake, while nitrogen-fixing bacteria (e.g., *Rhizobium*) supply biologically fixed nitrogen.
  - *Carbon sequestration*: Some microbial by-products contribute to stable humus formation, a long-term carbon pool.
  - *Disease suppression*: Beneficial microbes outcompete pathogens, reducing the need for chemical pesticides.

Climate link: Changes in temperature, moisture content, and substrate availability brought forth by climate change affect microbial activity. While methods that increase microbial diversity (such as organic additions and reduced tillage) can stabilize carbon in soils, rising temperatures can speed up decomposition and release CO_2.

Examples: Studies on maize systems in Africa have demonstrated that inoculating soil with nitrogen-fixing bacteria improves soil nitrogen status and boosts yields during drought stress. Similarly, more varied fungal communities that improve soil carbon stability are supported by conservation tillage in Latin America.

7.5.3 Soil and Water Conservation Practices

Soil and water conservation is integral to both productivity and climate resilience. Degradation through erosion, salinization, or compaction reduces the capacity of soils to sustain crops and store carbon.

- *Key practices include*:

  1. *Contour farming and terracing*—Reduces runoff and soil loss on slopes.
  2. *Cover cropping*—Protects soils from erosion, adds biomass, and enhances SOC.
  3. *Mulching*—Conserves moisture, moderates soil temperature, and feeds soil biota.
  4. *Reduced tillage*—Limits disturbance, maintaining aggregates and carbon pools.
  5. *Water harvesting*—Improves soil moisture availability, enhancing resilience in arid zones.

For instance, terracing has reduced soil erosion in Ethiopia's highlands by over 50%, increasing cereal yields and enhancing groundwater recharge. Mulching maize–wheat systems in India has increased SOC content while lowering irrigation needs by 20–30%.

Climate link: By stabilizing yields under climate stress, these approaches improve adaptation while simultaneously lowering emissions from damaged soils and conserving resources.

7.5.4 Challenges and Opportunities

Despite clear benefits, building soil health and increasing SOC face several challenges:

- *Short-term vs. long-term trade-offs*: Farmers often prioritize immediate yields over long-term soil building.
- *Measurement difficulties*: SOC changes are slow and spatially variable, making monitoring challenging.
- *Land degradation pressures*: Expanding populations and unsustainable practices accelerate soil loss.

However, there are significant opportunities:

- *Policy support*: Programs like India's National Mission on Sustainable Agriculture (NMSA) and the EU's Soil Mission emphasize SOC management.
- *Carbon markets*: Emerging soil carbon credit systems provide financial incentives for farmers to adopt soil-friendly practices.
- *Research innovations*: Advances in soil microbiome science and digital monitoring tools improve our capacity to manage soils for resilience.

Climate change and agricultural sustainability are closely related to soil health and carbon sequestration. Farmers may create resilient systems that can resist climatic extremes and help achieve global mitigation targets by increasing soil organic carbon, encouraging microbial diversity, and implementing soil and water conservation techniques.

Because they store carbon and provide ecosystem services that are necessary for sustained agriculture, healthy soils serve as organic climate friends. For food security and climate stability, soil management is therefore a worldwide necessity rather than just a local farm-level tactic.

7.6 Water Management in Sustainable Agriculture

Among the most important resources for agriculture is water, but it is also one of the most susceptible to climate change. Nearly 70% of freshwater withdrawals worldwide are used for agriculture, with irrigation being the main user (FAO 2017). By changing rainfall patterns, raising evapotranspiration, and making droughts and floods worse, climate change exacerbates the water crisis. Therefore, effective and sustainable water management is essential to climate adaptation and agricultural resilience.

Four interrelated features of water management in sustainable agriculture are examined in this section: water-saving technology, rainfall collection, groundwater recharge, and effective irrigation systems.

7.6.1 Efficient Irrigation Systems

The water-use efficiencies of traditional surface irrigation techniques, including flood irrigation, are frequently less than 40%, making them extremely wasteful. Runoff, seepage, and evaporation all result in significant volume loss. Effective irrigation systems present a substantial opportunity to conserve water while preserving or boosting production in the face of water scarcity.

- *Drip irrigation*: Drip systems reduce losses by using low-pressure emitters to feed water straight to the plant root zone. According to studies, drip irrigation can increase water output by up to 100% while using 30–70% less water than flood irrigation (Kang et al. 2017). It works particularly well for high-value crops, including cotton, fruits, and vegetables.
- *Sprinkler irrigation*: Sprinklers work well for a variety of crops and disperse water more uniformly than conventional techniques. Water-use efficiency is usually 20–40% higher than surface irrigation, notwithstanding the possibility of evaporation losses in hot or windy circumstances.
- *Subsurface irrigation*: By directly providing the root zone with water beneath the soil's surface, emerging systems lower evaporation. These are especially helpful in desert areas where evaporative demand is high.
- *Case example*: Drip irrigation has revolutionized agriculture in desert regions of Israel, allowing for high yields with limited water supplies. Micro-irrigation has also been encouraged by India's Pradhan Mantri Krishi Sinchai Yojana (PMKSY), which has been widely adopted in states like Gujarat and Maharashtra.

7.6.2 Rainwater Harvesting

Rainwater harvesting is an age-old practice that has regained attention in sustainable agriculture due to its role in mitigating drought and enhancing water availability.

- *Techniques*:
  - *Farm ponds*: Small reservoirs that capture rainwater for supplementary irrigation.
  - *Check dams*: Barriers across streams to slow runoff and increase infiltration.
  - *Contour bunds*: Earthen ridges constructed along contours to reduce runoff and capture rainwater.
  - *Percolation tanks*: Designed to recharge groundwater aquifers.

Advantages: Rainwater harvesting increases resilience during dry spells, decreases reliance on groundwater, and improves local water supply. Additionally, it reduces soil erosion by delaying runoff.

As an illustration, traditional Johads (earthen check dams) in Rajasthan, India, have restored groundwater levels, allowing for agriculture even in areas that are prone to drought. In a similar vein, zai pits—tiny planting basins that collect rainfall and concentrate organic matter—have been used by African Sahelian nations to repair degraded ground.

7.6.3 Groundwater Recharge and Conservation

Groundwater is a lifeline for agriculture, particularly in regions like South Asia, where tube wells support intensive cropping systems. However, over-extraction has led to alarming declines in water tables. Sustainable management requires both reducing demand and enhancing recharge.

- *Artificial recharge structures*: Techniques such as recharge wells, injection boreholes, and spreading basins channel excess runoff into aquifers.
- *Managed aquifer recharge (MAR)*: MAR integrates hydrological monitoring with engineering interventions to sustainably replenish groundwater.
- *Demand-side measures*: Groundwater conservation requires regulating pumping, promoting water-efficient crops, and incentivizing farmers to adopt micro-irrigation.
- *Case example*: The Sardar Patel Participatory Water Conservation initiative in Gujarat, India, has effectively built thousands of recharging structures and check dams, regulating water tables and boosting crop productivity.

7.6.4 Water-Saving Technologies and Practices

Beyond irrigation infrastructure, several agronomic practices and technologies can enhance water-use efficiency.

- Mulching: Adding plastic or organic mulch enhances water retention, moderates temperature, and decreases soil evaporation. Mulching has cut irrigation requirements in North Indian maize-wheat systems by 20–25%.

- Alternate Wetting and Drying: Rather than sustaining constant flooding, AWD in rice farming permits fields to dry out periodically. This lowers methane emissions and water use by up to 30% (Lampayan et al. 2015).
- Deficit irrigation: Water can be saved without causing a large loss in production by applying less water than the crop needs during non-critical growth phases. This method has worked well for growing wheat, cotton, and grapes.
- Crop varieties that can withstand drought: Plant breeding and biotechnology have created crop varieties that can withstand water stress or use less water. In Sub-Saharan Africa, the use of drought-resistant maize has increased harvests despite unpredictable rainfall.
- Precision technologies: Remote sensing, soil moisture sensors, and decision-support systems optimize irrigation scheduling, ensuring water is applied only when and where needed.

7.6.5 Integrated Water Resource Management (IWRM)

Sustainable water management requires a holistic, landscape-level approach. IWRM emphasizes coordinated development of land, water, and related resources to maximize economic and social welfare without compromising ecosystems.

- *Principles of IWRM*:
  - Integration across sectors (agriculture, industry, domestic).
  - Stakeholder participation, including farmers, local communities, and policymakers.
  - Ecosystem-based management that safeguards rivers, wetlands, and aquifers.

Case Example: The Murray–Darling Basin in Australia has implemented basin-wide water allocation and trading mechanisms to balance agricultural and ecological needs.

7.6.6 Challenges and Opportunities

While sustainable water management holds promise, several challenges persist:

- *High initial costs*: Drip and sprinkler systems require significant investment, which smallholders may find unaffordable.
- *Institutional gaps*: Weak governance and fragmented policies hinder integrated approaches.
- *Energy–water nexus*: Over-reliance on groundwater often coincides with subsidized electricity, encouraging over-pumping.
- *Climate uncertainty*: Unpredictable rainfall complicates water resource planning.

Yet opportunities are emerging:

- *Policy support*: Subsidies for micro-irrigation, water-user associations, and water-pricing reforms can drive change.
- *Digital technologies*: Mobile apps and IoT-based sensors improve farmer access to real-time water data.
- *Synergies with climate goals*: Efficient water management reduces emissions, improves adaptation, and aligns with SDGs.

7.7 Policy Frameworks and Institutional Support for Sustainable Agriculture

More than only creative farm-level methods are needed to make the shift to sustainable and climate-resilient agriculture; strong institutional structures and supportive legislative frameworks are also needed. Policies have an impact on risk management, practice incentives, and resource allocation. Institutions—from regional cooperatives to global organizations—are essential for expanding sustainable practices, encouraging information exchange, and facilitating farmer involvement.

The landscape of sustainable agriculture under climate change is shaped by national strategies, international frameworks, farmer participation, and institutional innovations, all of which are examined in this section.

7.7.1 International Policy Frameworks

Global agreements have recognized agriculture as both a contributor to climate change and a sector vulnerable to its impacts.

- *The Paris Agreement* (2015): Despite not specifically mentioning agriculture, the Agreement highlights mitigation and adaptation measures that are pertinent to the industry. Many of the Nationally Determined Contributions (NDCs) that nations submit include pledges to agroforestry, soil carbon sequestration, and climate-smart agriculture.
- *The Koronivia Joint Work on Agriculture (KJWA)*: KJWA is the first international mechanism devoted to agriculture and climate change, having been adopted under the UNFCCC in 2017. In addition to offering a forum for technical exchange and capacity building, it tackles soil health, nitrogen management, livestock systems, and water use (UNFCCC 2018).
- *Sustainable Development Goals (SDGs)*: Sustainable agriculture has strong ties to SDGs 2 ("Zero Hunger") and 13 ("Climate Action"). In order to accomplish these objectives, productivity must rise while ecosystems are preserved and resilience is strengthened.

- *Other frameworks*:
  - The FAO's Climate-Smart Agriculture (CSA) approach integrates productivity, adaptation, and mitigation.
  - The Convention on Biological Diversity (CBD) promotes agro-biodiversity conservation.
  - The Land Degradation Neutrality (LDN) targets of the UNCCD encourage sustainable land management.

Together, these frameworks provide global legitimacy for sustainable agriculture, though their effectiveness depends on national translation and implementation.

7.7.2 National Policies and Strategies

At the country level, policies define the enabling environment for sustainable practices.

- *India*:
  - *National Mission on Sustainable Agriculture (NMSA)*: Focuses on soil health management, water-use efficiency, and climate-resilient technologies.
  - *Pradhan Mantri Krishi Sinchai Yojana (PMKSY)*: Promotes micro-irrigation and watershed development.
  - *Soil Health Card Scheme*: Provides farmers with soil nutrient information to guide input use.
- *China*: Introduced the "Zero Growth in Fertilizer and Pesticide Use" policy (2015) and heavily invests in conservation agriculture and renewable energy for rural areas.
- *European Union*: The Common Agricultural Policy (CAP) now integrates "greening measures," rewarding farmers for crop diversification, maintaining permanent grasslands, and ecological focus areas. The EU Soil Mission seeks to ensure all soils are healthy by 2050.
- *Sub-Saharan Africa*: Initiatives such as the Comprehensive Africa Agriculture Development Programme (CAADP) emphasize climate-smart agriculture, while countries like Ethiopia and Kenya have mainstreamed CSA into national climate policies.

These strategies highlight how national governments are aligning agriculture with climate and sustainability goals, although implementation gaps remain.

7.7.3 Farmer Participation and Community-Based Approaches

Top-down policy measures alone cannot achieve sustainability; farmer participation and community empowerment are crucial.

- *Participatory extension systems*: Rather than merely receiving information, farmers are actively involved in its creation. In Asia and Africa, Farmer Field Schools (FFS) have been effective in advancing integrated pest management.
- *Collective action*: Farmers can share risks and pool resources with the support of community-based natural resource management programs, cooperative marketing groups, and water-user associations.
- *Indigenous and traditional knowledge*: Communities can adapt using centuries' worth of ecological understanding when policies acknowledge indigenous techniques (such as shifting cultivation, mixed cropping, or traditional irrigation systems).
- *Gender inclusion*: Women farmers, who make up about half of the agricultural workforce in developing nations, frequently do not have access to extension services, financing, or land. Household food security and resilience can be enhanced by gender-sensitive policies that encourage women's involvement (Dankelman 2010).

7.7.4 Institutional Innovations

Institutions, both formal and informal, are essential in bridging the gap between policy design and field-level action.

- *Research organizations*: Universities, international centers like CGIAR, and national agricultural research systems (NARS) produce climate-resilient technology, including organic inputs, precision irrigation equipment, and drought-tolerant crops.
- *Extension services*: Farmers are guaranteed access to sustainable techniques through efficient extension. Information distribution has changed as a result of ICT-enabled platforms, smartphone apps, and digital advisory services, especially in rural areas.
- *Financial establishments*: Farmers are less vulnerable when they have access to loans, crop insurance, and risk-sharing programs. In nations like Kenya and India, cutting-edge products like weather-indexed insurance are growing.
- *Carbon markets and certification programs*: Farmers can obtain credits for agroforestry and soil carbon sequestration through voluntary carbon markets. Market access and price advantages for sustainable products are offered by certification programs (such as Rainforest Alliance, Fairtrade, and organic).
- *Public–private partnerships (PPPs)*: PPPs bring together governments, NGOs, research bodies, and agribusinesses to scale up sustainable innovations. For example, partnerships in East Africa promote climate-smart dairy systems that enhance milk yields while lowering methane emissions.

7.7.5 Challenges in Policy and Institutional Support

Despite progress, several challenges persist in aligning policies and institutions with sustainability goals:

- *Fragmented policies*: Agricultural, environmental, and water policies often operate in silos, leading to contradictions.
- *Short-term political cycles*: Policymakers may prioritize immediate yield gains over long-term sustainability.
- *Limited funding*: Many developing countries face resource constraints to implement large-scale sustainability programs.
- *Weak institutional capacity*: Lack of coordination, transparency, and accountability undermines effectiveness.
- *Equity concerns*: Policies sometimes fail to address the needs of smallholders, landless labourers, and marginalized groups.

7.7.6 Opportunities for Strengthening Frameworks

The evolving global climate agenda provides opportunities to strengthen policy and institutional support for sustainable agriculture:

- *Integrating agriculture into climate strategies*: Ensuring agriculture is central to NDCs, adaptation plans, and climate financing mechanisms.
- *Decentralization*: Empowering local institutions, such as Panchayati Raj in India or village committees in Africa, to manage resources.
- *Digital governance*: Using satellite data, AI, and blockchain for monitoring, certification, and transparency.
- *Capacity building*: Investing in training programs for extension agents, farmer leaders, and policymakers.
- *International cooperation*: Cross-border knowledge exchange and funding partnerships can accelerate sustainable transitions.

The methods by which climate change can be addressed through sustainable agriculture are shaped by institutions and policies. Local institutions facilitate implementation, national policies specify strategies, and international frameworks offer legitimacy. However, innovations that connect science and practice, inclusive decision-making, and farmer participation are essential for effectiveness. Considering climate change, strengthening these frameworks is important for ensuring livelihoods, food security, and environmental sustainability, not just for safeguarding agriculture.

7.8 Case Studies in Sustainable Agriculture Under Climate Change

Case studies offer important insights into the global implementation of sustainable agriculture practices and their effects on environmental sustainability, productivity, and resilience (Table 7.2).

7.8.1 India: Conservation Agriculture in the Indo-Gangetic Plains

The Indo-Gangetic Plains (IGP) are one of the most intensively cultivated regions in the world, producing rice, wheat, and maize for hundreds of millions of people. However, the region faces multiple challenges: groundwater depletion, declining soil fertility, and increasing vulnerability to climate extremes.

- *Practice adopted*: Conservation agriculture (CA), including zero tillage, residue retention, and crop diversification.
- *Impact*:
  - Zero tillage reduced irrigation water use by 15–20%.
  - Retention of rice residues improved soil organic carbon by 0.2–0.3% over 5 years.
  - Yields of wheat increased by 10–15% compared to conventional tillage (Jat et al. 2014).

Climate link: Reduced tillage lowered fuel consumption and CO_2 emissions. Residue retention improved soil moisture buffering against droughts.

Lesson: While CA has proven benefits, adoption remains limited due to residue burning practices, machinery costs, and lack of awareness among smallholders.

Table 7.2 Case study summary of sustainable agriculture practices under climate change

Region/Country	Practice adopted	Climate benefits	Key lessons
India (IGP)	Conservation agriculture	Improved soil carbon, reduced water use	Needs policy support
Sub-Saharan Africa	Drought-tolerant maize + ISFM	Yield stability, food security	Input access is a barrier
Brazil	Agroforestry	Carbon sequestration, biodiversity	Long-term investment needed
Ethiopia	Terracing & watershed management	Soil erosion reduced	Community participation essential
Central America	Shade-grown organic coffee	Pest resistance, carbon storage	Certification costly
Bangladesh	Floating gardens	Adaptation to floods	Low-cost but training required

Policies promoting subsidies for zero-till seeders have helped but require further scaling.

7.8.2 Sub-Saharan Africa: Climate-Smart Maize in Drought-Prone Areas

Maize is the staple crop for much of Sub-Saharan Africa, but its production is increasingly threatened by recurrent droughts.

- *Practice adopted*: Climate-smart maize varieties (drought-tolerant hybrids) combined with integrated soil fertility management (ISFM).
- *Impact*:
  - Drought-tolerant maize increased yields by 20–30% under water stress conditions (Cairns et al. 2013).
  - ISFM practices (manure, compost, and small amounts of mineral fertilizers) improved soil fertility and yield stability.
  - Household food security improved as crop failures decreased.

Climate link: These practices reduced vulnerability to rainfall variability, directly enhancing adaptation.

Lesson: Access to improved seeds and inputs remains a barrier. Partnerships between research institutions (e.g., CIMMYT), governments, and seed companies have been critical in scaling adoption.

7.8.3 Brazil: Agroforestry in the Amazon and Cerrado

Brazil's agricultural frontier is expanding rapidly, often at the expense of forests and biodiversity. Agroforestry systems provide a balance between agricultural production and environmental conservation.

- *Practice adopted*: Agroforestry systems integrating crops, trees, and livestock. Examples include coffee with shade trees and silvo-pastoral systems.
- *Impact*:
  - Increased biodiversity and pollination services.
  - Higher soil organic carbon and improved water retention.
  - Farmers gained diversified income streams from timber, fruit, and livestock.
  - Silvo-pastoral systems reduced methane emissions per unit of beef produced.

Climate link: Agroforestry contributed to carbon sequestration in both soils and biomass while reducing pressure on deforestation.

Lesson: Scaling agroforestry requires long-term investments and supportive markets. Credit schemes and extension services have proven vital for farmer adoption.

7.8.4 Ethiopia: Sustainable Land Management in the Highlands

Ethiopia's highlands suffer from severe soil erosion due to steep slopes, deforestation, and intensive cultivation.

- *Practice adopted*: Terracing, reforestation, and integrated watershed management.
- *Impact*:
  - Soil erosion reduced by 50–60%.
  - Crop yields of teff and barley increased by 25–30%.
  - Groundwater recharge improved, reducing seasonal water stress.

Climate link: Restoration practices enhanced resilience against droughts and floods while contributing to long-term carbon sequestration.

Lesson: Strong community participation and government support were key to success. However, maintaining terraces requires ongoing collective effort.

7.8.5 Latin America: Organic Coffee in Central America

Central America is a major coffee-growing region, but climate change has increased the incidence of pests (e.g., coffee leaf rust) and reduced suitable growing areas.

- *Practice adopted*: Organic coffee production combined with shade-grown systems.
- *Impact*:
  - Reduced dependence on chemical inputs, lowering costs and environmental damage.
  - Shade trees moderated microclimates, reducing temperature extremes and pest outbreaks.
  - Farmers accessed premium markets through organic certification.

Climate link: Shade trees sequestered carbon while protecting coffee crops from heat and drought stress.

Lesson: Certification processes can be costly for smallholders. Cooperatives have played a vital role in providing collective certification and market access.

7.8.6 Bangladesh: Floating Agriculture for Flood-Prone Areas

Bangladesh faces recurrent flooding, which destroys crops and undermines food security. Communities have adapted with innovative practices.

- *Practice adopted*: Floating gardens made of water hyacinth and bamboo rafts for cultivating vegetables during floods.

- *Impact*:
  - Ensured year-round vegetable supply in flood-prone areas.
  - Improved household nutrition and provided income opportunities.
  - Low-cost and based on local materials.

Climate link: Floating agriculture is a direct adaptation strategy that transforms floods from a hazard into a resource.

Lesson: Scaling requires support for extension, training, and integration with broader flood management strategies.

7.9 Conclusion and Future Perspectives

Despite being essential to human life and advancement, agriculture is becoming more susceptible to the rapid effects of climate change. Global food security is being threatened by several factors, including soil erosion, rising temperatures, unpredictable rainfall, and the escalation of pests and diseases. This chapter has looked at how sustainable agriculture offers a vital means of striking a balance between ecological stewardship and productivity, emphasizing its function as a climate change adaptation and mitigation tactic.

7.9.1 Key Insights

From the discussion and case studies, several overarching insights emerge:

1. *Resilience via sustainability* Agroforestry, organic farming, conservation agriculture, and integrated nutrient management are examples of sustainable methods that improve biodiversity, soil fertility, and water usage efficiency. These enhancements boost resistance to climate-related stressors, including floods and droughts.
2. *The dual function of adaptation and mitigation* In addition to lowering greenhouse gas emissions (mitigation), many sustainable farming techniques also increase farming systems' resilience to shocks (adaptation). In conservation agriculture, for instance, residue retention enhances soil health and sequesters carbon.
3. *Integration of knowledge* Modern research, traditional wisdom, and farmer-led innovations are frequently combined in successful shifts to climate-resilient agriculture. This combination guarantees that procedures are both socially and technically acceptable.
4. *Institutional and policy support* Case studies revealed that adoption is most successful when accompanied by enabling environments: extension services, access to credit, favourable policies, and functioning markets. Without these, even proven practices struggle to achieve scale.

7.9.2 Challenges Ahead

Despite significant progress, several challenges remain:

Resource limitations: Most farmers in developing nations are smallholders, who frequently do not have access to high-quality inputs, capital, or technology.

Policy gaps: Short-term productivity is still prioritized over long-term sustainability in many agricultural programs. Sustainable practices may be discouraged by subsidies for water-intensive crops or artificial fertilizers.

Climate uncertainty: As unpredictability increases, planning becomes more challenging, and methods need to be resilient to a variety of future events.

Scaling innovations: Inadequate institutional frameworks, weak extension mechanisms, or a lack of funding are some of the reasons why successful pilot programs frequently fail to expand.

Equity and inclusivity: Despite their crucial role in farming, vulnerable groups—women, landless labourers, and indigenous communities—frequently encounter obstacles to their involvement in sustainable agriculture initiatives.

7.9.3 Opportunities for the Future

While the challenges are significant, the opportunities for transforming agriculture into a sustainable, climate-resilient system are equally compelling:

1. Climate-Smart Agriculture (CSA): Growing CSA can contribute to the "triple win" of improved mitigation, adaptation, and productivity. Customized CSA procedures that are adjusted by regional conditions will be crucial.
2. Digital agriculture: New technologies like AI-powered advisory tools, remote sensing, and precision farming can enhance climate predictions, maximize input use, and assist farmers in making decisions.
3. Carbon farming and markets: As carbon markets develop, farmers can profit from carbon-sequestering techniques (such as cover crops and agroforestry), which adds to the incentives for sustainability.
4. Agroecology and circular economies: Long-term sustainability can be achieved by shifting away from input-intensive systems and toward agroecological concepts, which include minimizing waste, recycling biomass, and closing nutrient loops.
5. Policy integration: Aligning agricultural, environmental, and climate policies can create synergies and remove contradictory incentives. Nationally Determined Contributions (NDCs) under the Paris Agreement provide a platform for integrating sustainable agriculture into global climate commitments.

One important takeaway is that sustainable agriculture is a socio-political endeavor that necessitates concerted action at all dimensions, from the farm to global frameworks, and is not only a technical solution. In the twenty-first century, agriculture, sustainability, and climate change are all intertwined. Even if climate change presents enormous obstacles, it also gives rise to the need to rethink agricultural systems in ways that are more resilient, fair, and ecologically responsible. A model for this kind of change, where food security is attained without endangering the health of the world, is provided by sustainable agriculture.

Looking ahead, how well humanity navigates this crucial crossroads will depend on how well scientific innovation, traditional wisdom, supporting policies, and farmer empowerment are integrated. A connected future where people, ecosystems, and economies all prosper together can be achieved by coordinating agricultural development with sustainability and climate resilience.

References

Bunn C, Läderach P, Rivera OO, Kirschke D (2015) A bitter cup: climate change profile of global production of Arabica and Robusta coffee. Clim Chang 129(1–2):89–101

Cairns JE, Hellin J, Sonder K, Araus JL, MacRobert JF, Thierfelder C, Prasanna BM (2013) Adapting maize production to climate change in sub-Saharan Africa. Food Secur 5:345–360

Campbell JR (2014) Climate-change migration in the Pacific. Contemp Pac 26(1):1–28

Challinor AJ, Watson J, Lobell DB, Howden SM, Smith DR, Chhetri N (2014) A meta-analysis of crop yield under climate change and adaptation. Nat Clim Chang 4:287–291

Dankelman I (2010) Gender and climate change: an introduction. Routledge, London

Derpsch R, Friedrich T, Kassam A, Hongwen L (2010) Current status of adoption of no-till farming in the world and some of its main benefits. Int J Agric Biol Eng 3(1):1–25

FAO (2011) The state of food and agriculture 2010–2011: women in agriculture—closing the gender gap for development. Food and Agriculture Organization of the United Nations, Rome

FAO (2013) Climate-smart agriculture sourcebook. Food and Agriculture Organization of the United Nations, Rome

FAO (2017) The future of food and agriculture—trends and challenges. Food and Agriculture Organization of the United Nations, Rome

FAO (2020) The state of world fisheries and aquaculture 2020: sustainability in action. Food and Agriculture Organization of the United Nations, Rome

Gerber PJ, Steinfeld H, Henderson B, Mottet A, Opio C, Dijkman J, Falcucci A, Tempio G (2013) Tackling climate change through livestock: a global assessment of emissions and mitigation opportunities. Food and Agriculture Organization of the United Nations, Rome

IFOAM (2018) Principles of organic agriculture. International Federation of Organic Agriculture Movements, Bonn

IPCC (2019) Climate change and land: an IPCC special report on climate change, desertification, land degradation, sustainable land management, food security, and greenhouse gas fluxes in terrestrial ecosystems. Intergovernmental Panel on Climate Change, Geneva

Jat ML, Gupta R, Saharawat Y, Khosla R, Humphreys E, Amor D, Sidhu HS (2014) Conservation agriculture in cereal systems of South Asia: nutrient management perspectives. Soil Tillage Res 143:50–62

Kang S, Hao X, Du T, Tong L, Su X (2017) Improving agricultural water productivity to ensure food security in China under changing environment: from research to practice. Agric Water Manag 179:5–17

Kassam A, Friedrich T, Shaxson F, Pretty J (2009) The spread of conservation agriculture: justification, sustainability and uptake. Int J Agric Sustain 7(4):292–320

Knox J, Hess T, Daccache A, Wheeler T (2012) Climate change impacts on crop productivity in Africa and South Asia. Environ Res Lett 7(3):034032

Lampayan RM, Rejesus RM, Singleton GR, Bouman BAM (2015) Adoption and economics of alternate wetting and drying water management for irrigated lowland rice. Field Crop Res 170:95–108
Lin BB (2011) Resilience in agriculture through crop diversification: adaptive management for environmental change. Bioscience 61(3):183–193
Meinzen-Dick R, Quisumbing A, Behrman J, Biermayr-Jenzano P, Wilde V, Noordeloos M, Ragasa C, Beintema N (2011) Engendering agricultural research, development and extension. International Food Policy Research Institute (IFPRI), Washington, DC
Minasny B, Malone BP, McBratney AB, Angers DA, Arrouays D, Chambers A, Chaplot V, Chen ZS, Cheng K, Das BS et al (2017) Soil carbon 4 per mille. Geoderma 292:59–86
Mwongera C, Shikuku KM, Twyman J, Läderach P, Ampaire E, Van Asten P, Winowiecki L (2017) Climate smart agriculture rapid appraisal (CSA-RA): a tool for prioritizing context-specific climate smart agriculture technologies. Agric Syst 151:192–203
Paris Agreement (2015) Adopted under the United Nations Framework Convention on Climate Change (UNFCCC), 21st Conference of the Parties (COP21), Paris, 12 December 2015
Rockström J, Falkenmark M, Karlberg L, Hoff H, Rost S, Gerten D (2010) Future water availability for global food production: the potential of green water for increasing resilience to global change. Water Resour Res 45(7):W00A12
Smith P, Bustamante M, Ahammad H, Clark H, Dong H, Elsiddig EA, Haberl H, Harper R, House J, Jafari M, Masera O, Mbow C, Ravindranath NH, Rice CW, Robledo Abad C, Romanovskaya A, Sperling F, Tubiello F (2014) Agriculture, Forestry and Other Land Use (AFOLU). In: Edenhofer O et al (eds) Climate change 2014: mitigation of climate change. Contribution of Working Group III to the Fifth Assessment Report of the Intergovernmental Panel on Climate Change. Cambridge University Press, Cambridge and New York, pp 811–922
Snapp SS, Blackie MJ, Gilbert RA, Bezner-Kerr R, Kanyama-Phiri GY (2010) Biodiversity can support a greener revolution in Africa. PNAS 107(48):20840–20845
Thornton PK, Ericksen PJ, Herrero M, Challinor AJ (2014) Climate variability and vulnerability to climate change: a review. Glob Chang Biol 20(11):3313–3328
Thornton PK, Ericksen PJ, Herrero M, Challinor AJ (2018) Climate change and vulnerability in African agriculture. Annu Rev Environ Resour 43:341–367
UNFCCC (2018) Koronivia joint work on agriculture: decision 4/CP.23. United Nations Framework Convention on Climate Change, Bonn
Van der Heijden MGA, Bardgett RD, van Straalen NM (2008) The unseen majority: soil microbes as drivers of plant diversity and productivity. Ecol Lett 11(3):296–310
Vorley B, Del Pozo-Vergnes E, Barnett A (2012) Small producer agency in the globalised market: making choices in a changing world. IIED and HIVOS, London
WCED (1987) Our common future. Oxford University Press, Oxford
Willer H, Trávníček J, Meier C, Schlatter B (eds) (2021) The world of organic agriculture: statistics and emerging trends 2021. FiBL and IFOAM—Organics International, Bonn
Zhao C, Liu B, Piao S, Wang X, Lobell DB, Huang Y, Huang M, Yao Y, Bassu S, Ciais P, Durand JL, Elliott J, Ewert F, Janssens IA, Li T, Lin E, Liu Q, Martre P, Müller C, Peng S, Peñuelas J, Ruane AC, Wallach D, Wang T, Wu D, Liu Z, Zhu Y, Asseng S (2017) Temperature increase reduces global yields of major crops in four independent estimates. PNAS 114(35):9326–9331

Technological Advances in Studying Plant Environment Interactions

8

Maneesh Lingwan, Yogesh Pant, and Manushree

Abstract

Understanding how plants perceive, respond, and adapt to their dynamic environments is fundamental to addressing the critical challenges of climate change, food security, and sustainable agriculture. Plant-environment interactions involve a complex network of molecular, biochemical, physiological, and ecological processes. Recent technological breakthroughs have revolutionized the study of these interactions, offering unprecedented insights into how plants detect and respond to diverse biotic and abiotic stimuli. Among these advances, metabolomics has emerged as a central and powerful tool, enabling comprehensive profiling of plant metabolites that act as both regulators and indicators of environmental responses. This chapter provides an in-depth overview of state-of-the-art metabolomics platforms, including mass spectrometry, nuclear magnetic resonance, and imaging-based technologies, as well as their integration with other omics, such as high-throughput phenotyping, genomics, transcriptomics, and proteomics. Both targeted and untargeted metabolomics methods are explored in the context of understanding complex metabolic networks that enable plants to adapt to environmental stresses such as drought, salinity, nutrient fluctuations, and pathogen attacks. The synergistic use of metabolomics with CRISPR/Cas systems, genome-editing tools, environmental simulation systems, and real-time biosensors is expanding the scope of plant-environment research. Through recent

M. Lingwan (✉)
Donald Danforth Plant Science Center, Saint Louis, MO, USA
e-mail: mlingwan@danforthcenter.org

Y. Pant
Indian Institute of Technology, Mandi, HP, India

Manushree
Metalytics, Inc., Cary, NC, USA

H. C. Joshi et al. (eds.), *Climate-Responsive Plant Biology*,
https://doi.org/10.1007/978-981-95-7354-7_8

case studies, we demonstrate how these integrative and multidisciplinary approaches are transforming our understanding of plant physiology and ecology, ultimately informing the development of climate-resilient crops and innovative, sustainable agricultural practices.

Keywords

Plants · Environments · Omics · Mass spectrometry · Interactions · Sustainable agriculture

8.1 Introduction

Plants live in dynamic environments where they must continuously perceive and respond to fluctuating signals such as light, temperature, water availability, salinity, nutrient status, pathogens, and herbivores (Yadav et al. 2020; Saini et al. 2020). Unlike animals, which can migrate to escape adverse conditions, plants are immobile and must adapt in place or risk failure. This constraint has driven the evolution of highly specialized sensory and signaling systems that convert environmental inputs into molecular, biochemical, and physiological responses, ultimately influencing plant growth, reproduction, and ecological interactions. Throughout the twentieth century, research on plant-environment interactions primarily focused on physiological and biochemical indicators, such as photosynthesis or enzyme activity, under stressful conditions. While this provided some insight into complex systems, it only offered a partial picture. The rise of molecular biology in the 1980s and 1990s led to the identification of specific genes and proteins linked to stress. Nevertheless, questions remained about how tissues respond differently, how systems work together, and how environments affect integration. Over the last two decades, however, advances in high-throughput omics, imaging technologies, genome editing, and computational biology have revolutionized the field, shifting it from descriptive analysis to predictive and translational research (Varadharajan et al. 2025; Araus et al. 2018).

Among multi-omics approaches, metabolomics has emerged as a compelling method. Because metabolite levels reflect the combined outcomes of gene regulation, enzyme activity, and environmental modulation, they provide a direct snapshot of cellular function. Unlike genomic or transcriptomic datasets, which reveal potential capacity, metabolomics captures real-time biochemical states. When integrated with genomics, transcriptomics, proteomics, ionomics, and next-generation phenomics, metabolomics enables the development of multiscale models that offer both diagnostic and predictive insights into plant stress responses. Single-cell and spatial transcriptomics enable the resolution of stress responses with cellular precision, uncovering specialized stress-responsive states in root and shoot tissues (Yin et al. 2023; Zhu et al. 2025; Lee et al. 2025; Adema et al. 2024; Lingwan et al. 2023a). When combined with spatial metabolomics, these approaches link local gene activity with chemical microenvironments, creating maps of stress-induced metabolic landscapes (Aratani et al. 2023; Rowe et al. 2025; Wagner and Weber 2025). Complementary progress in CRISPR-based genome editing, including base and prime editing, has expanded the toolkit for probing and manipulating stress response pathways. These technologies enable the precise functional validation of candidate genes and the rapid deployment of stress-resilient alleles into breeding pipelines (Vu et al. 2024; Jin et al. 2023; Pan et al. 2022; Chen et al. 2024).

At the field level, unmanned aerial vehicles (UAVs), multispectral and hyperspectral cameras, and robotic phenotyping platforms are producing vast amounts of data on canopy structure, pigment makeup, thermal signatures, and growth patterns. When paired with artificial intelligence (AI) and machine learning, these datasets are revolutionizing phenotyping into a predictive field that can detect early signs of stress, forecast yields, and speed up genotype selection in real-world conditions (Araus et al. 2018; Qiu et al. 2024; Kaur et al. 2025; Zhu et al. 2024). These breakthroughs, which include single-cell omics and field-level AI phenotyping, are revolutionizing our understanding of how plants interact with their environment. By revealing the underlying mechanisms of stress perception and adaptation, they are also closing the gap between basic research and practical applications. This combined approach provides a robust foundation for transforming fundamental science into climate-resilient and sustainable agricultural practices for the future.

8.2 Complexity of Plant Environment Interactions and Signaling

Plants have evolved complex adaptive traits and regulatory mechanisms that enable them to survive and reproduce under constantly changing environmental conditions. The primary strategies include changes in shape and structure, such as flexible roots and leaf shapes, as well as physiological and biochemical responses, like regulating stomata, adjusting osmotic pressure, accumulating antioxidants, and molecular adaptations that reprogram gene expression networks (Zhu 2016; Bailey-Serres et al. 2019). These environmental cues influence how plants function and their traits, such as the amount of biomass, reproductive efficiency, and yield. The genetics of the same species vary significantly when grown under different conditions or at different developmental stages. These variations underscore the complex relationship between genetics and environment, which has a profound impact on sustainable agriculture from a broader perspective (Varoquaux et al. 2019; Rasheed et al. 2020). Abiotic and biotic stresses not only influence crop yields but also affect the nutritional quality of edible tissues. Stress can reduce micronutrient accumulations such as iron, zinc, and carotenoids while increasing anti-nutritional compounds, such as phytates, oxalates, and cyanogenic glycosides (Sperotto et al. 2021).

Plants sense environmental signals, such as light, temperature, water levels, salt levels, nutrients, and interactions with other living things, through specialized receptors in their membranes, organelles, and nuclei. Specialized receptors allow them to convert external cues into responses that help them adapt to their environment, enabling them to manage stress effectively (Kushwaha et al. 2022). In plants, phytochromes detect red and far-red wavelengths, cryptochromes and phototropins respond to blue and UV-A light, and UVR8 senses UV-B. Phytochromes switch between inactive (Pr) and active (Pfr) forms to regulate plant growth, shade avoidance, and natural rhythm (Tripathi et al. 2019; Li et al. 2011; Paik and Huq 2019; Lingwan et al. 2024). On a biochemical level, plants facing environmental stress undergo significant changes in their metabolic processes to survive. Plants exposed to UV-B radiation, high light, drought, salinity, or pathogen infection exhibit characteristic metabolic signatures, including increases in osmoprotectants such as proline and trehalose, antioxidants (ascorbate and glutathione), and defense metabolites (flavonoids and alkaloids). These metabolites not only serve adaptive functions but also act as biomarkers of resilience. These changes often involve producing and using specific chemicals that help maintain water balance, regulate oxidation, and strengthen their defenses. When plants confront drought, salinity, or cold stress, they accumulate compatible solutes, including proline, glycine betaine, mannitol, and raffinose family oligosaccharides (Yadav et al. 2019a, b; Job et al. 2022).

8.3 Metabolic and Developmental Strategies of Plants in Response to Abiotic Stress

Temperature sensing involves disrupting proteins, which trigger the expression of chaperones to maintain protein stability, whereas cold changes the composition of membrane lipids and promotes the buildup of cryoprotectants. Osmosensors and mechanosensitive ion channels detect water deficit and salinity. Under heat stress, plants may also undergo structural adjustments in their thylakoid membranes and upregulate protective proteins such as heat shock proteins (HSPs) and Rubisco activase to stabilize photosynthetic function (Chaves et al. 2009; Lingwan 2024b). Drought and heat can accelerate flowering (drought escape) or induce floral abortion under severe conditions to conserve resources (Shavrukov et al. 2017). High temperatures during anthesis reduce pollen viability and stigma receptivity in cereals such as rice and wheat, while drought during grain filling limits assimilate transport, leading to reduced kernel number and weight (Zhao et al. 2016; Jagadish et al. 2021; Lingwan 2024a; Ali et al. 2025). Salicylic acid enhances resilience to heat and oxidative stress by stimulating antioxidant enzymes, such as catalase and ascorbate peroxidase, while other phytohormones, including ABA and auxin, refine stress responses and development in plants and bryophytes through their signaling pathways (Mohanasundaram et al. 2021; Ali et al. 2024). ABA is reported to interact with ethylene, jasmonates, salicylic acid, and auxin to regulate cold tolerance in tomatoes through osmolyte accumulation, modifying membrane composition, and antioxidant responses (Kim et al. 2024; Ding et al. 2025). These metabolites help reduce cellular damage, and studies have demonstrated that the accumulation of raffinose and galactinol in plants like Arabidopsis and various cereals can enhance their resistance to freezing. Wheat is particularly susceptible to terminal heat stress, which reduces grain filling and increases spike sterility. The integration of multi-omics approaches has demonstrated an upregulation of flavonoid biosynthesis and antioxidant pathways in heat-tolerant genotypes, offering reliable markers for selection (Kumar et al. 2023).

Drought responses rely heavily on abscisic acid (ABA) signaling. At the same time, salinity triggers the Salt Overly Sensitive (SOS) pathway, which includes CBL-interacting protein kinase and SOS1 (a sodium-hydrogen antiporter) to restore ionic balance (Ji et al. 2013). Root plasticity is a critical component of adaptation. Under drought, many crops exhibit deeper rooting or altered gravitropism to access subsoil water, often controlled by auxin and cytokinin signaling (Uga et al. 2013). One of the earliest responses to drought and salinity stress is stomatal closure, mediated by ABA signaling. This conserves water but restricts CO_2 uptake, forcing plants to shift toward photorespiration and alternative electron sinks such as the water–water cycle and cyclic electron transport (Lawson and Matthews 2020). Abscisic acid (ABA) plays a central role in abiotic stress tolerance. Under drought conditions, ABA signaling recruits Casein Kinase–Like Kinase 2 (CKL2) via PP2C–PYR/PYL/RCAR receptor complexes, leading to cytoskeletal rearrangements that reinforce stomatal closure (Mo et al. 2024). Arabidopsis guard cells accumulate sugars and osmolytes faster than mesophyll cells under dehydration. In rice, ABA is reported to increase ROS-scavenging enzymes and adjust root structure and leaf angle to boost drought resistance (Gul et al. 2024).

Plants detect nutrients through transporters and receptors that recognize nitrogen, phosphate, potassium, and micronutrients. When these nutrients are in short supply or in excess, they trigger changes in gene expression and metabolism (Pant et al. 2023). Nutrient deficiency triggers distinct architectural adjustments: phosphorus limitation induces lateral root proliferation and cluster root formation, while nitrogen scarcity promotes increased root hair elongation and nodulation, facilitating symbiosis with rhizobia or mycorrhizal fungi (Lynch 2019; Oldroyd and Leyser 2020). Under prolonged stress, resources are diverted from growth toward survival. At the same time, leaf expansion is inhibited, and senescence is accelerated to remobilize nutrients, particularly nitrogen and phosphorus, from older tissues to actively growing or reproductive organs. This process is strongly regulated by hormonal crosstalk involving ABA, ethylene, and jasmonic acid, which activate transcription factors such as NAC and WRKY to promote catabolic processes (Lim et al. 2007; Woo et al. 2019). Nutrient stress can delay flowering or alter seed set, further linking reproductive output to environmental plasticity. These responses demonstrate that adaptation encompasses not only short-term survival but also the strategic modulation of reproductive fitness under suboptimal conditions. These physiological and developmental strategies underscore the integrative nature of stress adaptation. By coordinating photosynthesis, resource allocation, root and leaf development, and reproductive processes, plants ensure survival while maximizing reproductive output in challenging environments.

8.4 Metabolic Reprogramming and Secondary Metabolite Accumulation Under Stress

Under stress, cells often switch their energy focus from building resources to breaking them down. This provides instant energy and the necessary building blocks for defense. The switch is characterized by increased activity in pathways such as glycolysis, the TCA cycle, and fatty acid breakdown. These pathways support ATP production and provide the components needed for other metabolic processes. Stress also triggers the activation of a mitochondrial enzyme called alternative oxidase. This enzyme helps maintain a healthy balance of electrons by preventing their overproduction in the electron transport chain. Exposure to stress typically leads to an increase in phenylpropanoids, flavonoids, terpenoids, and alkaloids. These substances act as antioxidants, UV protectors, signaling molecules, and antimicrobial agents. For example, flavonoids serve as antioxidants in leaves under drought conditions, while terpenoids are both direct defenses and indirect signals that attract predators of herbivores (Job et al. 2025). The production of stress-induced metabolites is controlled in specific areas, reflecting the plant's adaptive strategies. For example, anthocyanins accumulate in the outer layers of plant tissues when they are exposed to UV radiation, shielding the underlying cells from damage. In contrast, roots tend to store polyamines and compatible solutes, which help them adapt to changes in water and salt levels, thereby remaining viable. These biochemical responses demonstrate the adaptability of plant metabolism under stress and highlight how metabolic reprogramming provides an immediate defense mechanism before long-term developmental or genetic adaptations take effect (Fig. 8.1).

8.5 Analytical Advances Transforming Plant Metabolism and Interaction Studies

The metabolome, which encompasses the complete set of small molecules produced by enzymatic activity, reflects the most direct biochemical manifestation of these processes. Since metabolite levels combine genetic potential, transcriptional regulation, protein activity, and environmental factors, metabolomics is often viewed as the closest proxy to the actual phenotype (Fiehn 2022; Fernie and Tohge 2017; Saito and Matsuda 2010). It provides a real-time biochemical fingerprint that captures the outcome of gene-environmental interactions. For this reason, metabolomics has become a central part of plant stress biology and systems-level studies, bridging the gap between genotype, physiological processes, and observable traits (Fiehn 2022; Misra et al. 2021). The past two decades have been marked by a revolution in analytical technologies, which now enable us to interrogate plant responses at unprecedented resolution. While the foundational knowledge of plant physiology provided important insights into stress perception and adaptation, new methods have shifted the paradigm from descriptive biology to high-resolution, systems-based and predictive modeling. Different omics approaches are applicable in different aspects of plant interaction studies.

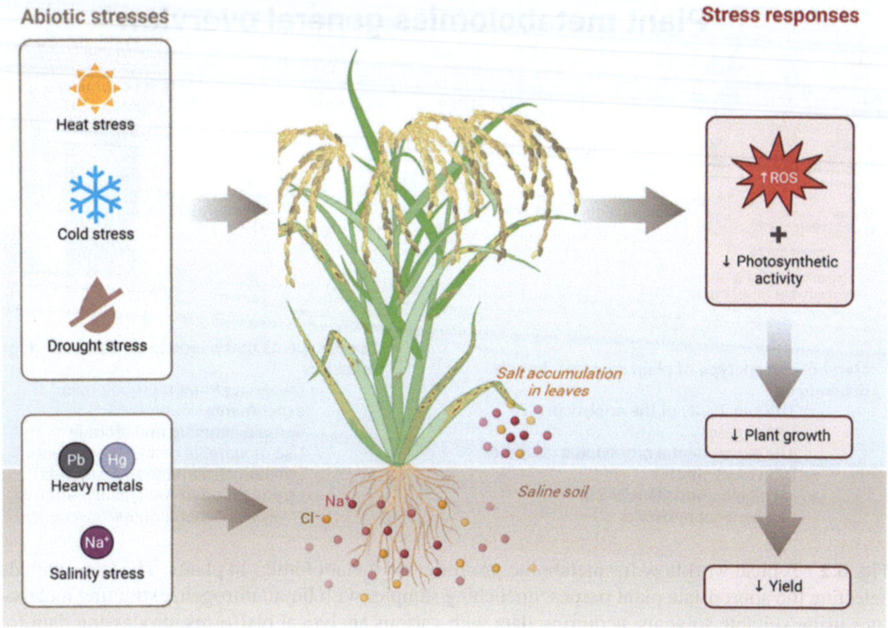

Fig. 8.1 Plants, being sessile in nature, must survive and reproduce in a wide range of environmental stresses, including drought, salinity, heat, and cold, as well as biotic pressures such as pathogenesis and herbivory

8.6 Technological Platforms in Plant Metabolomics

Metabolomics involves the comprehensive analysis of metabolites within a cell system. The identification of these small molecules highlights both differences and similarities in normal and disturbed physiological conditions. The principle of metabolite extraction primarily involves analyzing the cationic (amino acids, amino acid-related compounds, small peptides, amines, nucleobases, and nucleosides) and anionic (tricarboxylic acids and phosphorylated compounds) metabolome. Gas Chromatography-Mass Spectrometry (GC-MS) is highly sensitive to volatiles and derivatized compounds, which are commonly used in studies of central carbon metabolism (Shree et al. 2019). The process involves derivatizing the phytochemicals along with internal standards such as ribitol to make them volatile and detectable by GC-MS. The methoximation and silylation derivatization method is routinely employed to analyze a broad range of metabolites, including amino acids, alcohols, sugars, amines, sugar alcohols, and secondary metabolites (Lingwan and Masakapalli 2022). Liquid Chromatography-Mass Spectrometry (LC-MS) enables the profiling of semi-polar and specialized metabolites, such as flavonoids, stilbenes, and phytohormones. The Capillary Electrophoresis-Mass Spectrometry (CE-MS) technique is particularly suited for highly polar and charged molecules, including organic acids and amino acid derivatives. CE-MS-based profiling

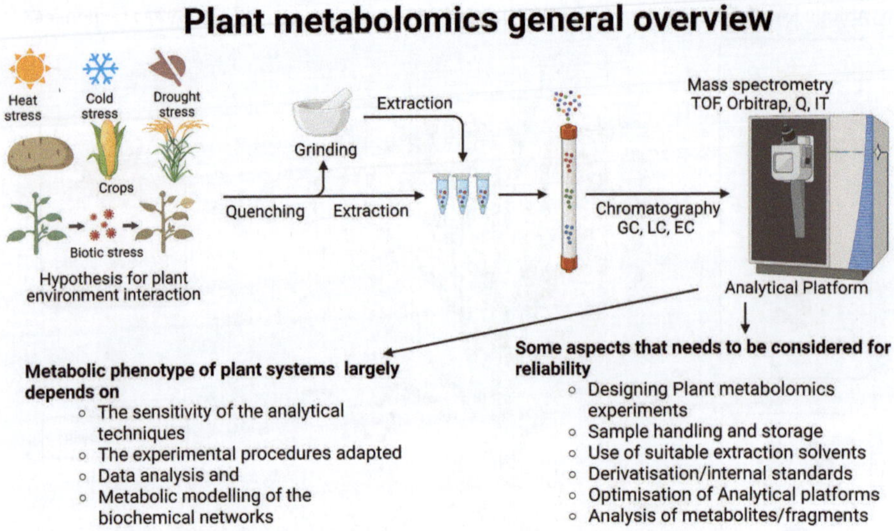

Fig. 8.2 Typical workflow for metabolite profiling and metabolomics in plants. The steps include selecting the appropriate plant tissues, quenching samples with liquid nitrogen, extracting metabolites using suitable solvents, acquiring data with various analytical platforms, processing data for metabolite profiling with specialized software, conducting multivariate statistical analysis for comprehensive metabolomics using statistical tools, and interpreting the biological results

facilitates the identification of differential metabolite responses in cadmium-accumulating soybean cultivars. However, this is not the most widely used approach for metabolome analysis. It offers a unique advantage over other omics in analyzing the polar metabolome (Fig. 8.2).

Nuclear Magnetic Resonance (NMR) spectroscopy offers reproducible, structural, and quantitative analysis of specific metabolites, including protein-bound metabolites (lipoprotein particles) and inorganic metabolites or ions (metal ions and H+ ions via pH), that cannot be done via LC-MS or GC-MS (Lingwan et al. 2023b; Shagun et al. 2024). Recent advances in cryoprobe technology have significantly improved NMR sensitivity, making it more competitive in large-scale plant studies. Breakthroughs allow metabolite profiling in single cells or even subcellular compartments. Methods such as microcapillary sampling coupled with nanoLC–MS or microprobe electrospray ionization enable interrogation at the picogram scale. Imaging mass spectrometry (IMS) provides spatially resolved information on metabolites without the need for extraction. MALDI-IMS has mapped the accumulation of phytoalexins in infection loci of Arabidopsis, highlighting the local defense nature of these compounds. Desorption electrospray ionization mass spectrometry (DESI–MS) enables in situ analysis of plant tissues with minimal preparation. Raman and FTIR imaging allow chemical fingerprinting of metabolites in intact tissues and even living roots. These approaches bridge molecular and ecological scales by uncovering how gradients in metabolites shape interactions with microbes, pests, and neighboring plants.

8.7 Metabolic Flux Analysis (MFA) for Quantitative Estimation of Central Metabolic Pathways

MFA is used in plants to quantify the rates of metabolic reactions within central and specialized pathways. Stable isotope labeling using 13C or 15N has enabled insights into metabolic fluxes under stress, revealing dynamic re-routing of carbon and nitrogen resources during acclimation (Koley et al. 2024). Fluxomics has been established to be the most relevant approach to define the metabolic phenotype of any cellular system, including plants. It deals with the determination of metabolic rates in central metabolism as fluxes. In vivo reaction rates or fluxes describe the rate at which one metabolite is converted into another within a metabolic pathway. Quantification of fluxes has already been proven to play an essential role in defining intracellular metabolism in various biological entities. MFA is a powerful approach for qualitative and quantitative estimation of the activities of central metabolic pathways. Qualitative estimation involves tracking the carbon flow from a labelled substrate into metabolic intermediates, including amino acids that retro-biosynthetically report on the central precursors. At the same time, quantitative estimation involves establishing a metabolic phenotype using the in vivo reaction rates of metabolic pathways (Cullen and Lingwan 2025). MFA is based on 13C labelling experiments, which employ various experimental and computational approaches (Lingwan et al. 2024). Its applications include mapping carbon and nitrogen flow through photosynthesis, respiration, and the synthesis of amino acids. MFA aids in identifying metabolic bottlenecks that limit growth, and understanding stress responses by tracking how plant stress alters fluxes (Lingwan and Yadav 2025).

8.8 Emerging Tools Beyond Omics: Gene Editing, Biosensors, Synthetic Biology

Beyond multi-omics, advances in gene editing, synthetic biology, and computational platforms are transforming the study of plant–environment interactions by shifting from correlation-based analyses to causal understanding and targeted modification of stress responses. Genome editing, particularly CRISPR–Cas systems, has revolutionized functional genomics by enabling the precise modification of stress-related genes and pathways. CRISPR–Cas9 editing of receptor PYL genes in rice created lines with improved drought tolerance without affecting growth. Editing the sodium transporter OsHKT1;5 enhanced sodium exclusion and salinity tolerance under field conditions (Meng et al. 2022; Zhang et al. 2023). In tomato, CRISPR-based reprogramming of carotenoid and flavonoid biosynthetic pathways increased both stress tolerance and fruit nutritional value, demonstrating how editing can provide dual benefits for resilience and food quality. Functional studies verified that CRISPR-mediated targeting of TaWRKY transcription factors enhances thermotolerance, thereby reducing reproductive sterility under field heat stress conditions (Singh et al., 2022; Luo et al., 2025).

Plant biosensors enable non-destructive, real-time monitoring of stress responses at the cellular, tissue, or whole-plant levels. Their high spatial and temporal resolution allows early detection of stress before visible symptoms appear, supporting timely interventions. Quantitative and sensitive measurements of signals, such as ROS, ions, and hormones, connect environmental cues to molecular signaling networks, providing mechanistic insights into how plants perceive and manage stress. To complement omics datasets that offer static snapshots of plant states, biosensors provide real-time, spatiotemporal insights into physiological and biochemical processes. By enabling dynamic monitoring of intracellular signals, metabolites, and stress markers, biosensors connect molecular discoveries with physiological outcomes. Genetically encoded biosensors (GEBs) are molecular tools that utilize fluorescent or luminescent proteins engineered to alter their shape or spectral properties upon binding to a target molecule. These sensors enable in vivo visualization of processes like hormone signaling, calcium fluxes, or metabolite accumulation (Toyota et al. 2018; Rowe et al. 2025; Lingwan and Yadav 2025). Engineered nanomaterials are emerging as versatile platforms for plant monitoring. Carbon nanotube–based biosensors, when introduced into leaves, can detect small molecules such as H_2O_2 or nitrate with high sensitivity, and they have been deployed in the field for real-time detection of stress markers. When combined with synthetic biology, biosensors can be integrated into engineered feedback systems, enabling crops to sense and respond to stress independently, thereby enhancing their resilience (Lew et al. 2020).

8.9 AI and Machine Learning in Plant Environment Interaction Studies

Artificial intelligence (AI) and machine learning (ML) are transforming the way plant–environment interactions are studied, providing tools to analyze large, complex datasets that include genomics, phenomics, metabolomics, and environmental monitoring. By extracting hidden patterns and building predictive models, AI/ML provides a data-driven and scalable framework to enhance agricultural decision-making and accelerate the development of climate-resilient crops. AI and ML techniques automate high-throughput plant phenotyping using image-based datasets that capture traits such as leaf area, pigmentation, and biomass. Convolutional Neural Networks (CNNs), a type of deep learning, can accurately classify disease symptoms and nutrient deficiencies in leaf images, surpassing the accuracy of human experts (Singh et al. 2023). AI models can detect, classify, and diagnose plant stress caused by abiotic factors such as drought, temperature extremes, and soil salinity, as well as biotic stressors including pathogens, pests, and weeds. ML algorithms analyze data streams from remote sensors, UAVs, or satellites to identify early signatures of stress before visible symptoms emerge (Araus et al. 2018; Chen et al. 2024).

AI enables the discovery of novel targets for crop breeding and supports genomic selection pipelines for climate-resilient varieties (Varoquaux et al. 2019). AI models are also being applied to simulate future climate scenarios, predicting shifts in plant growth, crop productivity, and pathogen dynamics. Messina et al. (2025) demonstrated that combining AI with crop modeling frameworks improves projections of yield responses under climate change, enabling the development of adaptive management strategies. The integration of AI and ML into plant sciences offers numerous benefits, transforming both research and agriculture. Their predictive capabilities enhance the accuracy of stress detection and classification, allowing researchers and farmers to identify problems before visible symptoms appear. Early detection supports timely interventions that help prevent yield losses and reduce reliance on excessive inputs, such as water, fertilizers, and pesticides.

8.10 Integrating Mutomics Techniques

By integrating multi-omics, genome editing, metabolite biomarkers, and high-throughput phenotyping, we are rapidly translating fundamental discoveries about plant–environment interactions into practical traits for agriculture. These methods collaborate to identify stress-response points, validate them through editing or transgenic techniques, and progress rapidly with field-relevant phenotyping, ultimately developing cultivars that are resistant to climate change and high in nutrition. Conventional breeding methods have struggled to keep pace with changing climates. A significant trend is the integration of metabolomics with other omics, including genomics, transcriptomics, and proteomics, to generate systems-level views. When combined with transcriptomics and metabolomics, it was revealed that drought-tolerant maize has elevated trehalose metabolism and sucrose partitioning as resilience hubs (Xu et al. 2020). High-throughput phenotyping linked with thermal and hyperspectral metabolic data enables prediction of canopy-level stress performance (Munns et al. 2020; Araus et al. 2018). Integrated metabolomics and transcriptomics in rice revealed trehalose, raffinose, and galactinol as biomarkers of drought tolerance, compounds that function in osmoprotection and carbon allocation (Wang et al. 2021). Overexpression of trehalose-6-phosphate phosphatase (OsTPP) enhanced drought resilience by increasing root vigor and improving sugar partitioning, resulting in yield increases under field stress (Nuccio et al. 2015). In maize, metabolomics-driven biomarker identification has reliably recognized proline, polyamines, and soluble sugars as indicators associated with drought tolerance. Furthermore, hyperspectral imaging facilitates early and non-destructive screening of these metabolites within breeding initiatives (Xu et al. 2020; Joshi et al. 2021). CRISPR-mediated promoter editing of ARGOS8, a negative regulator of ethylene signaling, resulted in the development of field-tested lines exhibiting increased grain yield under drought conditions, with no adverse effects observed under well-watered conditions (Shi et al. 2017).

8.11 Plant Microbiome Interactions and Rhizosphere Engineering for Sustainable Agriculture

Besides genetic factors in the host, metabolomic profiling of root exudates has identified compounds that attract beneficial microbes in the rhizosphere, such as stress-resilient strains of Pseudomonas and Bacillus, thus linking molecular breeding and microbiome engineering for resilience (Sharma et al. 2022; Afzal et al. 2025; Liu et al. 2025). Metabolomics profiling shows that drought and salinity stimulate the exudation of organic acids (malate, citrate), flavonoids, benzoxazinoids, and amino acids, which specifically attract beneficial microbes. In maize, drought-induced malate and citrate exudation recruited Pseudomonas fluorescens and Bacillus subtilis, microbes that produce ACC deaminase and indole-3-acetic acid (IAA), thereby reducing ethylene stress and encouraging root growth (Singh et al. 2023; Rolli et al. 2015).

Sustainable farming systems need to rely less on synthetic fertilizers and pesticides while maintaining high productivity. Plants accomplish this partly by reprogramming the chemistry of root exudates in response to stress, which then affects the composition of the rhizosphere microbiome. Modern omics platforms, including metabolomics, metagenomics, metatranscriptomics, and spatial metabolomics, have improved our understanding of how stress-induced exudates attract beneficial microbial taxa (Berendsen et al. 2012; Hu et al. 2018). In wheat, benzoxazinoid exudates promote the growth of Actinobacteria with siderophore and ROS-scavenging abilities, thereby enhancing drought resilience and iron absorption (Hu et al. 2018; Afzal et al. 2025). CRISPR–Cas9 modifications of OsF3H and OsCHS in rice increased flavonoid secretion and improved colonization by nitrogen-fixing bacteria, including Azospirillum and Herbaspirillum, thereby enhancing nitrogen use efficiency (Chen et al. 2024). In Arabidopsis, knocking out the ABC transporter PEN3 resulted in the production of indolic exudates and altered the rhizosphere microbiome (Badri et al. 2009). Synthetic microbial communities consisting of Pseudomonas, Bacillus, Streptomyces, and Rhizophagus irregularis (arbuscular mycorrhizal fungi), tailored to maize exudation patterns, enhanced biomass and phosphorus uptake under drought conditions (Compant et al. 2021). Similar metabolomics-informed inoculants enhanced water-use efficiency and yield stability in maize (Niu et al. 2017). In contrast, heat-modulated wheat exudates recruited Pseudomonas spp., which improved thermotolerance (Sharma et al. 2022; Liu et al. 2025). Overall, these findings show how omics-based exudate profiling, gene editing of metabolite pathways, and synthetic microbial inoculants create a strong foundation for incorporating microbiome engineering into sustainable agricultural practices (Fig. 8.3).

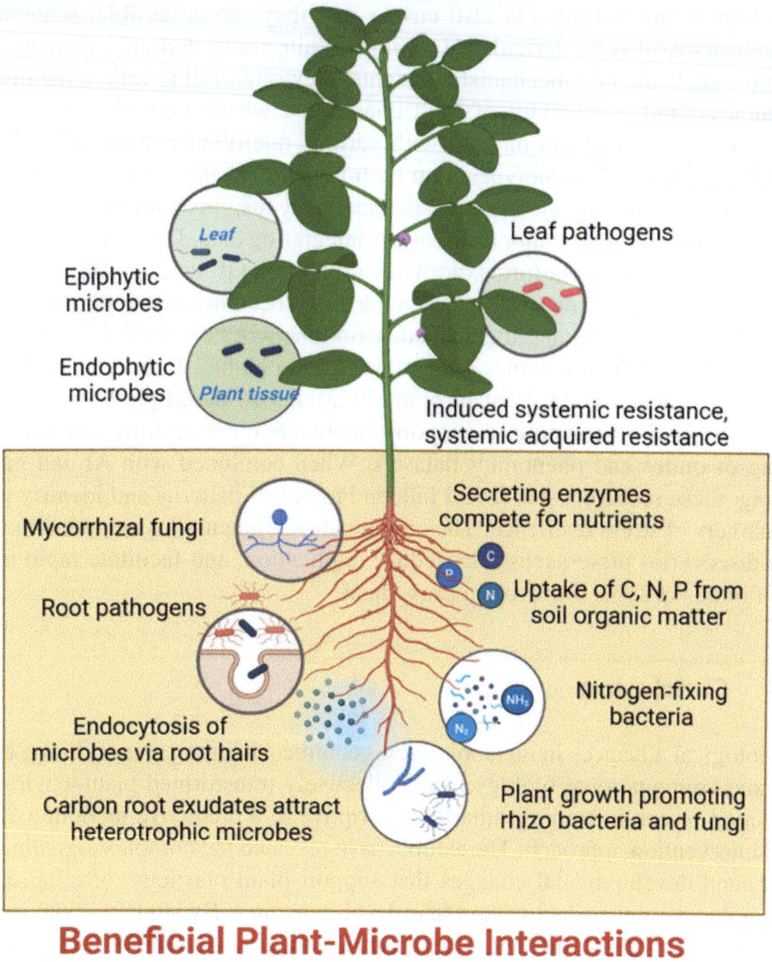

Fig. 8.3 Plants engage in diverse interactions with both below- and above-ground microbes that shape their growth and resilience. In the rhizosphere, carbon-rich root exudates attract heterotrophic microbes, while mycorrhizal fungi enhance nutrient exchange. Endocytosis of microbes via root hairs supports symbiosis, whereas pathogenic microbes challenge plant defenses. Nitrogen-fixing bacteria supply bioavailable nitrogen, while rhizobacteria and fungi promote plant growth. Microbial communities also secrete enzymes to compete for carbon, nitrogen, and phosphorus, improving nutrient cycling. In the phyllosphere and plant tissues, commensal microbes further support health and stress tolerance, together forming an integrated plant–microbe network

8.12 Future Perspectives and Technical Challenges and Socioeconomic Considerations

Despite significant progress in understanding plant–environment interactions and the use of advanced tools, numerous challenges remain in applying these laboratory findings to real-world field situations. Future research should not only enhance technological innovations but also ensure that they are accessible, scalable, and equitable across diverse agricultural settings. While controlled-environment studies have provided valuable mechanistic insights, they often fail to reflect the diversity, randomness, and complexity of field conditions, where multiple stresses occur simultaneously and plants interact with various microbial communities. Linking plant physiology and phenotypes with high-resolution omics datasets can provide an effective way to integrate mechanistic laboratory insights with the complex realities of the field. Another major challenge is integrating data. Different labs often use various metabolomics platforms, normalization methods, and approaches to annotate metadata. Comparing studies across the board becomes difficult. Standardizing protocols, data formats, and metadata descriptions will be essential for developing interoperable, FAIR (Findable, Accessible, Interoperable, Reusable) multi-omics databases (Lingwan 2025; Wishart et al. 2022). Cloud-based platforms and federated databases are beginning to support collaborative research by enabling global sharing of omics and phenomics datasets. When combined with AI and machine learning, these resources can reveal hidden biological patterns and identify predictive markers of stress resilience. The ongoing development of AI-enabled tools will make discoveries more accessible, reduce duplication, and facilitate rapid translation of omics insights into crop improvement.

8.13 Conclusion

Technological advances in metabolomics, genomics, imaging, gene editing, biosensors, and computational biology have collectively transformed plant–environment interaction research from a primarily descriptive to a predictive modeling and targeted intervention approach. These tools have revealed the complex signaling, metabolic, and developmental changes that support plant plasticity, offering insights from molecular networks to ecosystem-level dynamics. By studying both immediate cellular responses and long-term adaptive strategies, researchers can now link the basic mechanisms of stress perception to the observable traits that influence plant productivity and resilience. Adopting high-throughput technologies promises to increase yield and stress tolerance while also enhancing nutritional quality, environmental sustainability, and resilience to climate change. By integrating innovation within ethical, inclusive, and systems-level frameworks, plant science is positioned to provide the basis for a sustainable agricultural future capable of confronting the urgent challenges of global food security.

References

Adema K, Munnik T, Testerink C (2024) Lost in space: what single-cell RNA sequencing cannot tell plant biologists. Trends Plant Sci 29(12):1206–1218

Afzal M, Raza A, Hussain S et al (2025) Wheat root exudates under heat stress recruit beneficial *Pseudomonas* spp. Rhizosphere 25:100689

Ali A, Khoso MA, Mubeen M, Mehmood A, Khan MI, Zhou Y (2024) Salicylic acid: homeostasis, signalling and phytohormone crosstalk in abiotic stress tolerance. Biochem Biophys Res Commun 698:139–148

Ali A, Zulfiqar S, Riaz A, Lingwan M, Lianping SUN, Xianjun WU (2025) Molecular and functional roles of Tapetum organelles: a nursing staff for pollen development. Rice Sci 32(5):617–636

Aratani Y, Nakatani N, Nakajima K et al (2023) Green leaf volatile sensory calcium transduction in *Arabidopsis*. Nat Commun 14:6142

Araus JL, Kefauver SC, Zaman-Allah M, Olsen MS, Cairns JE (2018) Translating high-throughput phenotyping into genetic gain. Trends Plant Sci 23(5):451–466

Badri DV, Chaparro JM, Zhang R, Shen Q, Vivanco JM (2009) Application of natural blends of phytochemicals derived from root exudates reveals that phenolic compounds modulate the soil microbiome. J Biol Chem 284(22):15208–15219

Bailey-Serres J, Parker JE, Ainsworth EA, Oldroyd GED, Schroeder JI (2019) Genetic strategies for improving crop yields. Nature 575(7781):109–118

Berendsen RL, Pieterse CMJ, Bakker PAHM (2012) The rhizosphere microbiome and plant health. Trends Plant Sci 17(8):478–486

Chaves MM, Flexas J, Pinheiro C (2009) Photosynthesis under drought and salt stress: regulation mechanisms from whole plant to cell. Plant Physiol 149(2):439–448

Chen F, Li Y, Wang Y (2024) Recent advances of CRISPR-based genome editing for crop improvement under climate challenges. Front Plant Sci 15:1478398

Compant S, van der Heijden M, Sessitsch A et al (2021) Engineering synthetic microbial consortia based on root exudates to enhance plant growth under drought. Plant Cell Environ 44(4):1234–1245

Cullen E, Lingwan M (2025) Go with the flux: modeling accurately predicts phenotypes of Arabidopsis lipid mutants. Plant Physiol 197

Ding F, Zhang Y, Wang J, Li H (2025) Crosstalk of abscisic acid with other hormones and signaling molecules in tomato cold tolerance. Horticulturae 11(6):647

Fernie AR, Tohge T (2017) The genetics of plant metabolism. Annu Rev Genet 51:287–310

Fiehn O (2022) Metabolomics by gas chromatography–mass spectrometry: combined targeted and untargeted profiling. Nat Protoc 17(4):1033–1062

Gul RMS, Shah MA, Bacha S, Lee DJ (2024) Understanding abscisic acid-mediated stress signalling to sustain crop resilience. Front Sustain Food Syst 8:1477994

Hu L, Robert CAM, Cadot S et al (2018) Root exudate metabolites drive plant–soil feedbacks on growth and defense via rhizosphere microbiota. Nat Commun 9:2738

Jagadish SVK, Murty MVR, Quick WP (2021) Rice responses to rising temperatures—challenges, perspectives, and future directions. Plant Physiol 185(2):620–631

Ji H, Pardo JM, Batelli G, Van Oosten MJ, Bressan RA, Li X (2013) The Salt Overly Sensitive (SOS) pathway: established and emerging roles. Mol Plant 6(2):275–286

Jin S, Zong Y, Gao Q, et al (2023) Optimized prime editing in monocot plants using engineered pegRNAs and replicons. Unpublished preprint from GAO Lab site, later adapted into peer-reviewed reports

Job N, Lingwan M, Masakapalli SK, Datta S (2022) Transcription factors BBX11 and HY5 interdependently regulate the molecular and metabolic responses to UV-B. Plant Physiol 189(4):2467–2480

Job N, Dwivedi S, Lingwan M, Datta S (2025) BBX22 enhances the accumulation of antioxidants to inhibit DNA damage and promotes DNA repair under high UV-B. Physiol Plant 177(1):e70038

Joshi R, Wani SH, Singh B, Bohra A, Dar ZA (2021) Drought stress tolerance in maize: omics, breeding, and transgenic interventions. Front Plant Sci 12:698039

Kaur J, Kalantar M, Andrade-Sanchez P, White JW (2025) From sensors to insights: technological trends in image-based high-throughput plant phenotyping. Plant Phenomics 7:100148

Kim JS, Kim SY, Jeon J (2024) Regulatory networks in plant responses to drought and cold. Plant Physiol 195(1):170–186

Koley S, Jyoti P, Lingwan M, Allen DK (2024) Isotopically nonstationary metabolic flux analysis of plants: recent progress and future opportunities. New Phytol 242(5):1911–1918
Kumar RR, Goswami S, Singh K, Pathak H, Rai RD, Singh GP (2023) Multi-omics approaches reveal flavonoid-mediated thermotolerance and CRISPR-based *WRKY* editing in wheat. Plant Biotechnol J 21(6):1224–1239
Kushwaha AK, Dwivedi S, Mukherjee A, Lingwan M, Dar MA, Bhagavatula L, Datta S (2022) Plant microProteins: small but powerful modulators of plant development. Iscience 25(11):105400
Lawson T, Matthews J (2020) Guard cell metabolism and stomatal function. Annu Rev Plant Biol 71:273–302
Lee TA, Kim J, Brady SM (2025) A single-cell, spatial transcriptomic atlas of *Arabidopsis* development. Nat Plants 11:1234–1247
Lew TTS, Koman VB, Silmore KS, Seo JS, Gordiichuk P, Kwak SY et al (2020) Real-time detection of stress signaling in plants with carbon nanotube optical sensors. Nat Nanotechnol 15(7):618–623
Li J, Li G, Wang H, Deng XW (2011) Phytochrome signaling mechanisms. Arabidopsis Book 9:e0148
Lim PO, Kim HJ, Nam HG (2007) Leaf senescence. Annu Rev Plant Biol 58:115–136
Lingwan M (2024a) Damage and repair: how Poaceae plants fix DNA damaged by UV-B radiation. Plant Physiol 195(1):265–267
Lingwan M (2024b) Fats influencing flowering: pistil-derived lipids affect pollen tube growth and fertility in Arabidopsis thaliana. Plant Physiol 196(2):670–671
Lingwan M (2025) Metabolic modeling suggested noncanonical algal carbon concentrating mechanism in Cyanidioschyzon merolae. Plant Physiol 197(2):kiaf019
Lingwan M, Masakapalli SK (2022) A robust method of extraction and GC-MS analysis of Monophenols exhibited UV-B mediated accumulation in Arabidopsis. Physiol Mol Biol Plants 28(2):533–543
Lingwan M, Yadav A (2025) Sorbitol signaling: linker histone MdH1. 1 modulates malic acid buildup in apple. Plant Cell 37(1):koae332
Lingwan M, Pradhan AA, Kushwaha AK, Dar MA, Bhagavatula L, Datta S (2023a) Photoprotective role of plant secondary metabolites: biosynthesis, photoregulation, and prospects of metabolic engineering for enhanced protection under excessive light. Environ Exp Bot 209:105300
Lingwan M, Shagun S, Pahwa F, Kumar A, Verma DK, Pant Y et al (2023b) Phytochemical rich Himalayan Rhododendron arboreum petals inhibit SARS-CoV-2 infection in vitro. J Biomol Struct Dyn 41(4):1403–1413
Lingwan M, Yadav A, Datta S, Masakapalli SK (2024) Time course metabolomics and 13CO2 mapping establish BBX31/miP1b mediated metabolic readjustments conferring UV-B tolerance in Arabidopsis. bioRxiv, 2024-03
Liu Y, Chen Y, Wu W, Zhang X, Li P (2025) Heat stress reshapes wheat root exudates to recruit beneficial microbes. Rhizosphere 24:100678
Luo X, Dou Y, Lang Y, Zhao H, Liu X, Zhang X, Xia H (2025) CRISPR/Cas9-mediated editing of carotenoid biosynthesis genes alters carotenoid concentrations in kiwifruit. BMC Plant Biol, 25(1):1056
Lynch JP (2019) Root phenotypes for improved nutrient capture: an underexploited opportunity for global agriculture. New Phytol 223(2):548–564
Meng X, Yu H, Zhang Y, Zhou P, Li W (2022) CRISPR–Cas9-mediated editing of ABA receptor genes enhances drought tolerance in rice. Plant Physiol 190(3):1230–1243
Messina C, Garcia-Abadillo J, Powell O, Tomura S, Zare A, Ganapathysubramanian B, Cooper M (2025) Toward a general framework for AI-enabled prediction in crop improvement. Theor Appl Genet, 138(7):151
Misra BB, Langefeld C, Olivier M, Cox LA (2021) Integrated omics: tools, advances, and future approaches. J Mol Endocrinol 66(4):R21–R45
Mo W, Zheng Y, Xu Y (2024) Unveiling the crucial roles of abscisic acid in plant physiology. Front Plant Sci 15:1437184

Mohanasundaram B, Bhide AJ, Palit S, Chaturvedi G, Lingwan M, Masakapalli SK, Banerjee AK (2021) The unique bryophyte-specific repeat-containing protein SHORT-LEAF regulates gametophore development in moss. Plant Physiol 187(1):203–217

Munns R, Day DA, Fricke W, Watt M, Arsova B, Barkla BJ et al (2020) Energy costs of salt tolerance in crop plants. New Phytol 225(3):1072–1090

Niu B, Paulson JN, Zheng X, Kolter R (2017) Simplified bacterial communities of maize roots reveal functional contributions to drought resilience. PLoS Biol 15:e2001790

Nuccio ML, Wu J, Mowers R, Zhou HP, Meghji M, Primavesi LF et al (2015) Expression of trehalose-6-phosphate phosphatase in maize improves yield in drought conditions. Nat Biotechnol 33(8):862–869

Oldroyd GED, Leyser O (2020) A plant's diet, surviving in a variable nutrient environment. Science 368(6486):eaba0196

Paik I, Huq E (2019) Plant photoreceptors: multi-functional sensory proteins and their signaling networks. Semin Cell Dev Biol 92:114–121

Pan C, Sretenovic S, Qi Y (2022) Boosting plant genome editing with a versatile CRISPR-combo system. Nat Plants 8(5):513–525

Pant Y, Lingwan M, Masakapalli SK (2023) Metabolic, biochemical, mineral and fatty acid profiles of edible Brassicaceae microgreens establish them as promising functional food. Food Chemistry Advances 3:100461

Qiu Z, Zhang H, Guo W (2024) Integrating AI and high-throughput phenotyping for crop improvement. The Crop Journal 12(6):1543–1561

Rasheed A, Hao Y, Xia X, Khan A, Xu Y, Varshney RK, He Z (2020) Crop breeding chips and genotyping platforms: progress, challenges, and perspectives. Mol Plant 13(2):167–186

Rolli E, Marasco R, Vigani G et al (2015) Improved drought resistance is promoted by root-associated microbiomes as a water stress-dependent trait. Environ Microbiol 17:316–331

Rowe JH, Buckley CR, Frommer WB (2025) Quantifying plant biology with fluorescent biosensors. Annu Rev Plant Biol 76:1–29

Saini P, Bhatia S, Mahajan M, Kaushik A, Sahu SK, Kumar A et al (2020) ELONGATED HYPOCOTYL5 negatively regulates DECREASE WAX BIOSYNTHESIS to increase survival during UV-B stress. Plant Physiol 184(4):2091–2106

Saito K, Matsuda F (2010) Metabolomics for functional genomics, systems biology, and biotechnology. Annu Rev Plant Biol 61:463–489

Shagun S, Lingwan M, Masakapalli SK (2024) Mass spectrometry and Nuclear magnetic resonance spectroscopy profiles of red and pink Rhododendron flower petals establish them as rich sources of bioactive secondary metabolites. Sep Sci Plus 7(7):e202400007

Sharma R, Mishra M, Kumar S, Gupta A, Singh A (2022) Root exudates as modulators of plant–microbe interaction under heat stress in wheat. Rhizosphere 22:100517

Shavrukov Y, Kurishbayev A, Jatayev S, Shvidchenko V, Zotova L, Koekemoer F et al (2017) Early flowering as a drought escape mechanism in plants: how can it aid wheat production? Front Plant Sci 8:1950

Shi J, Gao H, Wang H, Lafitte HR, Archibald RL, Yang M, Habben JE (2017) ARGOS 8 variants generated by CRISPR-Cas9 improve maize grain yield under field drought stress conditions. Plant Biotechnol J, 15(2):207–216

Shree M, Lingwan M, Masakapalli SK (2019) Metabolite profiling and metabolomics of plant systems using 1H NMR and GC-MS. In: OMICS-based approaches in plant biotechnology, pp 129–144

Singh AK, Mishra P, Kashyap SP, Karkute SG, Singh PM, Rai N, Behera TK (2022) Molecular insights into mechanisms underlying thermo-tolerance in tomato. Front Plant Sci, 13:1040532

Singh A, Ganapathysubramanian B, Singh AK, Sarkar S (2023) Machine learning for plant stress phenotyping and genomic prediction. Plant J 114(5):1020–1040

Sperotto RA, Ricachenevsky FK, Waldow VA, Fett JP (2021) Iron biofortification in rice: a sustainable approach to crop improvement. Plant Sci 303:110760

Toyota M, Spencer D, Sawai-Toyota S, Jiaqi W, Zhang T, Koo AJ et al (2018) Glutamate triggers long-distance, calcium-based plant defense signaling. Science 361(6407):1112–1115

Tripathi S, Hoang QT, Han YJ, Kim JI (2019) Regulation of photomorphogenic development by plant phytochromes. Int J Mol Sci 20(24):6165

Uga Y, Sugimoto K, Ogawa S, Rane J, Ishitani M, Hara N et al (2013) Control of root system architecture by *DEEPER ROOTING 1* increases rice yield under drought conditions. Nat Genet 45(9):1097–1102

Varadharajan V, Thangavel G, Bharath P (2025) Multi-omics approaches against abiotic and biotic stress in plants. Plants 14(8):1785

Varoquaux N, Cole B, Gao C, Pierroz G, Baker CR, Patel D et al (2019) Transcriptomic analysis of field-droughted sorghum from seedling to maturity reveals biotic and metabolic responses. Proc Natl Acad Sci 116(52):27124–27132

Vu TV, Nguyen HN, Tran DK (2024) Prime editing: mechanism insight and recent applications in plants. Plant Biotechnol J 22(9):1721–1737

Wagner S, Weber APM (2025) Illuminating plant metabolism with genetically encoded biosensors. J Plant Physiol 320:154160

Wang X, Li Y, Fang C, Zhang Y, Zhu J, Chen H (2021) Multi-omics analysis of drought responses in rice identifies raffinose and trehalose as stress biomarkers. Plant Cell Environ 44(4):1234–1247

Wishart DS, Guo A, Oler E, Wang F, Anjum A, Peters H et al (2022) HMDB 5.0: the human metabolome database for 2022. Nucleic Acids Res 50(D1):D622–D631

Woo HR, Kim HJ, Nam HG, Lim PO (2019) Plant leaf senescence and death—regulation by multiple layers of control and implications for aging in general. J Cell Sci 132(1):jcs228005

Xu J, Yuan Y, Luo Y, Chen M, Huang Y, Li X (2020) Drought-induced metabolomic changes in maize reveal proline and polyamines as biomarkers. Plant J 103(4):1413–1427

Yadav A, Bakshi S, Yadukrishnan P, Lingwan M, Dolde U, Wenkel S et al (2019a) The B-box-containing microprotein miP1a/BBX31 regulates photomorphogenesis and UV-B protection. Plant Physiol 179(4):1876–1892

Yadav A, Lingwan M, Yadukrishnan P, Masakapalli SK, Datta S (2019b) BBX31 promotes hypocotyl growth, primary root elongation and UV-B tolerance in Arabidopsis. Plant Signal Behav 14(5):e1588672

Yadav A, Singh D, Lingwan M, Yadukrishnan P, Masakapalli SK, Datta S (2020) Light signaling and UV-B-mediated plant growth regulation. J Integr Plant Biol 62(9):1270–1292

Yin R, Schmid MW, Weigel D (2023) Spatial transcriptomics drives a new era in plant research. Plant J 116(4):840–856

Zhang Y, Liu J, Wang J, Zhao Y, Chen L, Zhou Q (2023) Base editing of *OsHKT1;5* improves salinity tolerance in rice field trials. Plant Biotechnol J 21(7):1301–1314

Zhao C, Liu B, Piao S, Wang X, Lobell DB, Huang Y et al (2016) Temperature increase reduces global yields of major crops in four independent estimates. Proc Natl Acad Sci 114(35):9326–9331

Zhu JK (2016) Abiotic stress signaling and responses in plants. Cell 167(2):313–324

Zhu X, Wang Y, Li J (2024) Yield prediction through UAV-based multispectral imaging and deep learning. Agric Syst 231:103881

Zhu M, Jiang L, Benfey PN (2025) Single-cell transcriptomics reveals how root tissues adapt to contrasting soils. Nature 626(8001):112–118

Comparative Evaluation on Current Progress of Global Crop Production Service in the Twenty-First Century and Its Future Prospects

9

Demus Matheus Huang, Widad Fadhullah, Weihong Hu, Pengqi Gao, Mohd Rafatullah, and Lv Wan

Abstract

Global crop production plays several vital provisioning roles to ensure the periodical sustainability of global human productivity. However, its recent progress in the twenty-first century is impeded by several pressuring challenges, mainly burgeoning population, climate change, arable land scarcity and economic instabilities. This book chapter comparatively evaluates the current progress of crop production service in some countries using the case studies from developed (United States of America or USA, Australia), developing (China, Brazil) and underdeveloped countries (Ghana, Myanmar) as comparison. Furthermore, this book chapter highlights the future prospects of global crop production services

D. M. Huang · W. Hu
Environmental Technology Division, School of Industrial Technology, Universiti Sains Malaysia, George Town, Pulau Pinang, Malaysia

W. Fadhullah (✉)
Environmental Technology Division, School of Industrial Technology, Universiti Sains Malaysia, George Town, Pulau Pinang, Malaysia

Renewable Biomass Transformation Cluster, School of Industrial Technology, Universiti Sains Malaysia, George Town, Pulau Pinang, Malaysia
e-mail: widad@usm.my

P. Gao
Department of Civil & Environmental Engineering, The Faculty of Engineering, The University of Auckland, Auckland, New Zealand

M. Rafatullah
Department of Mathematics and Natural Sciences, Prince Mohammad Bin Fahd University, Al-Khobar, Al-Khobar, Saudi Arabia

L. Wan
School of Life Sciences and Chemical Engineering, Jiangsu Second Normal University, Jiangsu, Nanjing, China

H. C. Joshi et al. (eds.), *Climate-Responsive Plant Biology*,
https://doi.org/10.1007/978-981-95-7354-7_9

and suggests solutions to solve its aforementioned challenges. The provisioning service provided by crop production service is vital to sustain human life in terms of health, food security, and economic growth. Currently, the global crop production service has been evaluated to become more demanding and complicated due to the aforementioned stressors. The crop production service trend may vary across the world depending on the countries, with developed and developing countries generally possessing more advanced crop production service than underdeveloped countries due to higher capital investment in the agricultural sector and lack of marginalisation experienced by the former. In the future, higher demand for the crop production is expected; hence, the integration of several crucial components such as advanced sciences and technologies, stricter enforcement of relevant policies and programs, better capital investment and other climate actions into crop agriculture is essential to preserve its sustainability.

Keywords

Global crop production · Ecosystem function and services · Provisioning ecosystem service · Sustainable food consumption · Climate action · Sustainable agricultural practices

9.1 Introduction

Crop production is a provisioning service under the SPRC (supporting, provisioning, regulating, and cultural services) ecosystem service framework (Millennium Ecosystem Assessment 2005). The crop provisioning products (i.e. food, health supplements and medicine) are not only meant to improve and sustain human vitality or well-being and ensure food security (Harbinson et al. 2021; Kumar et al. 2020; Pawlak and Kolodziejczak 2020; Poole et al. 2022; Roberts and Mattoo 2019), but also to contribute to better economic growth (Anand and Gupta 2020; Ahmed et al. 2019). The enhancement and preservation of crop production service are thus essential in promoting the fulfilment of sustainable development goals (SDGs) set by United Nations (UN), particularly SDG12 (responsible consumption and production) and SDG13 (climate action). As a matter of fact, agricultural crop production service has been vitally associated with human life since the ancient time, specifically around the earliest 12th millennium BCE (Wells and Stock 2020). At that time, the lifestyle of human populations experienced significant transition from nomadic to settled mode, and they developed a cognitive comprehension on crop production, coupled with several conventional agricultural methodologies and techniques which allowed them to be capable in planting, cultivating, harvesting and even processing the crops into edible foods or feasible consumables to sustain their life. Over time and until now, the development of agriculture has kept advancing, following the modernisation and advancement in science and technology (Parolini 2020).

Throughout the human history, crops and other energy resources have been indispensable energy sources of human productivity and survivability; hence, they are evidently inseparable from human life. However, since the early twenty-first century, their production has recently become more demanding due to intensification of human population growth. This factor, coupled with other drivers such as climate change and decrease in land availability, persistently depletes the availability of food and other energy resources over time, which subsequently induces global starvation, economic instability and even considerable human mortality (Chandio et al. 2022; Giller et al. 2021; Harbinson et al. 2021; Khan et al. 2021; Liu et al. 2020; Mitchell et al. 2020; Montoya et al. 2019; Pawlak and Kolodziejczak 2020; Sadigov 2022; Thiemann et al. 2022; Viana et al. 2022; Wang et al. 2021). In addition, the application of less sustainable conventional agricultural practices in crop production remains persistent, which significantly generates greater carbon footprint that highly promotes greenhouse gas (GHG) emission, thus, further exacerbating the ongoing climate change (Ozlu et al. 2022). Therefore, the implementation of several effective countermeasures is essential to prevent or minimise these undesirable consequences.

There have been many published research papers that extensively discussed the current progress and future prospects of overall global crop production service, but there is less attention focusing on discussing the comparison between crop production service progress in different countries and different socioeconomic backgrounds: developed or developing and underdeveloped countries. Hence, this book chapter attempts: (1) to elucidate the current progress of global crop production service in the recent twenty-first century (from year 2001 to 2024) by incorporating a brief comparative study on the crop production service in some countries based on their socioeconomic development category through literature review: developed (the United States of America or USA, Australia), developing (China, Brazil) and underdeveloped countries (Ghana, Myanmar); and (2) to highlight the future prospects of the aforementioned ecosystem service along with some suggested solutions to solve the associated challenges in the future.

9.2 Current Progress of Global Crop Production Service and Its Associated Environmental Issues

Global crop production as a provisioning service has recently become more demanding. This phenomenon is driven by several prominent factors, mainly an increase in global human population, the occurrence of climate change and a decrease in arable land availability, all of which further generate several instabilities and hardships correlated to the crop production service. The aforementioned factors are identified to be interconnected with each other to cause reduction in global crop production as well as jeopardisation of global food and other energy resource security.

Global population growth over time becomes more pronounced than before, with the current global population in this twenty-first century has reached over 8 billion (Sadigov 2022). As increase in population growth corresponds to increase in food and energy demand, depletion of global food and other energy resources will become more significant, which subsequently leads to the implementation of several measures to solve the aforementioned problems. These often include extensive land expansion for establishing several industries to sustain human daily productivity and needs, development of residential areas and urbanisation to possibly contain the population, over-exploitation of natural resources to meet the population needs and application of conventional agricultural technologies (i.e. usage of low-cost conventional pesticides, fertilisers and other relevant products) to increase agricultural crop yields (Chandio et al. 2022; Harbinson et al. 2021; Khan et al. 2021; Liu et al. 2020; Montoya et al. 2019; Pawlak and Kolodziejczak 2020; Sadigov 2022; Viana et al. 2022; Wang et al. 2021). These actions, however, are unsustainable and less environmentally friendly in long run, which tends to result in higher greenhouse gas (GHG) emissions, pronounced degradation of natural ecosystems (i.e. loss in biodiversity and polluted environmental compartments), arable land scarcity for crop agriculture, increase in mean temperature within the living environmental system and other detrimental consequences that are heavily associated with climate change. As shown in Fig. 9.1, it could be observed that the increase in world total population tends to be associated with declining total arable land areas across the world. The aforementioned consequences are synonymous with unstable trade-offs associated with the agricultural provisioning services, which eventually disrupt food and other energy resources security and reduction in crop production, and other disservices for long-term period. Hence, the current progress of global crop production service can be said to experience a mixture of improvements, hardships, and dilemmas, but it still attempts to achieve sustainability and to be capable in solving its associated issues.

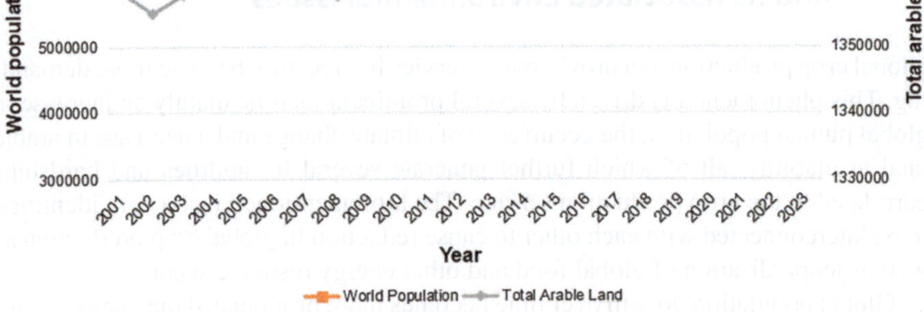

Fig. 9.1 The correlation between world total population and its total arable land area over time within the early twenty-first century. (Data source: FAO (2025) and World Bank (2025))

In general, the trends of global crop agriculture can be said as diverse in every country in the worldwide. According to Giller et al. (2021), these trends may vary depending on the main factors such as arable land availability, labour, and investment capital, which can be visualised as in Fig. 9.2. There are two major agricultural trends in this current era: intensification and expansion. The intensification mostly happens in several more developed and developing countries in Europe, Northern America and Asia with pronounced arable land scarcity but with medium-to-high investment capital which can be used to apply more advanced agricultural science and technology in improving their crop yield with absence or minimal increase in physical farm size. Conversely, the expansion mostly occurs in the developed countries like Australia with high arable land availability that allows them to expand their physical farm size and improve their crop yield in conjunction with high utilisation of invested technologies, although this trend also occurs in some developing countries such as Brazil and China (Abrahão and Costa 2018; Giller et al. 2021; Kopittke et al. 2019).

However, the agriculture in some less-developed countries (i.e. countries in sub-Saharan Africa) with very low gross domestic product (GDP) may undergo either marginalisation (restricted accessibility of farm labourers to the resources by capital owners) or extensification (expansion of farm size with minimal invested technologies) (Giller et al. 2021; Jellason et al. 2021). Besides, agricultural consolidation,

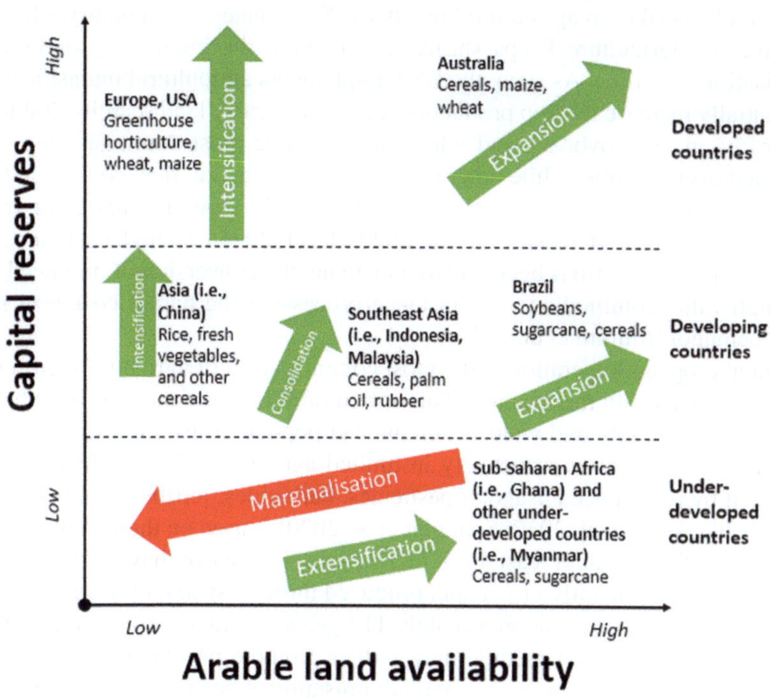

Fig. 9.2 The correlation between farm diversity and development trajectories and the availability of land and capital in different countries in the worldwide. (Adapted from: Giller et al. (2021))

which involves fusion of fragmented small lands or farms into larger farm to increase efficiency of crop production in meeting the population demanding needs, is identified to predominantly occur in some developing countries such as Indonesia, Malaysia, Philippines, and other countries in Southeast Asia (Giller et al. 2021; Vos 2018).

As a comparative study, this book chapter focuses on the current progress of agricultural crop production service in some developed (USA, Australia), developing (China, Brazil) and underdeveloped countries (Ghana, Myanmar). Table 9.1 summarises some important details on the crop production service of the six focused countries in terms of their principal agricultural crops and crop production method. Overall, there are significant differences between developed, developing and underdeveloped countries in terms of their current crop production service in the twenty-first century.

9.2.1 Developed Country: USA

The USA possesses very high overall capital resources and economic performance, with both key parameters having experienced dramatic increases since the beginning of twenty-first century. These parameters, coupled with the diverse climate and geographical features, provide the USA with higher degree of flexibility and feasibility of advanced crop agricultural practices. Nevertheless, the country's land area available for agriculture keeps shrinking following the intensive progression of urbanisation over time. As such, the USA implements agricultural intensification to substantially increase its crop production amidst the arable land scarcity (Table 9.1). Nationally, maize, soybeans and wheat are the three most dominant crops being cultivated over the time, while other minor crop variants (i.e. rice, cottons and potatoes) with varying proportions are also cultivated in several counties or regions within the country (Crossley et al. 2020; FAO 2025; Spangler et al. 2020). Furthermore, the USA has been widely practising the conversion of produced crops into high-value commodities, for instance processing the harvested maize kernel into bioethanol (Mumm et al. 2014).

Spatial crop agglomeration is the most common form of agricultural crop intensification practised in this country. This agricultural system functions by enhancing the cultivation quality and yield of suitable yet popular crops (i.e., wheats, maizes, soybeans) that can grow effectively in limited agricultural belts, which is further supported by pronounced usage of pesticides, herbicides, fertilisers and other technologies (Crossley et al. 2020; Spangler et al. 2020). Based on their study, Crossley et al. (2020) found that at least 4 out of 18 crops (i.e., maize, hay, soybean, wheat) were the most dominantly grown and produced through spatial crop agglomeration practice in the USA over approximately 117 years and more. To some extent, this implemented agricultural method may further promote profitability, easy agricultural management and monitoring, and minimisation of persistent widespread pests. Nevertheless, long-term implementation of spatial crop agglomeration may not be sustainable as it may induce suppression of crop biodiversity and agricultural belts,

Table 9.1 General overview of agricultural crop production service provided by the six countries focused on by the current study in the recent twenty-first century (2001–2024), along with the magnitude of their capital availability (indicated as gross fixed capital formation or GFCF), gross domestic product (GDP) per capita and arable land availability

Country	Socio-economic development background	Primary crops (top three)	Population as of 2024	GFCF (in millions of US$) as in 2024	GDP per capita (in US$) as of 2024	Arable land availability (in 1000 hectares) as of 2023	Predominant crop production trend	References
USA	Developed (High income)	Maize Soybeans Wheat	340 million **(+ 19.4% since 2001)**	6,293,963.00 **(+ 163.9% since 2001)**	85,809.90 **(+ 131.1% since 2001)**	151,563.53 **(− 13.6% since 2001)**	Intensification	Crossley et al. (2020), FAO (2025), Giller et al. (2021), Malone et al. (2022), Massigoge et al. (2023), Spangler et al. (2020)
Australia	Developed (High income)	Sugarcane Wheat Barley	27 million **(+ 41.1% since 2001)**	425,205.96 **(+ 382.7% since 2001)**	64,407.48 **(+ 227.0% since 2001)**	30,989.68 **(+ 28.1% since 2001)**	Expansion, intensification	Bell et al. (2014), FAO (2025), Giller et al. (2021), Greijdanus and Kragt (2014), Potapov et al. (2022), Thorburn et al. (2017)

(continued)

Table 9.1 (continued)

Country	Socio-economic development background	Primary crops (top three)	Population as of 2024	GFCF (in millions of US$) as in 2024	GDP per capita (in US$) as of 2024	Arable land availability (in 1000 hectares) as of 2023	Predominant crop production trend	References
China	Developing (Upper middle income)	Maize Rice Fresh vegetables	1.4 billion **(+ 10.8% since 2001)**	7,391,169.51 **(+ 1530.2% since 2001)**	13,303.15 **(+ 1148.6% since 2001)**	108,427.10 **(− 9.4% since 2001)**	Intensification, expansion	Chandio et al. (2022), FAO (2025), Giller et al. (2021), Li et al. (2019), Liu et al. (2020)
Brazil	Developing (Upper middle income)	Sugarcane Soybeans Maize (corn)	212 million **(+ 20.2% since 2001)**	371,217.22 **(+ 259.9% since 2001)**	10,280.31 **(+ 223.7% since 2001)**	55,642.00 **(+ 20.9% since 2001)**	Expansion, intensification	Abrahão and Costa (2018), FAO (2025), Giller et al. (2021), Martinelli and Filoso (2008)
Ghana	Underdeveloped (Lower middle income)	Cassava Yams Plantains	34 million **(+ 70.7% since 2001)**	8113.91 **(+ 462.9% since 2001)**	2405.79 **(+ 812.9% since 2001)**	4709.90 **(+ 24.2% since 2001)**	Extensification, marginalisation	Anaglo et al. (2020), Danquah et al. (2022), Dzomeku et al. (2011), FAO (2025), Giller et al. (2021), Jellason et al. (2021), Naab et al. (2017)

Myanmar	Underdeveloped (Lower middle income)	Rice Sugarcane Fresh vegetables	55 million **(+ 19.1% since 2001)**	*N/A*	1359.26 **(+860.4% since 2001)**	11,007.33 **(+ 10.2% since 2001)**	Extensification, marginalisation, expansion	Belton et al. 2021, FAO (2025), Giller et al. (2021); Herridge et al. (2019), Oo et al. (2023), Tun et al. (2024)

The data of population, GFCF and GDP per capita were acquired from World Bank (2025), whereas the data of arable land availability were acquired from FAO (2025) ecosystem degradation caused by excessive usage of the aforementioned chemical substances, increased vulnerability to climate change, and potential emergence of new pesticide-resistant pests that may endanger the crops (Crossley et al. 2020; Kopittke et al. 2019; Spangler et al. 2020).

Furthermore, most USA farmers have also been practising double-cropping agriculture as an effective means to increase crop productivity. Essentially, double cropping involves the cultivation and harvesting of two types of crops in the similar piece of land but at different seasons within one calendar year (Borchers et al. 2014). The implementation of this agricultural practice may help in optimising fertiliser usage, providing higher chance of additional economic profits, and protecting affected soil against climate or moisture erosion, although some concerns, such as potentially shorter growing time for the second crop and high susceptibility of the crops to climate change and pests, may also restrict its widespread application. Based on their case study, Massigoge et al. (2023) suggested the high feasibility of maize-soybean double cropping in the southern part of the USA. This result is mainly contributed by longer maize growth cycle during the first season with optimal solar irradiance and temperature, in contrast to soybean with slightly shorter growing season and less optimal growing parameters. Separately, trial study conducted by Malone et al. (2022) suggested high economical viability, crop productivity and biomass production in two-year soybean double-cropping with winter rye cover crops in the North Central USA region, which was further supported by 60 kg N ha^{-1} fertiliser input and high precipitation during that period. Additionally, they also revealed that proper fertilisation and harvest of ryes before planting soybeans may significantly improve soil quality by minimising nitrogen loss and excessive nitrate leaching within the soil environment, which further supports the potential ability of double cropping in mitigating environmental degradation.

9.2.2 Developed Country: Australia

Australia possesses a very unique climate and geographical profile compared to other countries or continents in the world. Overall, this continental country is known for its very dry climate, with most of its regions having only received less than approximately 600 mm mean annual rainfall in the latest 30 years (Bureau of Meteorology 2025). As a result, the largest proportion of Australia's landmass, mostly in its central region, is mainly constituted by arid and semi-arid desert, while some parts such as southern and northern regions are characterised by fertile landmass with temperate and tropical climates, respectively (Bell et al. 2014; Hochman et al. 2013).

Similar to USA, Australia is a developed country, yet there are some differences between these two countries in terms of their agricultural trends and economic features. As shown in Table 9.1, the capital reserve of Australia is lower than that of the USA but comparable in terms of GDP per capita. Additionally, Australia has observed slight increase in arable land compared to USA in recent decades (FAO 2025), although the size of total arable land area in Australia is insignificantly large due to its predominant dry land area and persistent shift in land use for urban development and other purposes (Grundy et al. 2016). In terms of agricultural practices, Australia mainly implements agricultural expansion of crops such as sugarcane, wheat and other cereals (Bell et al. 2014; Giller et al. 2021; Greijdanus and Kragt 2014; Thorburn et al. 2017). Nevertheless, its recent agricultural expansion rate was assessed to be moderate and nearly close to stable as the maximum arable land availability in that country has virtually been achieved (Bell et al. 2014; Potapov et al. 2022), which may likely indicate a gradual shift to agricultural intensification. Additionally, Australia has also been investing in a digital crop monitoring system (i.e., Graincast™) that can assist end-users, especially farmers, in generating essential details on cultivated crops (i.e., crop name, type and yield) and forecasting crop yield in the future (Lawes et al. 2023).

Sugarcane agriculture is dominant in Australia, especially near its eastern coastal region. Both previous research studies by Linnenluecke et al. (2020) and Thorburn et al. (2017) found that the majority of sugarcane plantations can be found in several catchments across coastal area within the vicinity of Great Barrier Reef Lagoon. Agricultural intensification of sugarcane production that involves high input of N fertiliser is common within these areas (Hochman et al. 2013; Linnenluecke et al. 2020; Thorburn et al. 2017). Based on their simulation study, Thorburn et al. (2017) reported that the nitrogen use efficiency (NUE) of sugarcane production within the last two decades ranged from approximately 0.3 to at least 5 Mg sugarcane (kg N)$^{-1}$, with mean NUE was approximately 0.91 Mg sugarcane (kg N)$^{-1}$. This mean NUE value was assessed to be slightly higher than the NUE range of the top eight sugarcane-producing countries (0.25–0.9 Mg sugarcane (kg N)$^{-1}$), which further corresponds to moderate N fertilisation application. Nevertheless, the aforesaid mean NUE value is considered to be less significantly high as there are many sugarcane-producing areas within the eastern region of Australia that tend to implement intensive fertilisation practices, which is attributed to the variability of seasonal climate and soil condition among those areas (Bell et al. 2014; Thorburn et al. 2017). The long-term implementation of this practice may further promote persistent nitrogen leaching and other environmental issues in the surrounding ecosystems, especially the Great Barrier Reef ecosystem (Hochman et al. 2013). Hence, there is an urgency to optimise the N fertiliser usage for sugarcane agriculture to achieve both socioeconomic and environmental sustainability in the long run.

Expanded and intensified agriculture of wheat and other grains is also very widespread across Australia. Out of six agroecological regions, the Central Western Australia region predominantly contributed to 17% of Australia's national grain production, whereas the South Australia Mid North, Lower Yorke and Eyre Peninsula region produced higher mean wheat yield (2.2 tonne/ha) in the latest decades (Greijdanus and Kragt 2014). Both of these regions are characterised by Mediterranean climate, low-to-moderate rainfall (approximately 300–600 mm) and varying soil textures (sandy loam to red and grey clay loam), which may significantly encourage the grain agricultural practice. Greijdanus and Kragt (2014) also noted that the grain cropland in certain agroecological regions such as Queensland Central has undergone expansion by 5.7% over the last two decades, although their recent crop production trend might have shifted to agricultural intensification with high fertiliser input due to limited arable land availability. Additionally, some methods of crop-livestock integration such as dual-purpose crops and sacrificial grazing are also very effective and commonly practised in the southern region of Australia (Bell et al. 2014). These methods may promote high productivity for both crops and livestock and diversified crop rotations, although future incorporation of perennial crops into these practices may further provide numerous benefits that are associated with sustainable agriculture, improved soil quality, and better crop yield and quality.

9.2.3 Developing Country: China

Recently, China has undergone intensive development in the aspects of sociocultural system and infrastructure, along with science and technology. As the world's most populous country, China only utilises 8% of global arable land for its national agriculture to fulfil the food requirement of its mega-population (Chandio et al. 2022). This is only made possible through implementation of agricultural intensification that depends on the utilisation of sophisticated technologies and usage of fertilisers to boost the crop yield. In comparison to USA, the capital availability of China is slightly higher but significantly lower in terms of GDP per capita and arable land availability (Table 9.1) (FAO 2025). The restricted arable land availability can be explained by the country's relatively high population growth over the years that leads to rapid urbanisation (Chandio et al. 2022; Yan et al. 2015), whereas the persistent socioeconomic inequality across that country mainly contributes to relatively lower national GDP per capita (Li et al. 2021).

China is regarded as world's biggest cereal-producing nation, contributing to approximately 28% of global production, with cereal grain yield of 148.5 million tonnes produced from its total planting area of 30,190 hectares by 2018. However, this yield is said to be decreasing in the upcoming decades due to the occurrence of climate change in China, mainly an increase in mean surface air temperature and less precipitation, which may significantly retard the aforementioned crop production (Giller et al. 2021; Liu et al. 2020). Additionally, rapid urbanisation that corresponds to the amplified water usage is considered as another key factor that contributes to yield suppression of water-dependent crops (i.e. rice and wheat) via pronounced reduction of crop irrigated areas and heavy reliance on rainfed agriculture (Yan et al. 2015). To overcome the aforementioned problems, several actions related to the crop intensification have been implemented in several provinces or regions in China.

Chandio et al. (2022) revealed the intensive cereal production in Sichuan Province of China. This province is well-known to be the sole cereal grain-producing province in the southwest part of China that produces local staple cereals such as rice, wheat and maize, making the province the major contributor to the overall national agriculture of China. As an approach to deal with climate variability, arable land scarcity and population overgrowth, the cereal agriculture in Sichuan and other provinces in China mainly utilises fertilisers with recommended doses, mechanisation technology, and other agricultural technologies innovated by the country to significantly improve the national cereal crop yield and quality. Conclusively, Chandio et al. (2022) found that fertilisation significantly contributes to higher rice, wheat, and maize yield in Sichuan Province, whereas the mechanisation technology is more favourable in increasing the rice and maize yield. Additionally, the maize production is said to be the most resilient compared to other two crops in facing the climate change in China, as the physiology of maize allow it to withstand high mean temperatures and less precipitation for certain periods. Their study also emphasised the importance of stricter enforcement of policy and programmes to combat against climate change while improving the crop production service. Therefore, it could be said that the combination of proper agricultural technologies, selection of crops with resilience to climate change and high-yield capability, stricter policy enforcement along with intensive comprehension of the impacts of climate change factors on the crop yield on both annual and seasonal basis can be said as some solutions to improve the sustainability of crop production in this current era.

Li et al. (2019) highlighted the overusage of nitrogen-enriched fertilisers in the intensive annual rotation planting of summer wheat-winter maize within North China Plain (NCP) region. This region generally consists of several provinces such as Henan, Hebei, Shandong, Tianjin and Beijing that collectively contribute to approximately 71% of total national wheat production and 55% of total national maize production (Shao et al. 2016). Nevertheless, the crop production service within that region is evaluated to be less sustainable as it involves the excessive usage of fertilisers and water supply to significantly boost the crop yield. Li et al. (2019) estimated the mean *N* input for both wheat and maize cropping within that region to be approximately 204 kg N ha^{-1} and 225 kg N ha^{-1}, respectively. Their simulations further revealed two major findings: (1) Optimum yields of wheat and maize after the application of optimum *N* input level with consideration of environmental costs (216 kg N ha^{-1} for wheat and 206 kg N ha^{-1} for maize) within the studied region were approximately 9769 kg ha^{-1} and 10,912 kg ha^{-1}, respectively; (2) insignificant increase in wheat-maize crop yield was evident following the increased N input application that exceeds the aforementioned optimum *N* input level for both respective crops. Moreover, Li et al. highlighted that only few numbers of wheat-maize farmers within NCP region achieved optimally high yields by accounting for optimum fertiliser management. Overall smaller farm size and low economic profits of grain production, along with deficient manpower in some rural areas, were assessed to be the main factors that induce inefficient and unsustainable fertiliser application in rural wheat-maize agriculture within that region.

9.2.4 Developing Country: Brazil

Brazil is widely known for its extensive land, abundant natural resources, and three climate zones (tropical, equatorial and subtropical). These beneficial traits allow the country to serve as feasible medium for agriculture of various high-value crops, most predominantly sugarcane, maize and soybeans (Abrahão and Costa 2018; Cherubin et al. 2022; FAO 2025; Martinelli and Filoso 2008; Rausch et al. 2019). Similar to Australia, Brazil mainly implements agricultural expansion due to its relatively extensive arable land availability (Giller et al. 2021; Martinelli and Filoso 2008; Potapov et al. 2022), although there has been gradual shift to agricultural intensification following the decrease in land availability caused by rapid urbanisation. Crop agricultural system in Brazil has experienced remarkable progress since last few decades. Cherubin et al. (2022) noted that Brazil has been slowly investing in the application of precision agriculture and irrigation to enhance national crop agricultural system, although the former is relatively new as it was recently introduced during the late 1990s and has not been widely implemented across that country.

Similar to Australia, sugarcane expansion is considered one of the most popular and feasible agricultural practices in Brazil. Being the largest producer of sugarcane in the world, Brazil's annual mean production has been recorded to be at least 566 million tonnes of sugarcane as of today (FAO 2025). Among five major regions in Brazil, the southeastern region is the most dominant contributor to national sugarcane production, accounting for at least a half of total national sugarcane cropland area (Martinelli and Filoso 2008). Based on their research study, Martinelli and Filoso (2008) highlighted that sugarcane is highly valued by the country as it can be further processed into bioethanol as a renewable energy to fulfil the country's high energy demand. Additionally, sugarcane possesses high carbon sequestration capability, which may further contribute to reduced carbon footprints in the surrounding environments. Nevertheless, long-term expansion of sugarcane cropland that is intended to further amplify sugarcane bioethanol yield may not be environmentally sustainable. The persistence of this practice has been assessed to potentially induce disturbance and degradation of surrounding natural ecosystems, soil quality degradation and potential nitrogen leaching caused by high fertiliser demand by sugarcane agriculture (Martinelli and Filoso 2008; Thorburn et al. 2017).

The agriculture of soybeans and other grains in Brazil had been experiencing rapid expansion prior to twenty-first century, but then the rate became slower and stable since the beginning of twenty-first century. Based on their research on periodical cropland extent and change, Abrahão and Costa (2018) revealed that the intensive soybean agriculture was initially established and more predominant in the southern region of Brazil, and then slowly experienced rapid expansion to the northward following introduction of advanced mechanisation and technologies from 1974 until 2000s. This is similar to the findings of Potapov et al. (2022), in which they revealed the recent cropland expansion rate in the South America region, including Brazil, has declined by virtually 50% compared to the years in the early twenty-first century. From 2000s onwards, the expansion rate becomes lower, with rainfed double-cropping (i.e., soybean-maize double cropping) and other agricultural intensification practices becoming more significant. Nevertheless, in the upcoming decades, the feasibility of this double-cropping has been estimated to be reduced due to the potential shortening of rainfall period, which is primarily induced by persistent climate change (Abrahão and Costa 2018).

9.2.5 Underdeveloped Country: Ghana

Ghana is one of tropical Sub-Saharan countries in Africa that heavily depend on agricultural production to encourage its national socioeconomic performance. Among all countries in Africa continent, this country is well-regarded as one of largest producers of many types of crops. Cassava, yam, and plantain are top three crops that are widely cultivated as well as largely exported by Ghana (FAO 2025). These harvested crops are commonly further processed into commodities with higher market value that can further promote the socioeconomic condition of the smallholder farmers and national market in the country (Akrofi-Atitianti et al. 2018; Anaglo et al. 2020; Danquah et al. 2022; Dzomeku et al. 2011). Ghana's national agriculture is predominantly contributed by smallholder farmers, mainly through agricultural extensification, which is often characterised by cropland size expansion with minimal technological inputs (Anaglo et al. 2020; Giller et al. 2021; Jellason et al. 2021).

Nevertheless, Ghana's recent agricultural progress is significantly hindered by farm marginalisation. This phenomenon is primarily driven by less viable policy frameworks, minimal capital availability and investments, labile market conditions and persistent climate change (Ahmed et al. 2019; Akrofi-Atitianti et al. 2018; Anaglo et al. 2020; Asante and Amuakwa-Mensah 2015; Dzomeku et al. 2011; Giller et al. 2021; Jellason et al. 2021; Naab et al. 2017). This can be further supported with the relevant data provided in Table 9.1: despite the fact that arable land availability in Ghana has observed periodical increases over the recent decades, the country's GDP per capita and capital growth are underperforming (FAO 2025), which further impedes its national agricultural advancement. Three respective research studies also suggested the potential threat of declining crop production and food security jeopardisation in Ghana in the future due to the ongoing farm marginalisation (Anaglo et al. 2020; Danquah et al. 2022; Dzomeku et al. 2011). The persistent farm marginalisation may further cause weakening commitment by the smallholder farmers to invest in the crop production for long term. This can be deduced from the results of analytical study by Ahmed et al. (2019), in which they revealed that even though the objective well-being parameters (i.e., household income, fulfilment of daily consumption) of the smallholder farmers may be improved through the smallholder crop farming, their subjective well-being (i.e., happiness, life satisfaction, worthwhileness) may not be necessarily improved. This implication is mainly resulted by discriminated public accessibility to different crop markets, national economic instability, climate variability (i.e., droughts, floods), discrimination and anxiety among different Ghana communities that are involved in different crop agricultural practices. Hence, it is crucial for Ghana government to seriously consider creation of stable and profitable crop markets, strong regulation of climate change in Ghana, revamping the policies related to crop agricultural practice for the local communities to ensure the fulfilment of both subjective and objective well-being of the smallholder farmers as well as to promote sustainable crop agriculture.

9.2.6 Underdeveloped Country: Myanmar

Similar to Brazil, Myanmar has very abundant natural resources and tropical climate. These conditions further encourage the agricultural activities of various crops such as rice, fresh vegetables, sugarcane, pulses and other crops (Aung et al. 2022; FAO 2025; Gumma et al. 2018; Herridge et al. 2019; Oo et al. 2023). In terms of its crop agricultural practice trend, extensification and marginalisation are common in Myanmar (Giller et al. 2021; Gumma et al. 2018), followed by expansion with mechanisation, which is significant in certain particular regions within Myanmar (Tun et al. 2024; Zhang et al. 2018), although shift to agricultural intensification may become more significant in long run (Aung et al. 2022; Gumma et al. 2018). Among all agroecological regions present in Myanmar, Central Dry Zone is well-known as the biggest contributor to overall national crop agriculture (Gumma et al. 2018; Herridge et al. 2019; Oo et al. 2023), which contributes to at least three-fifths of the national crop production.

Nevertheless, similar to Ghana, the overall performance of Myanmar's agriculture has been evaluated to be below average, which is caused by several factors. Climate change is one of the major factors that cause underperformance of the country's crop agriculture. Both research studies by Herridge et al. (2019) and Oo et al. (2023) revealed that the climate condition of Myanmar has become labile, with persistent increases in temperatures caused by GHG emissions, irregular rainfall, and unpredictable and intense natural disasters (i.e., floods, drought and cyclones) have been associated with the climate change within that country. These climate change factors further cause stress and damage to the cultivated crops, which subsequently induce reduced crop yield, jeopardised food security, degraded ecosystems and imbalance in human livelihood. Political instability is another main cause of the impeded agricultural progress in Myanmar (Oo et al. 2023; Zhang et al. 2018). The persistent political instability, which is often associated with weak governance and frequent conflicts, may significantly hinder effective and consistent implementation of climate and agriculture policies, thus, decelerating transition to sustainable agriculture. Lastly, poor socioeconomic condition is also significantly detrimental to the agricultural progress in Myanmar (Oo et al. 2023; Zhang et al. 2018). From Table 9.1, it could be seen that the GDP per capita for Myanmar is lower than Ghana's even though the arable land availability has recently experienced an increase due to ongoing cropland expansion or extensification (FAO 2025). In terms of its current capital availability, there is no data provided by FAO, although the magnitude could be assumed to be equivalent to that of Ghana. This may further suggest the poor financial and social livelihood quality of most Myanmar farmers, which could be due to market price instability and political instability, hence, resulting in the occurrence of cropland marginalisation and extensification (Oo et al. 2023; Herridge et al. 2019; Zhang et al. 2018). In the end, similar to the Ghana's case, it is crucial for government of Myanmar to seriously consider the implementation of consistent and effective policy-making to resolve the aforementioned issues to promote better socioeconomic quality for the farmers and sustainable agriculture.

9.3 Future Prospects of Global Crop Production Service and Solutions

In the future, the increase in demand on crops and other energy resources is highly expected following the projected increase in the global population growth. According to the projection made by United Nations Department of Economic and Social Affairs: Population Division (2017), global human population may increase to approximately 8.6 billion by 2030, 9 billion by 2050, and then approximately over 10 billion by 2100. Thus, to meet the food demand by 2050, agricultural crop production areas must double the crop production to provide more accessible food supply or better food security instead of maintaining approximately 2 billion tonnes of crop production, which should be over 4 billion tonnes in the future time (Khan et al. 2021; Viana et al. 2022). Nevertheless, the intensification of climate change and land scarcity which are associated with the increasing population in the future is inevitable as these factors significantly pressurise the crop agriculture over time. Additionally, current progress of agriculture in most developing countries is still dropping behind in terms of crop yield and quality compared to other developed countries (Khan et al. 2021). As such, several ongoing attempts related to application of smart and sustainable agricultural practices, as listed in Table 9.2, are being made to achieve sustainable global crop agriculture provisioning service and to combat the climate change in the future.

Progressive advancement in science and technology along with renewable energy sources plays an integral part in enhancing the crop agricultural system to possibly achieve greater crop production efficiency and sustainable agriculture in long term. For instance, the development of artificial intelligence (AI) and machine learning (ML) and other smart technologies can assist in researching and inspecting any suitable genes that may potentially provide several beneficial traits to the crops, including high-crop yield and crop resilience to climate change (Khan et al. 2021). The integration of AI and ML into crop agricultural system, as demonstrated in precision agriculture, also offers several benefits in form of optimisation of fertiliser and other nutrition supplement consumption for crop cultivation along with effective monitoring of crop growth through simulations, which further allows eco-sustainable agricultural management, mitigation of excessive harmful chemical usage and better crop yield (Cherubin et al. 2022; Greijdanus and Kragt 2014; Lawes et al. 2023). Other smart technologies such as distributed energy sources (DER) system that comprise smart grids and microgrids, and other alternative rechargeable devices powered by renewable energy may also help in resolving excessive power consumption by the agricultural sector due to its heavy reliance on machinery and irrigation operational activities (Khan et al. 2021), thus, promoting sustainable energy consumption. Additionally, there are several potential green technologies that are capable of stabilising the growth and yield quality of climate-sensitive crops under certain restricted conditions. For instance, Widmer et al. (2024) explored the significant potential of sunlight-powered agrivoltaics in promoting the growth of light-dependent crops (i.e. maize and tomatoes) amidst climate change. Another emerging green technology such as CRISPR/Cas9 genome editing technology, that relies on

Table 9.2 Some examples of sustainable and novel agricultural practices that have been recently implemented in some countries across the world

Type of crop agricultural practice	Details	References
Ponic (soilless) agriculture	Crop agriculture using media other than soil, such as water (hydroponic), air (aeroponic) and other substrates	Khan et al. (2021)
Perennial cropping	Crop agriculture that incorporates crops that can last longer for at least 2 years with minimal treatments of fertiliser, tillage and others	Chapman et al. (2022), Greijdanus and Kragt (2014), Ryan et al. (2018)
Precision agriculture	Utilising AI and ML technologies to monitor crop growth and yield progress and to assist the users in decision-making on optimal fertiliser or nutrient inputs for the cultivated crops	Cherubin et al. (2022), Greijdanus and Kragt (2014), Khan et al. (2021), Lawes et al. (2023)
Agrivoltaics farming	Incorporation of sunlight-powered facilities into crop cultivation to provide the crops with sufficient light amidst limited light condition	Widmer et al. (2024)
Minimal or no-tillage farming	Less or zero reliance on tillage activities in cropping to ensure better soil quality and stability	Greijdanus and Kragt (2014), Naab et al. (2017), Shao et al. (2016)

the concept of gene modification, may potentially enhance the physiology and resilience of certain types of crops against climate change, diseases and other threats, thus, assuring long-term food security and sustainability (Haque et al. 2018).

The introduction and application of novel agricultural practices and relevant programmes can also be regarded as another factor that contributes to sustainable crop agriculture. The adoption of vertical farming or ponic (soilless) farming that is not dependent on land media can be regarded as effective, feasible and environmentally friendly method, considering that the intense urbanisation and arable land scarcity become more pronounced over time (Khan et al. 2021). Minimal or no-tillage farming is also beneficial for agriculture, especially for the soil quality, as it may minimise or mitigate potential occurrence of soil erosion and soil nutrient leaching, which may otherwise degrade soil quality and negatively affect crop yield and quality (Greijdanus and Kragt 2014; Naab et al. 2017; Shao et al. 2016). Another sustainable crop agriculture practice is perennial grain crop (e.g., intermediate wheatgrass, rye, sorghum, rice, etc.) agriculture, which is feasible, low-cost and environmentally friendly (Chapman et al. 2022; Greijdanus and Kragt 2014; Ryan et al. 2018). Due to the fact that perennial crops reproduce vegetatively and possess robust and extensive rooting systems that allow highly-effective carbon sequestration, they are capable of growing steadily for several seasons (typically more than 2 years) with minimal input of fertilisers, tilling and other treatments. The implementation of perennial crop agriculture may not only help in reducing the necessity of annual replanting and high-cost expenditure but also may significantly contribute to reduced occurrence of GHG emission and other environmental problems.

Additionally, the perennial crops can be intercropped with legumes, other different perennial crop species, and other crops to minimise nitrogen requirements, to improve high-crop yield stability, to improve pollination and forage quality and for other beneficial services (Ryan et al. 2018). The flexibility and resilience of perennial grain crops make them possible to be cultivated on sloped land terrain, which further helps in improving water quality and preventing soil erosion. On the other hand, the ongoing innovation and development of sustainable future-proof crops within the framework of proposed CropBooster Programme can be regarded as another sustainable approach to ensure sustainable crop agriculture, stable food and energy security and sustainable human productivity in the future (Harbinson et al. 2021).

In the end, the most crucial factor that determines the sustainability of future crop agriculture is the initiative of government, decision-makers and public communities across the world. It is essential for them to work collaboratively in urgently resolving this matter by revamping relevant policies, strategies, programmes and management practices to prevent the future socioeconomic and agricultural crises that may further induce several worst impacts on the society's well-being and its surrounding ecosystems. In other words, there must be an increase in the awareness and real actions of government, decision-makers and communities along with their cooperative actions in preserving every important aspect of life that contributes to the sustainable future of the world by minimising the aforementioned crises. Several suggested actions that can be done by the government and the associated parties to ensure sustainable future of crop agriculture are such as dedicating capital investment to agricultural science and technologies and relevant research studies, launching and supporting socialisation programmes about the importance of sustainable agriculture, revamping crop market policies to provide the crop markets and more market options, improving the global farmers' adaptability in conducting agricultural practices, revamping climate change policies and having strong commitment to launch and support several programmes and policies to preserve global ecosystems and their ecosystem services. By considering the implementation of these measures, it is highly possible to achieve the sustainable crop agriculture in the future, to preserve healthy well-being of society and environment and to sustain global food and energy security, as well as fulfil the relevant SDGs.

9.4 Conclusion and Future Perspectives

Overall, the current progress of global crop production provisioning service is evaluated to be rigorous due to the increase in the pressure driven by several challenges, such as increasing population growth, climate change, arable land scarcity, economic instabilities and other associated issues. There is a trend of diversity in crop agriculture worldwide, which mainly depends on socioeconomic and climate conditions of every country in the world. The advancement in crop agriculture service is more pronounced in developed and developing countries due to high capital investment in agricultural science and technologies and government commitment, whereas there is still underdevelopment of the service in several less-developed countries due to thc marginalisation and lack of capital investment which negatively affect the well-being of the locals.

There have been several efforts and measures being implemented to promote the sustainability of crop agriculture in the future, including the integration of IoT, AI, ML, green technologies and other advanced technologies into current crop agriculture practices, development of future-proof crops and application of new and alternative agricultural practices. In the long run, these efforts are expected to improve the productivity and efficiency of crop agriculture, to act as climate actions in minimising GHG emission and other impacts associated with climate change and to counteract the current global challenges. Nonetheless, there are some constraints that hinder their progress, such as lack of investment from government and other affiliated decision-makers and less awareness from the society of these actions. Therefore, it is essential for the government, decision-makers and the society around the world to be more cooperative and more open-minded in investing and supporting these efforts to ensure sustainable future for crop agriculture and other essential aspects that contribute to sustainable future of the world.

Eventually, the strength and limitation of this book chapter are highlighted. The strength of this book chapter is attributed to the incorporation of a comparative study as a means to supplement the discussion of current global crop production progress, which focuses on discussing the comparison between crop production service progress in different countries with different socioeconomic backgrounds using real case studies. However, the main limitation of this book chapter may be the number of case studies that are incorporated in the comparative study, with only six countries selected as the representative case studies for the discussion of current global crop production trends. This is so, as this book chapter is only intended to conduct comparative study of the global crop agricultural progress between developed, developing and underdeveloped countries in a less exhaustive manner. Therefore, for future studies that would like to provide an exhaustive discussion pertaining to this topic, the inclusion of higher number of relevant case studies is highly recommended.

References

Abrahão GM, Costa MH (2018) Evolution of rain and photoperiod limitations on the soybean growing season in Brazil: the rise (and possible fall) of double-cropping systems. Agric For Meteorol 256-257:32–45. https://doi.org/10.1016/j.agrformet.2018.02.031

Ahmed A, Dompreh E, Gasparatos A (2019) Human wellbeing outcomes of involvement in industrial crop production: evidence from sugarcane, oil palm and jatropha sites in Ghana. PLoS One 14(4):e0215433. https://doi.org/10.1371/journal.pone.0215433

Akrofi-Atitianti F, Speranza CI, Bockel L, Asare R (2018) Assessing climate smart agriculture and its determinants of practice in Ghana: a case of the cocoa production system. Land 7(1):e30. https://doi.org/10.3390/land7010030

Anaglo JN, Antwi G, Manteaw SA, Kwapong NA (2020) Influence of agricultural information sources on the practices and livelihood outcomes of cassava farmers in eastern region of Ghana. J Sustain Develop 17(1):2–10

Anand S, Gupta S (2020) Provisioning ecosystem services: multitier bibliometric analysis and visualisation. Environ Sustain Indic 8:e100081. https://doi.org/10.1016/j.indic.2020.100081

Asante FA, Amuakwa-Mensah F (2015) Climate change and variability in Ghana: stocktaking. Climate 3(1):78–99. https://doi.org/10.3390/cli3010078

Aung NN, Khaing EE, Mon YY (2022) History of sugarcane breeding, germplasm development and related research in Myanmar. Sugar Tech 24:243–253. https://doi.org/10.1007/s12355-021-01079-y

Bell LW, Moore AD, Kirkegaard JA (2014) Evolution in crop-livestock integration systems that improve farm productivity and environmental performance in Australia. Eur J Agron 57:10–20. https://doi.org/10.1016/j.eja.2013.04.007

Belton B, Win MT, Zhang X, Filipski M (2021) The rapid rise of agricultural mechanization in Myanmar. Food Policy 101:e102095. https://doi.org/10.1016/j.foodpol.2021.102095

Borchers A, Truex-Powell E, Wallander S, Nickerson C (2014) Multi-cropping practices: recent trends in double-cropping. United States Department of Agriculture: Economic Research Service. https://doi.org/10.22004/ag.econ.262122

Bureau of Meteorology (2025) Average annual, seasonal and monthly rainfall maps. Rainfall. https://www.bom.gov.au/climate/maps/averages/rainfall/

Chandio AA, Nasereldin YA, Anh DLT, Tang Y, Sargani GR, Zhang H (2022) The impact of technological progress and climate change on food crop production: evidence from Sichuan—China. Int J Environ Res Public Health 19(16):e9863. https://doi.org/10.3390/ijerph19169863

Chapman EA, Thomsen HC, Tulloch S, Correia PMP, Luo G, Najafi J, DeHaan LR, Crews TE, Olsson L, Lundquist P-O, Westerbergh A, Pedas PR, Knudsen S, Palmgren M (2022) Perennials as future grain crops: opportunities and challenges. Front Plant Sci 13:e898769. https://doi.org/10.3389/fpls.2022.898769

Cherubin MR, Damian JM, Tavares TR, Trevisan RG, Colaço AF, Eitelwein MT, Martello M, Inamasu RY, de Castro Pias OH, Molin JP (2022) Precision agriculture in Brazil: the trajectory of 25 years of scientific research. Agriculture 12(11):e1882. https://doi.org/10.3390/agriculture12111882

Crossley MS, Burke KD, Schoville SD, Radeloff VC (2020) Recent collapse of crop belts and declining diversity of US agriculture since 1840. Glob Chang Biol 27(1):151–164. https://doi.org/10.1111/gcb.15396

Danquah EO, Danquah FO, Frimpong F, Dankwa KO, Weebadde CK, Ennin SA, Asante MOO, Brempong MB, Dwamena HA, Addo-Danso A, Nyamekye DR, Akom M, Opoku AY (2022) Sustainable intensification and climate-smart yam production for improved food security in West Africa: a review. Front Agron 4:e858114. https://doi.org/10.3389/fagro.2022.858114

Dzomeku BM, Dankyi AA, Darkey SK (2011) Socioeconomic importance of plantain cultivation in Ghana. J Anim Plant Sci 21(2):269–273

Food and Agriculture Organization of the United Nations (2025) FAOSTAT. https://www.fao.org/faostat

Giller KE, Delaune T, Silva JV, Descheemaeker K, van de Ven G, Schut AGT, van Wijk M, Hammond J, Hochman Z, Taulya G, Chikowo R, Narayanan S, Kishore A, Bresciani F, Mancini Teixeira H, Andersson JA, van Ittersum MK (2021) The future of farming: who will produce our food? Food Secur 13:1073–1099. https://doi.org/10.1007/s12571-021-01184-6

Greijdanus A, Kragt ME (2014) The grains industry: an overview of the Australian broad-acre cropping (Working Paper 1402). School of Agricultural and Resource Economics, University of Western Australia. https://doi.org/10.22004/ag.econ.164256

Grundy MJ, Bryan BA, Nolan M, Battaglia M, Hatfield-Dodds S, Connor JD, Keating BA (2016) Scenarios for Australian agricultural production and land use to 2050. Agric Syst 142:70–83. https://doi.org/10.1016/j.agsy.2015.11.008

Gumma MK, Thenkabail PS, Deevi KC, Mohammed IA, Teluguntla P, Oliphant A, Xiong J, Aye T, Whitbread AM (2018) Mapping cropland fallow areas in Myanmar to scale up sustainable intensification of pulse crops in the farming system. GIScience Remote Sens 55(6):926–949. https://doi.org/10.1080/15481603.2018.1482855

Harbinson J, Parry MAJ, Davies J, Rolland N, Loreto F, Wilhelm R, Metzlaff K, Lankhorst RK (2021) Designing the crops for the future; the cropbooster program. Biology 10(7):e690. https://doi.org/10.3390/biology10070690

Haque E, Taniguchi H, Hassan MM, Bhowmik P, Karim MR, Śmiech M, Zhao K, Rahman M, Islam T (2018) Application of CRISPR/Cas9 genome editing technology for the improvement of crops cultivated in tropical climates: recent progress, prospects, and challenges. Front Plant Sci 9:e617. https://doi.org/10.3389/fpls.2018.00617

Herridge DF, Win MM, Nwe KMM, Kyu KL, Win SS, Shwe T, Min YY, Denton MD, Cornish PS (2019) The cropping systems of the Central Dry Zone of Myanmar: productivity constraints and possible solutions. Agric Syst 169:31–40. https://doi.org/10.1016/j.agsy.2018.12.001

Hochman Z, Carberry PS, Robertson MJ, Gaydon DS, Bell LW, McIntosh PC (2013) Prospects for ecological intensification of Australian agriculture. Eur J Agron 44:109–123. https://doi.org/10.1016/j.eja.2011.11.003

Jellason NP, Robinson EJZ, Chapman ASA, Neina D, Devenish AJM, Po JYT, Adolph B (2021) A systematic review of drivers and constraints on agricultural expansion in Sub-Saharan Africa. Land 10(3):e332. https://doi.org/10.3390/land10030332

Khan N, Ray RL, Sargani GR, Ihtisham M, Khayyam M, Ismail S (2021) Current progress and future prospects of agriculture technology: gateway to sustainable agriculture. Sustainability 13(9):e4883. https://doi.org/10.3390/su13094883

Kopittke PM, Menzies NW, Wang P, McKenna BA, Lombi E (2019) Soil and the intensification of agriculture for global food security. Environ Int 132:e105078. https://doi.org/10.1016/j.envint.2019.105078

Kumar D, Kumar S, Shekhar C (2020) Nutritional components in green leafy vegetables: a review. J Pharmacogn Phytochem 9(5):2498–2502

Lawes R, Hochman Z, Jakku E, Butler R, Chai J, Chen Y, Waldner F, Mata G, Donohue R (2023) Graincast™: monitoring crop production across the Australian grainbelt. Crop Pasture Sci 74:509–523. https://doi.org/10.1071/CP21386

Li S, Lei Y, Zhang Y, Liu J, Shi X, Jia H, Wang C, Chen F, Chu Q (2019) Rational trade-offs between yield increase and fertilizer inputs are essential for sustainable intensification: a case study in wheat–maize cropping systems in China. Sci Total Environ 679:328–336. https://doi.org/10.1016/j.scitotenv.2019.05.085

Li W, Cai Z, Cao S (2021) What has caused regional income inequality in China? Effects of 10 socioeconomic factors on per capita income. Environ Dev Sustain 23:13403–13417. https://doi.org/10.1007/s10668-020-01218-7

Linnenluecke MK, Zhou C, Smith T, Thompson N, Nucifora N (2020) The impact of climate change on the Australian sugarcane industry. J Clean Prod 246:e118974. https://doi.org/10.1016/j.jclepro.2019.118974

Liu Y, Tang L, Qiu X, Liu B, Chang X, Liu L, Zhang X, Cao W, Zhu Y (2020) Impacts of 1.5 and 2.0°C global warming on rice production across China. Agric For Meteorol 284:e107900. https://doi.org/10.1016/j.agrformet.2020.107900

Malone RW, O'Brien P, Herbstritt S, Emmett BD, Karlen DL, Kaspar TC, Kohler K, Radke A, Lence SH, Wu H, Richard TL (2022) Rye-soybean double-crop: planting method and N fertilization effects in the North Central US. Renew Agric Food Syst 37:445–456. https://doi.org/10.1017/S1742170522000096

Martinelli L, Filoso S (2008) Expansion of sugarcane ethanol production in Brazil: environmental and social challenges. Ecol Appl 18(4):885–898. https://doi.org/10.1890/07-1813.1

Massigoge I, Carcedo A, de Borja Reis AF, Mitchell C, Day S, Oliverio J, Truong SH, McCormick RF, Rotundo J, Lira S, Ciampitti I, Messina CD (2023) Exploring avenues for agricultural intensification: a case study for maize-soybean in the southern US region. Agric Syst 204:e103539. https://doi.org/10.1016/j.agsy.2022.103539

Millennium Ecosystem Assessment (2005) Ecosystems and human well-being: synthesis. Island Press

Mitchell MGE, Chan KMA, Newlands NK, Ramankutty N (2020) Spatial correlations don't predict changes in agricultural ecosystem services: a Canada-wide case study. Front Sustain Food Syst 4:e539892. https://doi.org/10.3389/fsufs.2020.539892

Montoya D, Haegeman B, Gaba S, de Mazancourt C, Bretagnolle V, Loreau M (2019) Trade offs in the provisioning and stability of ecosystem services in agroecosystems. Ecol Appl 29(2):e01853. https://doi.org/10.1002/eap.1853

Mumm RH, Goldsmith PD, Rausch KD, Stein HH (2014) Land usage attributed to corn ethanol production in the United States: sensitivity to technological advances in corn grain yield, ethanol conversion, and co-product utilization. Biotechnol Biofuels 7:e61. https://doi.org/10.1186/1754-6834-7-61

Naab JB, Mahama GY, Yahaya I, Prasad PVV (2017) Conservation agriculture improves soil quality, crop yield, and incomes of smallholder farmers in North Western Ghana. Front Plant Sci 8:e996. https://doi.org/10.3389/fpls.2017.00996

Oo AT, Boughton D, Aung N (2023) Climate change adaptation and the agriculture-food system in Myanmar. Climate 11:e124. https://doi.org/10.3390/cli11060124

Ozlu, E., Arriaga, F. J., Bilen, S., Gozukara, G., & Babur, E. (2022). Carbon footprint management by agricultural practices. Biology, 11 (10), Article e1453. https://doi.org/10.3390/biology11101453

Parolini G (2020) Then and now: re-positioning the history of agriculture within the history of science and technology. Cahiers François Viète 3(9):53–77. https://doi.org/10.4000/cahierscfv.508

Pawlak K, Kolodziejczak M (2020) The role of agriculture in ensuring food security in developing countries: considerations in the context of the problem of sustainable food production. Sustainability 12(13):e5488. https://doi.org/10.3390/su12135488

Poole N, Donovan J, Erenstein O (2022) Continuing cereals research for sustainable health and well-being. Int J Agric Sustain 20(5):693–704. https://doi.org/10.1080/14735903.2021.1975437

Potapov, P., Turubanova, S., Hansen, M. C., Tyukavina, A., Zalles, V., Khan, A., Song, X.-P., Pickens, A., Shen, Q., & Cortez, J. (2022). Global maps of cropland extent and change show accelerated cropland expansion in the twenty-first century. Nature Food, 3 , 19–28. https://doi.org/10.1038/s43016-021-00429-z

Rausch LL, Gibbs HK, Schelly I, Brandão A Jr, Morton DC, Filho AC, Strassburg B, Walker N, Noojipady P, Barreto P, Meyer D (2019) Soy expansion in Brazil's Cerrado. Conserv Lett 12(6):e12671. https://doi.org/10.1111/conl.12671

Roberts DP, Mattoo AK (2019) Sustainable crop production systems and human nutrition. Front Sustain Food Syst 3:e72. https://doi.org/10.3389/fsufs.2019.00072

Ryan MR, Crews TE, Culman SW, Dehaan LR, Hayes RC, Jungers JM, Bakker MG (2018) Managing for multifunctionality in perennial grain crops. Bioscience 68(4):294–304. https://doi.org/10.1093/biosci/biy014

Sadigov, R. (2022). Rapid growth of the world population and its socioeconomic results. The Scientific World Journal, 2022 , Article e8110229. https://doi.org/10.1155/2022/8110229

Shao Y, Xie Y, Wang C, Yue J, Yao Y, Li X, Liu W, Zhu Y, Guo T (2016) Effects of different soil conservation tillage approaches on soil nutrients, water use and wheat-maize yield in rainfed dry-land regions of North China. Eur J Agron 81:37–45. https://doi.org/10.1016/j.eja.2016.08.014

Spangler K, Burchfield EK, Schumacher B (2020) Past and current dynamics of U.S. agricultural land use and policy. Front Sustain Food Syst 4:e98. https://doi.org/10.3389/fsufs.2020.00098

Thiemann M, Riebl R, Haensel M, Schmitt TM, Steinbauer MJ, Landwehr T, Fricke U, Redlich S, Koellner T (2022) Perceptions of ecosystem services: comparing socio-cultural and environmental influences. PLoS One 17:e0276432. https://doi.org/10.1371/journal.pone.0276432

Thorburn PJ, Biggs JS, Palmer J, Meier EA, Verburg K, Skocaj DM (2017) Prioritizing crop management to increase nitrogen use efficiency in Australian sugarcane crops. Front Plant Sci 8:e1504. https://doi.org/10.3389/fpls.2017.01504

Tun PT, Safi Z, Nguyen TT, Wiehle M (2024) Cropping systems and production assessment in southern Myanmar: informing strategic interventions. Open Agric 9(1):e20220318. https://doi.org/10.1515/opag-2022-0318

United Nations Department of Economic and Social Affairs: Population Division (2017) World population projected to reach 9.8 billion in 2050, and 11.2 billion in 2100. https://www.un.org/development/desa/en/news/population/world-population-prospects-2017.html

Viana CM, Freire D, Abrantes P, Rocha J, Pereira P (2022) Agricultural land systems importance for supporting food security and sustainable development goals: a systematic review. Sci Total Environ 806(3):e150718. https://doi.org/10.1016/j.scitotenv.2021.150718

Vos R (2018) Agricultural and rural transformations in Asian development: past trends and future challenges (Working Paper No. 2018/87). The United Nations University–World Institute for Development Economics Research (UNU–WIDER). https://doi.org/10.35188/UNU-WIDER/2018/529-9

Wang Y, Cai Y, Liu G, Zhang P, Li B, Li B, Jia Q, Huang Y, Shu T (2021) Evaluation of sustainable crop production from an ecological perspective based emergy analysis: a case of China's provinces. J Clean Prod 313:e127912. https://doi.org/10.1016/j.jclepro.2021.127912

Wells JCK, Stock JT (2020) Life history transitions at the origins of agriculture: a model for understanding how niche construction impacts human growth, demography and health. Front Endocrinol 11:e325. https://doi.org/10.3389/fendo.2020.00325

Widmer, J., Christ, B., Grenz, J., & Norgrove, L. (2024). Agrivoltaics, a promising new tool for electricity and food production: A systematic review. Renewable and Sustainable Energy Reviews, 192 , Article e114277. https://doi.org/10.1016/j.rser.2023.114277

World Bank (2025) World Bank Open Data. https://data.worldbank.org/

Yan T, Wang J, Huang J (2015) Urbanization, agricultural water use, and regional and national crop production in China. Ecol Model 318:226–235. https://doi.org/10.1016/j.ecolmodel.2014.12.021

Zhang Y, Prescott GW, Tay RE, Dickens BL, Webb EL, Htun S, Tizard RJ, Rao M, Carrasco LR (2018) Dramatic cropland expansion in Myanmar following political reforms threatens biodiversity. Sci Rep 8:e16558. https://doi.org/10.1038/s41598-018-34974-8

Role of Epigenetics in Plant Adaptation to Environmental Stress

10

Apurva Malik, Kishan Kumar, Hukum Singh, and Santan Barthwal

Abstract

Plants are continuously exposed to a wide range of changing environmental and climatic stresses, including drought, salinity, extreme temperatures, nutrient imbalances, greenhouse gases, heavy metal toxicity, and pathogen attacks. Unlike animals, plants are immobile and cannot escape these adverse conditions, making adaptive resilience and physiological responses critical for their survival and growth. Recent advances in plant science, particularly in the field of epigenetics, have revealed key mechanisms that modulate gene expression without altering the underlying DNA sequence. Epigenetic processes, including DNA methylation, histone modifications, and small RNA-mediated regulation, play significant roles in modulating plant responses to environmental and climatic stressors. These mechanisms contribute to stress memory, enabling plants to "remember" previous stress events and exhibit enhanced resilience when exposed to similar conditions in the future. This chapter explores the pivotal role of epigenetic modifications in plant adaptive resilience, with a focus on stress-induced epigenetic changes, their reversibility, and heritability. Understanding these epigenetic processes holds immense potential to transform plant breeding and improvement strategies, offering sustainable solutions for agriculture under a changing climate. Furthermore, leveraging epigenetic knowledge can help in developing stress-tolerant crop varieties, thereby contributing to food security and achieving the Sustainable Development Goals.

A. Malik (✉) · K. Kumar · H. Singh · S. Barthwal
ICFRE-Forest Research Institute, Dehradun, India

H. C. Joshi et al. (eds.), *Climate-Responsive Plant Biology*,
https://doi.org/10.1007/978-981-95-7354-7_10

Keywords

Epigenome plasticity · Plant adaptation mechanisms · Plant-environment interactions · Stress memory · Stress-induced epigenetic changes · Sustainable agriculture

10.1 Introduction

The changing environmental and climatic conditions represent a major global challenge. Under such scenarios, plants encounter multiple and severe threats to their survival, growth, reproduction, and ecological functions. These environmental and climatic stresses include drought, soil salinity, elevated temperatures, nutrient deficiencies, heavy metal toxicity, increased greenhouse gas concentrations, and biotic pressures such as pathogen attacks and pest infestations (Singh et al. 2017; Nawaz et al. 2023). In urban regions, plants face additional challenges from elevated air pollutants, including particulate matter, nitrogen oxides, sulfur dioxide, volatile organic compounds, urban heat islands, limited water availability, and soil degradation (Singh et al. 2022). Because plants are immobile and cannot relocate to avoid these stresses, they rely on highly dynamic physiological, morphological, biochemical, and molecular strategies to withstand and acclimate over time (Kumar et al. 2025).

Epigenetics has emerged as one of the most intriguing and promising mechanisms by which plants achieve adaptive resilience to diverse stresses, primarily through modulation of gene expression without altering the underlying DNA sequence (Gayathiri et al. 2025). Epigenetic mechanisms include DNA methylation, histone modifications, chromatin remodeling, and the regulation mediated by non-coding RNAs. These processes enable plants to fine-tune the activation of stress-responsive genes in a cell-specific, temporal, and quantitative manner (Kumari et al. 2022). Such regulation imparts short-term adaptation and stress memory, allowing plants to "remember" prior exposure and respond more rapidly, strongly, or efficiently upon subsequent encounters. This stress memory may be expressed through somatic modifications within the same generation. In some instances, there is evidence of intergenerational or even transgenerational inheritance, wherein offspring exhibit modified responses as a result of their parents' exposure to stress (Bowers and Yehuda 2016).

In urban environments, where environmental stressors often occur in combination, such as heat and drought or pollution and temperature extremes, plant resilience becomes crucial. These synergistic conditions may exacerbate stress responses, profoundly altering plant morphology and physiology. The adaptations that confer tolerance to multiple stressors or allow rapid acclimation help maintain green cover, urban biodiversity, air quality, and ecosystem services (Kumar and Rani 2023). Understanding epigenetic responses is thus not only of theoretical and ecological importance but also has potential applications, such as breeding or genetically modifying plants to better suit urban and changing climatic conditions. In this chapter, we provide an in-depth exploration of the role of molecular and epigenetic mechanisms through which plants respond to environmental stresses, examining how epigenetic modifications contribute to plant resilience, stress memory, and possibly transgenerational inheritance. This chapter also highlights the potential applications of epigenetic insights in plant breeding, crop improvement, and the broader pursuit of climate-resilient agriculture aligned with Sustainable Development Goals.

10.2 Plant Stress Response Mechanisms: Conventional Versus Epigenetic

The continuous exposure of plants to environmental stress and pathogen attacks has led them to evolve intricate response mechanisms at physiological, biochemical, and molecular levels. Plant stress responses have been associated with conventional defense strategies. Epigenetic regulation of gene expression under stress conditions has evolved as a mechanism that offers both immediate resilience and long-term adaptive benefits (Bashir and Setyawati 2025).

Conventional plant stress responses operate through rapid biochemical and physiological adjustments, which include

10.2.1 Antioxidant Defense Systems

Under oxidative stress, plants produce reactive oxygen species (ROS) such as hydroxyl radicals, superoxide radicals, and hydrogen peroxide, and to mitigate these ROS-induced stresses, enzymatic antioxidants like superoxide dismutase, catalase, and peroxidases, and non-enzymatic antioxidants like ascorbic acid, flavonoids, glutathione, etc., get activated (Malik et al. 2024).

10.2.2 Hormonal Signaling

Stress conditions trigger the biosynthesis and signaling of phytohormones, such as salicylic acid, abscisic acid, and ethylene, which coordinate the activation of stress-responsive genes (Salvi et al. 2021). For example, abscisic acid helps mediate responses to drought and salt stress by regulating stomatal closure and osmoprotectant accumulation (Bharath et al. 2021).

10.2.3 Stomatal Regulation

Stomata act as gatekeepers, controlling the exchange of gases and the transpiration process. Under water-deficit conditions, abscisic acid-mediated stomatal closure reduces water loss, a key adaptation during drought (Liu et al. 2022).

These conventional mechanisms offer immediate and efficient protection, and they are often transient. In contrast, epigenetic regulation allows plants to store stress memory and pass adaptive traits across generations without genetic mutations. Epigenetics is the study of heritable and reversible modifications that influence gene activity without altering the underlying DNA sequence. These modifications allow plants to rapidly adjust gene expression in response to environmental cues, thereby enhancing plasticity and adaptation (Zhang et al. 2018).

The brief account of key epigenetic mechanisms in plants includes:

10.2.4 DNA Methylation

This mechanism involves the addition of a methyl group to the fifth carbon of cytosine residues. This modification leads to transcriptional silencing or activation, depending on the genomic context and location (Kumari et al. 2022). Under environmental stress, methylation processes can be dynamically altered, either through hypermethylation, which often suppresses transposable elements and some gene promoters, or through hypomethylation, which activates stress-responsive genes. These methylation changes are partially heritable, enabling progeny to exhibit improved tolerance to stresses encountered by the parent plant (Kinoshita and Seki 2014).

10.2.5 Histone Modifications

Histones are the core proteins around which DNA is wrapped to form chromatin. Chromatin structure and gene accessibility are regulated by post-translational modifications, such as acetylation, methylation, and phosphorylation (Bannister and Kouzarides 2011). Histone acetylation is associated with gene activation and the loosening of chromatin, facilitating transcription. In contrast, histone methylation can either activate or suppress gene expression, depending on the specific residue modified (Lawrence et al. 2016).

10.2.6 Small RNAs (sRNAs)

It includes microRNAs (miRNAs) and small interfering RNAs (siRNAs), which play a crucial role in post-transcriptional gene regulation by guiding RNA-induced silencing complexes to target mRNAs for degradation or translational repression (Matzke and Mosher 2014). MiRNAs regulate the expression of transcription factors and other genes involved in stress responses, whereas siRNAs are involved in RNA-directed DNA methylation, reinforcing transcriptional silencing of target loci (Erdmann and Picard 2020). Environmental stresses modulate sRNA expression profiles, thereby orchestrating a precise and reversible regulation of the plant's response machinery.

10.3 Stress-Induced Epigenetic Changes, Stress Memory, and Heritability

Environmental stressors induce specific epigenetic modifications, including DNA methylation, histone tail modifications, and small RNA activity, which collectively help plants modulate gene expression. One of the most remarkable features of such epigenetic regulation is the ability to establish stress memory, where plants remember prior exposure and respond more efficiently upon re-exposure to similar stress. Plants exposed to stress often exhibit priming, which is a state of enhanced preparedness for subsequent stress events (Avramova 2015). This memory effect has been documented in several contexts. For example, drought exposure in Arabidopsis thaliana leads to changes in DNA methylation that prime the plant for faster stomatal regulation during future droughts (Rahman et al. 2025). Similarly, salinity stress in rice has been linked to histone acetylation changes that facilitate enhanced ion transport and osmotic balance. Additionally, pathogen challenges often trigger the RNA-directed DNA methylation pathway, which silences transposable elements and activates defense-related genes, thereby increasing pathogen resistance (Yung et al. 2021).

Stress-induced epigenetic changes also directly regulate the detoxification of reactive oxygen species and the activation of antioxidant enzymes, such as superoxide dismutase, catalase, and peroxidases. Since ROS accumulation is a common consequence of abiotic and biotic stresses, epigenetic control over detoxification pathways provides plants with a crucial defense mechanism to maintain cellular homeostasis (Nadarajah 2020; Jangra et al. 2024). Interestingly, in urban environments, roadside trees exposed to high vehicular pollution display altered methylation patterns and histone modifications that correlate with improved antioxidant activity, providing real-world evidence of epigenetic adaptation outside laboratory conditions.

Stress memory can be either somatic, affecting only the current life cycle and not being inherited by the next generation, or transgenerational, which involves stable modifications transmitted through gametes or seeds to the next generation, thereby preparing the next generation to withstand similar stresses without direct prior exposure (Nguyen et al. 2022). This phenomenon has been demonstrated *in Arabidopsis*, where exposure to heat or drought resulted in progeny with enhanced tolerance, and in rice, where salinity stress-induced methylation changes were inherited across multiple generations (Ou et al. 2012). Not all epigenetic modifications are stable or heritable; many are reset during the reproductive process. However, those that persist serve as a natural mechanism for evolutionary adaptation. These modulations enable plant populations to adapt to changing climates much more rapidly than would be possible through genetic mutations alone.

Taken together, stress-induced epigenetic changes and their heritability form a powerful adaptive toolkit for plants. They not only provide short-term resilience by reprogramming stress responses within a single generation but also contribute to long-term adaptation by passing beneficial modifications to offspring. This opens new opportunities for crop improvement, climate-resilient agriculture, and the management of urban ecosystems, where plants are regularly exposed to multiple overlapping stresses. Harnessing epigenetic insights in breeding and ecological management strategies has the potential to enhance food security, sustainability, and biodiversity conservation under rapidly changing environmental conditions (Rajpal et al. 2022).

10.4 Methodologies for Studying Epigenetic Modulations

Epigenetic regulation studies have advanced with the advent of next-generation sequencing (NGS) technologies. These provide detailed and comprehensive insights into the epigenetic modifications of various cell types, unraveling their impact on gene regulation. The techniques available for epigenetic profiling include RNA sequencing (RNA-seq) for small RNAs and long non-coding RNAs (lncRNAs), chromatin immunoprecipitation sequencing (ChIP-seq), assay for transposase-accessible chromatin with sequencing (ATAC-seq), and methylated DNA immunoprecipitation sequencing (MeDIP-seq) (Selvamani et al. 2025).

RNA-seq provides a comprehensive and quantitative view of gene expression, offering a deeper understanding of epigenetic regulation by regulatory RNAs, including long non-coding RNAs (lncRNAs), small interfering RNAs (siRNAs), and microRNAs (miRNAs) (Kumar et al. 2018). ChIP-seq analysis in epigenetics enables the identification of genome-wide histone marks and transcription factor binding sites (TFBSs) (Mundade et al. 2014). ATAC-seq defines chromatin accessibility through the mapping of Tn5 transposase activity, facilitating transcription factor binding analysis, nucleosome mapping, novel enhancer identification, comparative epigenomics, and biomarker discovery. This makes it a significant technique in epigenetics analysis (Ma and Zhang 2020). MeDIP-seq is a genome-wide DNA methylation profiling technique that enriches methylated DNA fragments using antibodies against 5-methylcytosine for subsequent sequencing (Neary and Carless 2020).

The analysis of nuclear protein interactions through epigenetic studies helps develop personalized medicines and construct the transcriptional regulatory network, which significantly reduces noise in epigenetic analyses (Selvamani et al. 2025). The decision between untargeted and targeted epigenetic techniques is based on the study environment and the level of resolution required. Untargeted approaches, such as whole-genome bisulfite sequencing (WGBS), chromatin Immunoprecipitation (ChIP) sequencing (ChIP-seq), and RNA sequencing (RNA-seq), are critical during the exploratory stages of research.

Targeted approaches, such as locus-specific bisulfite polymerase chain reaction (PCR), methylation-sensitive qPCR, targeted ChIP, or microRNA (miRNA) panels, are better suited for hypothesis-driven research, marker validation, and clinical applications that require precision and reproducibility (Benčik et al. 2025). Methylation-specific PCR (MSP) is a highly sensitive and gene-targeted method for detecting CpG methylation, particularly in promoter regions. Herman et al. (1996) introduced this technique, which utilizes bisulfite-treated DNA to distinguish between methylated and unmethylated cytosines through the use of primer pairs designed to bind sequence variants resulting from differential conversion. MSP marked a major advancement in early epigenetic research by enabling the reliable detection of methylation in small or degraded DNA samples, such as those from formalin-fixed, paraffin-embedded (FFPE) tissues, where traditional methods, like Southern blotting, proved ineffective (Yoshioka et al. 2018).

A typical MSP workflow consists of DNA extraction, bisulfite conversion, primer design, PCR amplification, and post-PCR analysis. Bisulfite conversion is central to the assay and must be optimized to ensure complete deamination of unmethylated cytosines without excessive DNA degradation (Simons et al. 2025). Primer design is critical, as primers must target CpG-rich promoter regions, ideally near transcription start sites, and avoid CpG islands, which can reduce amplification efficiency. Notably, methylated primers are generally shorter and richer in G/C content, whereas unmethylated primers are longer and A/T-rich, allowing for the conversion of uracil (Benčik et al. 2025).

10.4.1 Limitations and Challenges in Epigenetics Research

Research on plant epigenetics has primarily focused on DNA methylation, although other mechanisms are also gaining attention. For instance, histone acetylation has been linked with root growth, developmental processes, and photosynthesis in several poplar species. At the same time, microRNAs (miRNAs) have been shown to influence plant growth and phosphorus uptake (Schönberger et al. 2016). Poplar is often studied because it serves as a model tree species, possessing a relatively small genome (0.5 Gb), few repetitive sequences, and a fully sequenced and annotated reference genome. This makes it an excellent resource for genomic and epigenomic research. By contrast, knowledge about histone modifications and non-coding RNAs in conifers remains very limited, presenting a significant challenge for advancing forest tree epigenomics.

The first complete plant methylome was reported for *Arabidopsis thaliana* using bisulfite sequencing, while conifer epigenomic mapping began later with Picea abies in 2016 (Ausin et al. 2016). Studies have examined methylation during normal development, heteroblasty, phenotypic plasticity, and stress memory. However, many studies have focused on model plants, such as Arabidopsis (Markus et al. 2021), crops (Mercé et al. 2020), and fruit trees (Rothkegel et al. 2021). In forest species, limited progress includes studies on epigenetic memory during embryogenesis in *Picea abies, Picea glauca, and Picea sylvestris* (Lafon-Placette et al. 2018). Moreover, studies have revealed stress-related methylation changes in poplars under drought and in Pinus radiata exposed to heat stress (Lamelas et al. 2020), again highlighting the prominence of poplar studies due to its genomic advantages.

10.4.2 Need for Integrated Multi-omics and Long-Term Studies

Plants face various stresses, which often coincide. Studying these complex interactions requires more than single-layer molecular approaches. Integrated multi-omics studies, which combine genomics, transcriptomics, epigenomics, proteomics, and metabolomics, offer a holistic view of plant adaptation by linking genetic information with gene expression, protein activity, and metabolite accumulation. Genomics provides the foundation, but it cannot explain phenotypic plasticity, which is strongly influenced by epigenetic modifications and post-transcriptional regulation (Yang et al. 2023). Merging multiple omics datasets may help identify functional genes, pathways, and networks that control stress tolerance in trees. For example, combined methylation and transcriptome analyses have revealed how DNA methylation influences stress-related gene expression, while system-level integration has identified key regulators of lignin biosynthesis in Populus (Balmant et al. 2020).

Long-term studies are equally important, as they track changes in epigenetic and molecular profiles over time, providing insights into stress memory and resilience across seasons or generations. Despite challenges in data integration, clustering, and interpretation, the combination of multi-omics with long-term monitoring can build predictive models for plant responses in urban ecosystems (Pierre-Jean et al. 2020). Such approaches will be crucial for breeding resilient species, guiding urban forest management, and strengthening climate adaptation strategies.

10.5 Applications in Plant Breeding, Agriculture, and Urban Forestry

Epigenetics provides a robust framework for enhancing plant adaptation to environmental stresses without altering the underlying DNA sequence. Understanding and manipulating epigenetic modifications, such as DNA methylation, histone modifications, and small RNAs, may help develop new strategies to enhance crop resilience and sustain urban forests in a changing climate.

10.5.1 Plant Breeding and Agriculture

Epigenetic marks contribute to phenotypic diversity and stress tolerance. Unlike mutations, these modifications are often reversible and can be inherited across generations. Epigenetic recombinant inbred lines (epiRILs) in *Arabidopsis* have revealed novel stress-tolerant traits that are independent of DNA sequence changes (Kim et al. 2022). Exposing plants to mild stress can trigger epigenetic "memory," enabling them to respond more effectively to future stress. This concept is being explored in cereals, rice, and legumes for drought and salinity tolerance. DNA methylation or histone modification signatures serve as biomarkers for stress tolerance. The integration of these biomarkers into marker-assisted selection accelerates the identification of resilient genotypes. CRISPR/dCas9-based tools allow targeted editing of DNA methylation or histone marks. This can modulate the expression of stress-responsive genes without the need for transgenic modification, making it a promising approach in the era of climate-smart agriculture (Yang et al. 2025).

10.5.2 Urban Forestry Applications

Urban trees provide vital ecosystem services, including shade, cooling, carbon sequestration, pollution mitigation, and biodiversity support. However, they face combined stresses, including heat islands, drought, and pollution. Epigenetic studies in forestry open new avenues for sustainable urban green infrastructure. Epigenetic markers can help identify urban tree species that maintain stable growth under chronic stresses, such as high ozone, particulate matter, and irregular water supply. *Populus* and *Quercus* species have exhibited stress-induced methylation changes that may inform species selection in urban planning (Klupczyńska and Ratajczak 2021). Since trees are long-lived, epigenetic memory plays a crucial role in their survival. Understanding how trees "remember" stresses can help urban planners select genotypes capable of thriving under varying urban conditions. Monitoring epigenetic stability ensures that beneficial stress adaptations are transmitted to subsequent generations. Additionally, the early detection of stress via methylation or small RNA changes can serve as diagnostic tools to predict decline before visible symptoms appear. Epigenetic data, combined with thermal imaging, hyperspectral sensing, and UV-based monitoring, can provide holistic assessments of tree health in urban landscapes.

10.5.3 Broader Implications

In climate-smart agriculture and forestry, epigenetics provides strategies for developing resilient crops and trees that can withstand future climate scenarios. In sustainable urban ecosystems, epigenetic-based management can enhance the longevity and services of urban forests, thereby contributing to human well-being and ecological balance. Furthermore, translating lab findings on epigenetics into breeding pipelines and forestry management will be crucial for real-world applications (Pakhira et al. 2024).

10.6 Implications for Sustainable Development Goals (SDGs)

Epigenetic modulations reveal how plants cope with environmental stressors and provide innovative solutions to some of the most pressing Sustainable Development Goals (SDGs). Regulating gene expression in response to environmental cues may contribute directly to some of the SDGs, such as climate resilience (SDG 13), food security (SDG 2), and sustainable agriculture (SDG 12 & 15).

10.6.1 Climate Resilience (SDG 13: Climate Action)

Plants exposed to heat waves, drought, salinity, and pollution can develop epigenetic "stress memory," enabling them to respond more effectively to recurring stresses. Long-term adaptation strategies in crops and trees rely on heritable epigenetic modifications that increase tolerance without altering genetic diversity. Furthermore, epigenetic biomarkers can guide breeding programs aimed at developing climate-resilient crops and urban trees, strengthening adaptation to climate change. For instance, some crops exhibit enhanced drought tolerance through epigenetic priming, enabling more stable yields under changing temperatures (Suruthi et al. 2025).

10.6.2 Zero Hunger and Food Security (SDG 2: Zero Hunger)

Epigenetic mechanisms enhance crop productivity under stress conditions, ensuring stable yields in water-limited and degraded soils. These plants adapt more quickly through epigenetic mechanisms, suggesting that breeding programs can incorporate traits like stress tolerance without requiring genetic mutations. Moreover, stress priming through epigenetic memory (e.g., drought priming in cereals) provides a low-cost and sustainable approach to securing food production. Epigenome-assisted breeding can enhance the nutritional quality and resilience of staple crops, thereby contributing to food and nutritional security (Sun et al. 2021).

10.6.3 Sustainable Agriculture and Biodiversity (SDG 12: Responsible Consumption and Production & SDG 15: Life on Land)

Epigenetics reduces reliance on chemical inputs by enhancing natural stress tolerance and pest resistance, thereby supporting eco-friendly agricultural practices. In forestry, epigenetic studies in long-lived trees help design sustainable urban and rural ecosystems, enhancing ecosystem services such as carbon sequestration and biodiversity conservation. By promoting plant varieties that can thrive under low-input conditions, epigenetics supports sustainable land management and reduces environmental degradation (Guarino et al. 2022).

10.7 Future Directions and Research Needs

Although plant epigenetics has shown great promise, there are still many underlying gaps that need attention to translate plant epigenetics knowledge into resilient, equitable agriculture and conservation. Currently, most studies are concentrated in developed countries, while expanding research to underrepresented regions and agro-ecological contexts is crucial for understanding how local crops and native plants utilize epigenetic changes to survive stress (Auge et al. 2024). Long-term monitoring and well-designed transgenerational studies are also necessary to confirm whether stress-induced epigenetic marks can be stably passed on to the next generation or are only transient, distinguishing between stress memory and heritable changes (Bird 2024).

Furthermore, integrating epigenetic studies with modern tools such as remote sensing, environmental monitoring, and field-based observations allows for linking molecular changes with large-scale ecosystem responses, enabling the detection of epigenetically linked phenotypes across space and time, and improving predictive models of crop and habitat responses to climate change (Thackeray and Hampton 2020). Moreover, for practical benefits, interdisciplinary collaborations among molecular biologists, ecologists, agronomists, and policymakers are needed to enable capacity building in developing countries, promote open data sharing, and establish clear policies. This will further ensure that epigenetic knowledge is applied safely and fairly for sustainable agriculture and biodiversity conservation.

10.8 Conclusion

Epigenetics has emerged as a key driver of plant adaptation to environmental stress, enabling plants to modulate their responses without permanently altering their DNA sequences. Plants can rapidly adapt to varying environmental conditions and pass these adaptive traits on to the next generation through reversible, heritable, and flexible regulatory mechanisms, such as DNA methylation, histone modifications, and small RNAs. This unique adaptive capacity enhances plant resilience and holds significant promise for crop improvement, where stress-tolerant varieties can contribute to more sustainable and climate-resilient agriculture. Moreover, epigenetic insights can also be applied to sustainable urban ecosystem management, where resilient roadside plantations, urban forests, and green infrastructure are increasingly crucial for mitigating the impacts of pollution and climate change. Integrating molecular research with practical land-use strategies may offer a powerful approach to achieving agricultural sustainability and resilient urban ecosystems, aligning with global sustainability goals.

References

Auge G, Sunil RS, Ingle RA, Rahul PV, Mutwil M, Estevez JM (2024) Current challenges for plant biology research in the Global South. New Phytol 244(4):1168–1174

Ausin I, Feng S, Yu C, Liu W, Kuo HY, Jacobsen EL, Wang H (2016) DNA methylome of the 20-gigabase Norway spruce genome. Proc Natl Acad Sci 113(50):E8106–E8113

Avramova Z (2015) Transcriptional 'memory'of a stress: transient chromatin and memory (epigenetic) marks at stress-response genes. Plant J 83(1):149–159

Balmant KM, Noble JD, Alves FC, Dervinis C, Conde D, Schmidt HW et al (2020) Xylem systems genetics analysis reveals a key regulator of lignin biosynthesis in Populusdeltoides. Genome Res 30(8):1131–1143

Bannister AJ, Kouzarides T (2011) Regulation of chromatin by histone modifications. Cell Res 21(3):381–395

Bashir H, Setyawati A (2025) Decoding the mechanisms of plant epigenetic adaptations to stress. Rev Agric Sci 13(2):1–19

Benčik I, SaftićMartinović L, Mladenić T, Ostojić S, DevićPavlić S (2025) From DNA methylation and histone modifications to non-coding RNAs: evaluating tools for epigenetic research. Appl Sci 15(18):9940

Bharath P, Gahir S, Raghavendra AS (2021) Abscisic acid-induced stomatal closure: an important component of plant defense against abiotic and biotic stress. Front Plant Sci 12:615114

Bird A (2024) Transgenerational epigenetic inheritance: a critical perspective. Front Epigenet Epigenom 2:1434253

Bowers ME, Yehuda R (2016) Intergenerational transmission of stress in humans. Neuropsychopharmacology 41(1):232–244

Erdmann RM, Picard CL (2020) RNA-directed DNA methylation. PLoS Genet 16(10):e1009034

Gayathiri E, Viswanathan D, Prakash P, Pandiaraj S, Kondapavuluri BK, Singh CD, Thiruvengadam M (2025) A critical review on epigenetics and epigenomics in plant development and stress resilience. Rev Bras Bot 48(1):51

Guarino F, Cicatelli A, Castiglione S, Agius DR, Orhun GE, Fragkostefanakis S, Martinelli F (2022) An epigenetic alphabet of crop adaptation to climate change. Front Genet 13:818727

Herman JG, Graff JR, Myöhänen SBDN, Nelkin BD, Baylin SB (1996) Methylation-specific PCR: a novel PCR assay for methylation status of CpG islands. Proc Natl Acad Sci U S A 93(18):9821–9826

Jangra A, Kumar K, Maikhuri S, Bhandari MS, Pandey S, Singh H, Barthwal S (2024) Unveiling stress-adapted endophytic bacteria: characterizing plant growth-promoting traits and assessing cross-inoculation effects on *Populus deltoides* under abiotic stress. Plant Physiol Biochem 210:108610

Kim EY, Kim KD, Cho J (2022) Harnessing epigenetic variability for crop improvement: current status and future prospects. Genes & Genomics 44(3):259–266

Kinoshita T, Seki M (2014) Epigenetic memory for stress response and adaptation in plants. Plant Cell Physiol 55(11):1859–1863

Klupczyńska EA, Ratajczak E (2021) Can forest trees cope with climate change?—effects of DNA methylation on gene expression and adaptation to environmental change. Int J Mol Sci 22(24):13524

Kumar M, Rani K (2023) Epigenomics in stress tolerance of plants under the climate change. Mol Biol Rep 50(7):6201–6216

Kumar V, Khare T, Shriram V, Wani SH (2018) Plant small RNAs: the essential epigenetic regulators of gene expression for salt-stress responses and tolerance. Plant Cell Rep 37(1):61–75

Kumar K, Malik A, Singh H, Meena RK, Barthwal S (2025) Foliar application of silicon and salicylic acid on morpho-physiological and biochemical characteristics of poplar under varying water regimes. Plant Physiol Rep:1–12

Kumari P, Khan S, Wani IA, Gupta R, Verma S, Alam P, Alaklabi A (2022) Unravelling the role of epigenetic modifications in development and reproduction of angiosperms: a critical appraisal. Front Genet 13:819941

Lafon-Placette C, Le Gac AL, Chauveau D, Segura V, Delaunay A, Lesage-Descauses MC et al (2018) Changes in the epigenome and transcriptome of the poplar shoot apical meristem in response to water availability affect preferentially hormone pathways. J Exp Bot 69(3):537–551

Lamelas L, Valledor L, Escandón M, Pinto G, Cañal MJ, Meijón M (2020) Integrative analysis of the nuclear proteome in Pinusradiata reveals thermopriming coupled to epigenetic regulation. J Exp Bot 71(6):2040–2057

Lawrence M, Daujat S, Schneider R (2016) Lateral thinking: how histone modifications regulate gene expression. Trends Genet 32(1):42–56

Liu H, Song S, Zhang H, Li Y, Niu L, Zhang J, Wang W (2022) Signaling transduction of ABA, ROS, and Ca2+ in plant stomatal closure in response to drought. Int J Mol Sci 23(23):14824

Ma S, Zhang Y (2020) Profiling chromatin regulatory landscape: insights into the development of ChIP-seq and ATAC-seq. Mol Biomed 1(1):9

Malik A, Singh H, Kumar K, Kumari G, Sojitra A (2024) Understanding the impact of climate change and environments on urban forests: insights from morphological, physiological, and molecular perspectives. In: Urban forests, climate change and environmental pollution: physio-biochemical and molecular perspectives to enhance urban resilience. Springer Nature Switzerland, Cham, pp 161–183

Markus C, Pecinka A, Merotto A Jr (2021) Insights into the role of transcriptional gene silencing in response to herbicide-treatments in Arabidopsis thaliana. Int J Mol Sci 22(7):3314

Matzke MA, Mosher RA (2014) RNA-directed DNA methylation: an epigenetic pathway of increasing complexity. Nat Rev Genet 15(6):394–408

Mercé C, Bayer PE, Tay Fernandez C, Batley J, Edwards D (2020) Induced methylation in plants as a crop improvement tool: progress and perspectives. Agronomy 10(10):1484

Mundade R, Ozer HG, Wei H, Prabhu L, Lu T (2014) Role of ChIP-seq in the discovery of transcription factor binding sites, differential gene regulation mechanism, epigenetic marks and beyond. Cell Cycle 13(18):2847–2852

Nadarajah KK (2020) ROS homeostasis in abiotic stress tolerance in plants. Int J Mol Sci 21(15):5208

Nawaz M, Sun J, Shabbir S, Khattak WA, Ren G, Nie X, Sonne C (2023) A review of plants strategies to resist biotic and abiotic environmental stressors. Sci Total Environ 900:165832

Neary JL, Carless MA (2020) Methylated DNA immunoprecipitation sequencing (MeDIP-seq): principles and applications. In: Epigenetics methods. Academic Press, pp 157–179

Nguyen NH, Vu NT, Cheong JJ (2022) Transcriptional stress memory and transgenerational inheritance of drought tolerance in plants. Int J Mol Sci 23(21):12918

Ou X, Zhang Y, Xu C, Lin X, Zang Q, Zhuang T et al (2012) Transgenerational inheritance of modified DNA methylation patterns and enhanced tolerance induced by heavy metal stress in rice (Oryza sativa L.). PLoS One 7

Pakhira R, Biswas S, Manoj C, Singh H (2024) Biological insights into forest adaptation to climate change: ecosystem resilience and adaptation. In: Urban forests, climate change and environmental pollution: physio-biochemical and molecular perspectives to enhance urban resilience. Springer Nature Switzerland, Cham, pp 307–329

Pierre-Jean M, Deleuze JF, Le Floch E, Mauger F (2020) Clustering and variable selection evaluation of 13 unsupervised methods for multi-omics data integration. Brief Bioinform 21(6):2011–2030

Rahman MM, Keya SS, Bulle M, Ahsan SM, Rahman MA, Roni MS et al (2025) Past trauma, better future: how stress memory shapes plant adaptation to drought. Funct Plant Biol 52(6)

Rajpal VR, Rathore P, Mehta S, Wadhwa N, Yadav P, Berry E, Raina SN (2022) Epigenetic variation: a major player in facilitating plant fitness under changing environmental conditions. Front Cell Dev Biol 10:1020958

Rothkegel K, Espinoza A, Sanhueza D, Lillo-Carmona V, Riveros A, Campos-Vargas R, Meneses C (2021) Identification of DNA methylation and transcriptomic profiles associated with fruit mealiness in Prunuspersica (L.) Batsch. *Frontiers*. Plant Sci 12:684130

Salvi P, Manna M, Kaur H, Thakur T, Gandass N, Bhatt D, Muthamilarasan M (2021) Phytohormone signaling and crosstalk in regulating drought stress response in plants. Plant Cell Rep 40(8):1305–1329

Schönberger B, Chen X, Mager S, Ludewig U (2016) Site-dependent differences in DNA methylation and their impact on plant establishment and phosphorus nutrition in Populustrichocarpa. PLoS One 11(12):e0168623

Selvamani SB, Selvapandian U, Shihabdeen MNS, Muthusamy K, Nallusamy S (2025) Epigenetics: RNA-seq, ChIP-seq, ATAC-seq, and MeDIP-seq. In: Next-generation sequencing. CRC Press, pp 17–47

Simons RB, Karkala F, Kukk MM, Adams HH, Kayser M, Vidaki A (2025) Comparative performance evaluation of bisulfite-and enzyme-based DNA conversion methods. Clin Epigenetics 17(1):56

Singh H, Verma A, Kumar M, Sharma R, Gupta R, Kaur M et al (2017) Phytoremediation: a green technology to clean up the sites with low and moderate level of heavy metals. Austin Biochem 2(2):1–8

Singh H, Bandyopadhyay S, Kumar A, Singh M, Malik A, Kumari G, Rawat PS (2022) Understanding the physiological and biophysical response of urban roadside plantations for assessing adaptation and mitigation mechanisms toward vehicular emissions. Urban Clim 44:101183

Sun C, Ali K, Yan K, Fiaz S, Dormatey R, Bi Z, Bai J (2021) Exploration of epigenetics for improvement of drought and other stress resistance in crops: a review. Plants 10(6):1226

Suruthi GS, Anand AV, Rajesh M, Begum TN, Begum SB, Priya A, Robert AA (2025) Epigenetics and epigenomics for climate-smart and sustainable agriculture: unlocking potential for crop improvement. In: Epigenetics for climate-smart and sustainable agriculture. CABI, GB, pp 24–40

Thackeray SJ, Hampton SE (2020) The case for research integration, from genomics to remote sensing, to understand biodiversity change and functional dynamics in the world's lakes. Glob Chang Biol 26(6):3230–3240

Yang L, Yang Y, Huang L, Cui X, Liu Y (2023) From single-to multi-omics: future research trends in medicinal plants. Brief Bioinform 24(1)

Yang X, Sohail H, Noor I, Costa FCL, Zhong S, Zhang L, Chen X (2025) Epigenetic crop improvement: integrating ENCODE strategies into horticultural breeding. Hortic Res 12:uhaf213

Yoshioka M, Matsutani T, Hara A, Hirono S, Hiwasa T, Takiguchi M, Iwadate Y (2018) Real-time methylation-specific PCR for the evaluation of methylation status of MGMT gene in glioblastoma. Oncotarget 9(45):27728

Yung WS, Li MW, Sze CC, Wang Q, Lam HM (2021) Histone modifications and chromatin remodelling in plants in response to salt stress. Physiol Plant 173(4):1495–1513

Zhang H, Lang Z, Zhu JK (2018) Dynamics and function of DNA methylation in plants. Nat Rev Mol Cell Biol 19(8):489–506

Plant Microbe Interaction in the Context of Climate Change 11

Vivek Kumar, Kapil Khulbe, Sushma Tamta,
Rashmi Srivastava, and A. K. Sharma

Abstract

Plant-microbe interactions are pivotal in shaping terrestrial ecosystem processes, especially as climate change accelerates. Rising atmospheric CO_2 and global warming are altering ecosystem dynamics, making it essential to understand how plants and soil microbes collaborate to influence carbon sequestration, nutrient cycling, and ecosystem resilience. This chapter highlights the significance of mutualistic partnerships, such as those with mycorrhizal fungi and plant growth-promoting rhizobacteria, alongside specialized processes like the oxalate–carbonate pathway (OCP), in which plant-derived oxalate is transformed by microbes into stable soil carbonates. Field studies in Indian forests, particularly under oxalogenic trees like Terminalia alata, reveal that these interactions can enhance soil pH, calcium carbonate content, and fertility, while supporting diverse communities of beneficial bacteria and fungi. Climate change-induced stresses, such as increased temperature and altered rainfall, can disrupt plant-microbe relationships, potentially affecting ecosystem functions. Nevertheless, microbial diversity and functional redundancy offer resilience, helping maintain critical processes like nutrient cycling and long-term carbon storage. Practical applications, including agroforestry, reforestation, and microbial inoculants, leverage these plant-microbe partnerships to boost soil carbon sequestration,

V. Kumar (✉)
Department of Botany, IPGGPGCC Haldwani, Kumaun University,
Nainital, Uttarakhand, India

K. Khulbe · S. Tamta
Department of Botany, D.S.B Campus, Kumaun University, Nainital, Uttarakhand, India

R. Srivastava · A. K. Sharma
Department of Biological Sciences, College of Basic Sciences & Humanities, G. B. Pant University of Agriculture and Technology, Pantnagar, Uttarakhand, India

H. C. Joshi et al. (eds.), *Climate-Responsive Plant Biology*,
https://doi.org/10.1007/978-981-95-7354-7_11

restore degraded lands, and promote sustainable agriculture in a changing climate.

In summary, plant-microbe interactions are a vital but underutilized tool for climate change mitigation. Advancing our knowledge of these relationships through field research, molecular techniques, and biotechnology can unlock new strategies for sustaining ecosystem services, improving soil health, and enhancing carbon sequestration in terrestrial environments.

Keywords

Climate change · Carbon sequestration · Mycorrhizal fungi · Plant growth-promoting rhizobacteria (PGPR) · Oxalate–carbonate pathway (OCP), Oxalogenic trees

11.1 Introduction

Climate change, marked by rising greenhouse gas concentrations, shifting weather patterns, and more frequent extreme events, is exerting cascading effects on terrestrial ecosystems, including the intricate interactions between plants and soil microorganisms (Sharma et al. 2022). Elevated temperatures, changes in precipitation, and abiotic stresses such as drought and salinity directly influence plant physiology and productivity while also reshaping the composition, diversity, and functioning of soil microbial communities (Bardgett 2011).

Within the rhizosphere, the soil zone immediately surrounding plant roots, these changes are particularly profound, as this dynamic interface underpins critical processes such as nutrient cycling, plant stress tolerance, and disease suppression. The rhizosphere harbors a network of beneficial microbes that facilitate nutrient acquisition, produce growth-promoting hormones, and help plants adapt to adverse environmental conditions (e.g., by synthesizing antioxidants, osmoprotectants, and regulatory metabolites) (Chen et al. 2024). Climatic stresses can disrupt these mutualistic associations, potentially reducing essential functions like nitrogen fixation, phosphate solubilization, and mycorrhizal colonization, thereby impacting plant health and ecosystem resilience (Bhattacharyya et al. 2021). Nevertheless, some soil microbes exhibit remarkable adaptive plasticity and resilience, with the ability to shift community composition toward taxa better able to support ecosystem functions under challenging conditions (Sharma et al. 2022).

Recent advances in molecular and field-based research have illuminated the genetic and chemical dialogues between plant roots and their microbiomes, revealing mechanisms mediating plant responses to climate-driven stresses. Such insights are being leveraged to develop sustainable management approaches such as the application of plant growth-promoting rhizobacteria, arbuscular mycorrhizal fungi, and tailored bioinoculants that reinforce both crop yields and ecosystem stability under ongoing environmental fluctuations (Saini and Sharma 2019). By harnessing these plant-microbe partnerships, it may be possible to mitigate some adverse impacts of climate change, enhance soil health, maintain productivity, and contribute to climate-smart agriculture at regional and global scales (Sharma et al. 2022).

A distinctive example of microbial adaptation to environmental change is found in oxalotrophic bacteria, microorganisms capable of using oxalate as their sole carbon and energy source. These bacteria, widely distributed in soils and closely associated with plant roots, play key roles in the oxalate–carbonate pathway, where they degrade plant-derived oxalic acid, promoting soil carbon cycling, calcium availability, and mineral weathering (Bravo et al. 2015). Through their metabolic activities, oxalotrophic bacteria help regulate rhizosphere microbial communities, suppress fungal pathogens (many of which rely on oxalic acid for infection), and contribute to the formation of stable inorganic carbon sinks (via carbonates), with implications for global CO_2 sequestration and soil organic matter restoration. Their ecological significance is especially pronounced in arid, saline, or nutrient-poor ecosystems, where their functions can enhance soil health, mitigate abiotic stresses, and support sustainable agriculture under the pressures of ongoing climate change. Harnessing the metabolic diversity and environmental resilience of oxalotrophic bacteria therefore represents a promising strategy for improving both ecosystem stability and natural climate mitigation solutions (Syed et al. 2020).

11.2 Climate Change and Its Impact on Terrestrial Ecosystems

- *Rising Greenhouse Gases:* Global environmental change has ushered the planet into a new geological epoch known as the Anthropocene, a term popularized by Paul Crutzen to describe the profound dominance of human activity over Earth's systems since the industrial revolution (Crutzen 2016). Over the past 150 years, atmospheric CO_2 concentrations have surged from 280 ppm to nearly 400 ppm due to human activities such as fossil fuel burning and deforestation, which are well-documented as the primary drivers of the recent sharp rise in greenhouse gas levels and global warming. This persistent increase in CO_2 has intensified the greenhouse effect, leading to warming temperatures, altered precipitation patterns, more droughts, and frequent extreme weather events (Geeks for Geeks 2025). The US EPA highlights how these climate alterations cause direct ecological effects, shifting vegetation zones, disrupting wildlife migration, and stressing various species. As a consequence, many terrestrial plants and animals now face increased extinction risks if they cannot adapt to rapid environmental changes—a challenge further detailed by leading reviews of terrestrial ecosystem vulnerability (U.S. Environmental Protection Agency 2025). Regional studies emphasize that in nations like India, climate change is reducing agricultural yields and limiting water resources, thereby threatening both food security and human livelihoods. Collectively, these processes disrupt ecosystem structure, weaken biodiversity, and undermine ecological stability, revealing the urgent necessity of greenhouse gas mitigation and adaptation measures to sustain terrestrial ecosystems, as also recognized by international and national science agencies (Conradi et al. 2024).

- *Terrestrial Carbon Sinks:* Forests and soils serve as major terrestrial carbon sinks by capturing and storing large amounts of atmospheric carbon dioxide (CO_2), thereby playing a crucial role in mitigating climate change. Forests sequester carbon primarily through photosynthesis, storing it in biomass such as trunks, branches, leaves, and roots. This carbon can remain stored for decades or even centuries, depending on forest type and age. Soils, on the other hand, act as significant carbon reservoirs by accumulating organic carbon derived from plant litter, root exudates, and microbial biomass. The capacity of these carbon sinks is influenced by biotic factors like vegetation type, microbial diversity, and plant-microbe interactions, as well as abiotic factors such as climate, soil type, temperature, and moisture availability. For example, cooler and moist climates tend to support higher soil organic carbon levels compared to arid regions. Additionally, processes like the oxalate-carbonate pathway, where some plants, bacteria, and fungi facilitate carbon transfer from atmospheric CO_2 to calcium carbonate, enable long-term mineral carbon storage in soils (Verrecchia et al. 2006; Cailleau et al. 2011). The biological oxidation of calcium oxalate not only alters local soil pH but also leads to the precipitation of calcium carbonate, acting as a stable and persistent carbon sink (Cailleau et al. 2004). Understanding these interactions, both biological and chemical, is essential to enhance carbon sequestration potential and develop effective climate mitigation strategies.

11.3 Plant-Microbe Interactions: Mechanisms and Functions

11.3.1 Symbiotic and Mutualistic Relationships

- *Rhizosphere Microbiome:* The rhizosphere is the narrow zone of soil directly influenced by plant roots, hosting a highly diverse microbial community comprised of bacteria, fungi, actinomycetes, and archaea. These organisms play a crucial role in facilitating nutrient uptake for the plant by solubilizing phosphate, fixing atmospheric nitrogen, and producing essential phytohormones such as auxins and cytokinins (Luo et al. 2025). The composition of this microbiome is actively shaped by root exudates and plant species, with beneficial microbes recruited from the surrounding bulk soil to the root-soil interface. In addition to promoting plant growth, rhizosphere microbes enhance tolerance to environmental stressors by inducing plant immune responses and outcompeting pathogens (Duret et al. 2024). The interplay between plant signaling and microbial activity ensures dynamic cycling of carbon and nitrogen, supporting robust ecosystem functioning and sustainable agriculture (Ling et al. 2022). Overall, the rhizosphere microbiome is a key driver of soil fertility, plant health, and ecosystem resilience.
- *Mycorrhizal Fungi:* Arbuscular mycorrhizal fungi (AMF) form highly beneficial symbiotic associations with the roots of most terrestrial plants, significantly enhancing plant nutrition and resilience. These fungi colonize root cortical cells, developing intricate arbuscule structures that serve as the primary sites for nutri-

ent and water exchange between the plant and fungus. By expanding the effective root surface area through a dense hyphal network in the soil, AMF improve the uptake of essential minerals, especially phosphorus, but also nitrogen, potassium, and micronutrients that are otherwise difficult for plant roots to access (Bennett and Groten 2022). Moreover, this symbiosis increases plant tolerance to abiotic stresses, including drought and high temperature, by promoting water acquisition and activating plant defense mechanisms (Begum et al. 2019). AMF also contribute to improved soil structure and fertility, indirectly supporting broader ecosystem health. Through these multiple mechanisms, arbuscular mycorrhizal fungi play a critical role in sustaining plant growth and productivity, especially under challenging environmental conditions (Chen et al. 2018).
- *Plant Growth-Promoting Rhizobacteria (PGPR):* Plant Growth-Promoting Rhizobacteria (PGPR) are soil bacteria that actively colonize plant root surfaces and interiors, where they exert numerous beneficial effects on plant growth and health. These microbes produce siderophores iron-chelating molecules that enhance iron bioavailability for plants, especially in iron-limited soils. PGPR also solubilize phosphates, making essential phosphorus more accessible in the rhizosphere (Glick 2012), which is crucial for root development and energy transfer. Another key function is the induction of systemic resistance, where PGPR stimulate plant defense mechanisms, thereby helping plants fend off pathogens and reducing disease incidence. Through direct mechanisms like nutrient mobilization and phytohormone production, and indirect protection against biotic stresses, PGPR contribute to improved crop yield, resilience to environmental stresses, and reduced reliance on chemical fertilizers and pesticides (Backer et al. 2018). Their multifaceted role is fundamental for sustainable agriculture, soil health, and plant productivity (Vejan et al. 2016).

11.3.2 The Oxalate–Carbonate Pathway (OCP): A Model for Climate-Resilient Interactions

- *Oxalogenic Plants:* Oxalate is a widespread and crucial component of soils, originating from plant root exudates, decomposition of organic matter, animal residues, and as a metabolite from fungi and microorganisms. Its concentration is notably higher in the rhizosphere, the region closely associated with plant roots, due to ongoing biological activity (Fox and Comerford 1990). One of oxalate's key roles in the soil environment is its ability to chelate and solubilize metals, thereby assisting in metal transport and detoxification processes for both plants and soil organisms (Gadd 1999). In plants, oxalic acid is produced through metabolic pathways often facilitated by ascorbic acid as a precursor and can accumulate as insoluble calcium oxalate crystals in specialized cells called crystal idioblasts (Franceschi and Nakata 2005). Oxalic acid and oxalate minerals are the secondary product of plants and microbes present in soil (Tamer and Aragno 1980). After plant death and decay, this oxalate is released into the soil, where it is persistently present due to its low solubility. Fungi also contribute by producing oxalate, which participates in metal speciation and nutrient cycling. While oxalate is not readily decomposed in litter, its fate is determined by microbial oxidation, mainly by oxalotrophic bacteria. This microbial process converts oxalate into carbon dioxide and hydroxide ions, enabling the precipitation of stable calcium carbonate (Fig. 11.1) and forming a long-lived inorganic carbon sink as part of the oxalate-carbonate pathway (OCP).

- The oxalate-carbonate pathway (OCP) is regulated primarily by three factors: oxalate, calcium ions (Ca^{2+}), and oxalate-degrading microbes (Dumas et al. 1995; Makela et al. 2002; Tuason and Arocena 2009). Studies on Milicia excelsa trees in Ivory Coast and Cameroon have demonstrated significant pedogenic carbonate accumulation in rhizospheric soils, resulting from oxalate metabolism by oxalotrophic bacteria. For instance, 170-year-old M. excelsa rhizospheric soil contained about 8 tons of calcium carbonate, equating to 1 ton of stable carbon, acting as an important carbon sink (Braissant et al. 2002, 2004; Cailleau et al. 2005). The residence time of this mineral carbon in soil can reach up to 10^6 years, far exceeding that of soil organic carbon. This stability highlights the value of the OCP as a natural carbon sequestration mechanism. Identification of oxalogenic systems includes biomineralization in plants, soil alkalinization, carbonate presence, pH stability, and oxalotrophic bacteria presence. Upon plant death, oxalate is released from plant cells and oxidized by microbes, producing carbonate the final product of the oxalate-carbonate, pathway, and contributing to long-term carbon storage in soils (Cailleau et al. 2004).

Fig. 11.1 Decomposition of $CaCO_3$ below plant

- *Oxalotrophic Bacteria:* Oxalic acid and oxalate minerals are prominent secondary products produced by plants, animals, fungi, and bacteria in soils. Calcium oxalate is widespread, present in over 215 angiosperm and gymnosperm families, sometimes accounting for over 50% of dry plant weight (Franceschi and Nakata 2005; Libert and Franceschi 1987). Oxalate plays a vital role in increasing phosphorus and micronutrient availability in soils (Sahin 2003). Oxalotrophic bacteria metabolize oxalate, facilitating calcium carbonate ($CaCO_3$) precipitation in acidic tropical soils, linking calcium and carbon cycles, and offering a potential long-term CO_2 sink (Braissant et al. 2002; Cailleau et al. 2004). The oxalate-carbonate pathway requires calcium ions, oxalate, and oxalate-degrading organisms, mainly bacteria, as spontaneous chemical oxidation of calcium oxalate is energetically unfavorable (Dumas et al. 1995; Verrecchia et al. 2006). Bacterial oxalate degradation raises soil pH, converting a strong acid into a weaker one, fostering carbonate precipitation (Braissant et al. 2004). Enzymes like formyl coenzyme A transferase (frc) and oxalyl-CoA decarboxylase (oxc) catalyze this catabolism. Fungi are also key players in oxalate cycling, but their role in oxidation is less clear (Dutton and Evans 1996).
- *Role of Fungi:* Fungi play a critical role as decomposers in terrestrial ecosystems, breaking down complex plant litter (Osono 2020), such as lignocellulose, by secreting extracellular enzymes, including cellulases, hemicellulases, pectinases, and phenol oxidases. Through this process, oxalate is released from calcium oxalate crystals present in plant tissues into the surrounding soil. Oxalate then acts as an essential substrate for oxalotrophic bacteria, linking plant litter degradation to microbial metabolic activities within the oxalate-carbonate pathway (OCP) (Braissant et al. 2004; Cailleau et al. 2005; Rowley et al. 2017; Durand et al. 2018), which promotes carbon sequestration by facilitating calcium carbonate precipitation in soils (Cailleau et al. 2011, 2014). Besides carbon stabilization, fungal decomposition significantly influences soil nutrient cycling, metal speciation, and microbial community structure, fundamentally shaping ecosystem functioning. The mutualistic interaction between fungi and oxalotrophic bacteria highlights the integrated role of soil microbial communities in converting plant-derived organic substrates into mineral carbon sinks, emphasizing fungi's role as key biological agents in maintaining long-term soil carbon storage and ecosystem resilience (Braissant et al. 2004; Cailleau et al. 2005). This interplay is vital for sustaining soil health and mitigating atmospheric CO_2 through natural biogeochemical cycles.
- *Ecological and Climate Significance:* The oxalate-carbonate pathway (OCP) holds significant ecological and climatic importance due to its ability to convert atmospheric CO2 into stable inorganic carbon in soils. This process involves the oxidation of plant-produced calcium oxalate by oxalotrophic bacteria, resulting in the precipitation of calcium carbonate (CaCO3) with a residence time spanning up to 100,000 years-vastly longer than the turnover time for soil organic carbon, which ranges from decades to millennia. Such mineral carbon sequestration provides a highly durable carbon sink, contributing to the regulation of atmospheric CO2 and mitigating climate change impacts. Studies of the tropical

tree Milicia excelsa in Africa reveal pedogenic carbonate accumulation in the rhizosphere as a direct outcome of the OCP, quantifying significant amounts of stable carbon stored in soils. The pathway's sustainability depends on the interaction of oxalogenic plants, oxalotrophic microbes, and soil chemistry conditions, particularly soil pH and calcium availability. While research remains limited, the OCP represents a promising natural mechanism for enhancing carbon sequestration in terrestrial ecosystems, especially in acidic tropical soils. Its quantitative contributions to the global carbon cycle and potential for integration into land management strategies offer vital prospects for climate mitigation and soil health restoration (Syed et al. 2020; Cailleau et al. 2011; Braissant et al. 2002).

11.4 Effects of Climate Change on Plant-Microbe Interactions

11.4.1 Abiotic Stress and Microbial Community Shifts

- *Temperature and Drought:* Climate change, particularly elevated temperatures and altered precipitation patterns, profoundly disrupts soil microbial communities, impacting plant-microbe interactions. Recent global meta-analyses reveal that experimental warming reduces soil bacterial and fungal diversity by 16% and 20%, respectively, with even greater losses projected under prolonged warming scenarios (Qiao et al. 2024; Fu et al. 2023; Jayaramaiah et al. 2025). These changes are driven by thermal filtering, competitive exclusion, and nutrient limitations, especially in nitrogen-poor soils (Qiao et al. 2024; Fu et al. 2023). Drought and heat stress further favor drought-tolerant and thermophilic taxa, reducing overall microbial abundance and functional potential. Such shifts weaken beneficial plant-microbe relationships, including those involved in nutrient cycling and disease suppression, ultimately compromising plant health and ecosystem resilience (Jayaramaiah et al. 2025; Qiao et al. 2024). The decline in microbial diversity and abundance also diminishes soil organic carbon sequestration, threatening long-term ecosystem functioning and stability (Qiao et al. 2024; Fu et al. 2023). These findings underscore the urgent need to conserve soil microbial diversity to sustain plant productivity and ecosystem services under ongoing climate change.
- *Greenhouse Gas Emissions:* Soil microbes play a pivotal role in regulating greenhouse gas (GHG) emissions, including carbon dioxide (CO_2), methane (CH_4), and nitrous oxide (N_2O), through processes such as respiration, methanogenesis, and denitrification (Das and Mandal 2022; Muhammad et al. 2022; Hao et al. 2025). Climate change-induced stressors such as elevated temperatures and altered precipitation significantly impact microbial metabolism and community structure. Warming accelerates microbial respiration and organic matter decomposition, increasing CO_2 emissions, while waterlogged or anaerobic conditions favor methanogenic archaea and denitrifying bacteria, leading to higher CH_4 and N_2O outputs (Hao et al. 2025). Soil moisture is a critical regulator, determining whether aerobic (CO_2-producing) or anaerobic (CH_4- and N_2O-producing)

microbes dominate. Droughts can suppress methanotrophs, reducing CH_4 mitigation potential, while optimal moisture enhances microbial activity and carbon sequestration (Hao et al. 2025). These shifts disrupt carbon and nitrogen cycles, potentially creating positive feedback loops that exacerbate climate change. Conversely, managing microbial communities such as promoting methanotrophs or optimizing soil moisture can mitigate GHG emissions and enhance ecosystem resilience (Rastogi and Sani 2011). Thus, understanding soil microbial responses is crucial for regulating GHG fluxes and sustaining ecosystem functions.

- *Resilience Through Microbial Diversity:* Diverse soil microbial communities provide functional redundancy, ensuring that key ecosystem processes such as decomposition and nutrient cycling persist despite environmental fluctuations (Boyle et al. 2024; Jayaramaiah et al. 2025). This redundancy means that if some microbial taxa are lost or suppressed by climate-induced stress, others with similar functions can compensate, buffering the ecosystem against disruptions (De Vries and Shade 2013). High microbial diversity enhances ecosystem resilience by supporting rapid recovery and maintaining soil functions under drought, heatwaves, or other climate extremes (Boyle et al. 2024). For example, soils with greater fungal-to-bacterial ratios and higher microbial biomass show stronger resistance and faster functional recovery after heat stress (Beugnon et al. 2025). This resilience sustains plant-microbe relationships, supporting plant health, nutrient uptake, and productivity even under adverse conditions. Ultimately, conserving and promoting microbial diversity is critical for maintaining ecosystem stability, carbon storage, and nutrient cycling in the face of ongoing climate change (Boyle et al. 2024; Jayaramaiah et al. 2025; De Vries and Shade 2013).

11.4.2 Feedbacks to Plant Health and Productivity

- *Enhanced Stress Tolerance:* Beneficial microbes play a crucial role in enhancing plant stress tolerance under climate change by modulating hormone levels, improving nutrient uptake, and inducing resistance to pathogens. These microbes, including plant growth-promoting rhizobacteria (PGPR) and mycorrhizal fungi, produce phytohormones such as auxins, gibberellins, and cytokinins, which regulate plant growth and help mitigate the negative effects of abiotic stresses like drought and heat (Ali et al. 2023; Tripathi et al. 2025). Additionally, microbes secrete enzymes and metabolites that solubilize nutrients, enhance root growth, and maintain membrane stability, thereby improving water and nutrient acquisition (Khawula et al. 2025). Some bacteria produce ACC deaminase, lowering ethylene levels and supporting root development during stress (Meza et al. 2025). Microbial consortia can also activate plant defense pathways, increasing resistance to pathogens and oxidative stress. These interactions enable plants to adapt to climate-induced stress while sustaining productivity and ecosystem functioning. Harnessing beneficial microbiomes offers a sustainable strategy for improving crop resilience and maintaining soil health in the face of ongoing climate change (Ali et al. 2023; Tripathi et al. 2025).

- *Altered Disease Dynamics:* Climate change fundamentally alters the balance between beneficial and pathogenic microbes, resulting in increased plant disease incidence and severity. Rising temperatures accelerate pathogen life cycles, enabling more infection cycles per season, while altered precipitation patterns create favorable conditions for fungal and bacterial pathogens such as Phytophthora infestans and Xanthomonas species (Singh et al. 2023). Elevated temperatures and extreme weather events weaken plant defense mechanisms, including salicylic acid and jasmonic acid pathways, making crops more susceptible to opportunistic pathogens (Singh et al. 2023). Simultaneously, climate stress disrupts beneficial plant-microbe relationships, reducing the protective capacity of disease-suppressive microbiomes and microbial antagonists (Singh et al. 2023). This microbial dysbiosis facilitates pathogen invasion, as the plant's "first line of defense" becomes compromised. Consequently, agricultural productivity faces significant threats, with yield losses exceeding 50% documented during severe pathogen epidemics under climate stress (Singh et al. 2023). These shifts also threaten ecosystem stability by disrupting natural plant communities and biodiversity. Effective management requires integrated approaches combining climate-smart practices, resistant varieties, and microbiome restoration strategies (Lahlali et al. 2024; Nega 2025).
- *Carbon Sequestration Potential:* Plant-microbe interactions, particularly those involving oxalotrophic bacteria and mycorrhizal fungi, play a crucial role in soil carbon sequestration and climate change mitigation. Oxalotrophic bacteria oxidize plant-derived oxalate, leading to the precipitation of calcium carbonate ($CaCO_3$) through the oxalate-carbonate pathway (OCP), which forms a stable inorganic carbon sink with residence times up to 100,000 years (Syed et al. 2020; Cowan et al. 2024). This process not only traps atmospheric CO_2 but also improves soil structure and fertility, especially in arid and nutrient-poor environments (Prasanna et al. 2025; Mason et al. 2023). Mycorrhizal fungi further enhance carbon stabilization by promoting plant growth, increasing root biomass, and facilitating the transfer of organic carbon into soil aggregates (Sun et al. 2019). The synergy between oxalogenic plants, saprophytic fungi, and oxalotrophic bacteria maintains the oxalate pool and drives biomineralization, supporting long-term ecosystem sustainability (Syed et al. 2020; Cowan et al. 2024). These interactions underpin a natural, low-cost strategy for atmospheric CO_2 mitigation and highlight the importance of conserving soil microbial diversity for ecosystem resilience and climate regulation.

11.5 Case Study: Plant-Microbe Interactions in Indian Forest Soils

11.5.1 Field Investigations in Uttarakhand

- *Biomineralizing Trees:* Field investigations in Uttarakhand's forest soils highlight the role of biomineralizing tree species like Terminalia alata in facilitating the oxalate–carbonate pathway (OCP) and soil carbon stabilization. Terminalia alata and other oxalogenic plants produce substantial calcium oxalate in idioblast cells, which is released to soil upon litter decomposition (Fig. 11.2). In alkaline rhizospheric zones beneath these trees, oxalotrophic bacteria oxidize soil oxalate, resulting in an increased pH and precipitation of calcium carbonate—a stable inorganic carbon sink with residence times far exceeding organic carbon, enhancing long-term sequestration (Kumar et al. 2016). Molecular diversity studies from Uttarakhand forests reveal abundant oxalotrophic bacterial genera such as Stenotrophomonas, Pseudomonas, and Mesorhizobium, confirmed by unique 16S rDNA and frc gene markers (Kumar et al. 2016). Arbuscular mycorrhizal fungi, also present in these soils, support rhizospheric interactions, nutrient cycling, and microbial diversity. The OCP not only traps atmospheric CO_2 but also improves soil fertility and structure, supporting ecosystem services and agricultural productivity. These synergistic plant-microbe interactions offer a

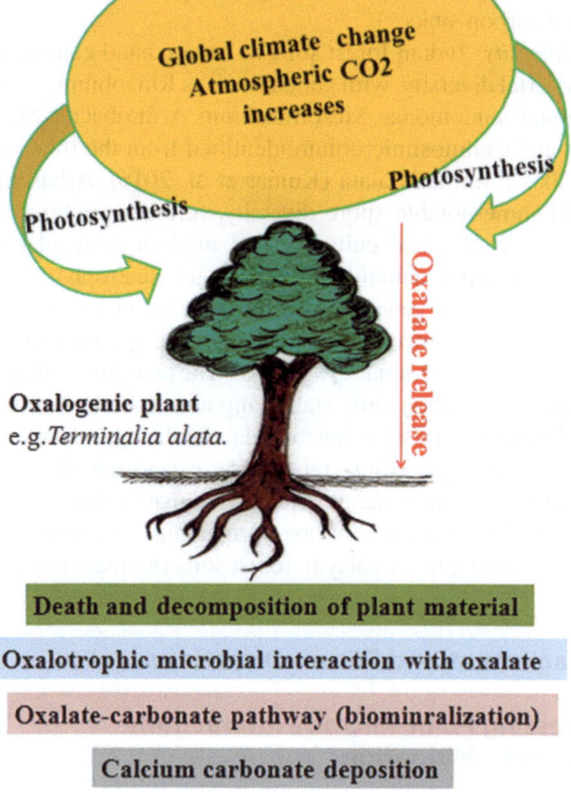

Fig. 11.2 Schematic representation of oxalate-carbonate pathway showing the main biotic and abiotic factors

scalable, low-cost strategy for carbon sequestration and ecosystem resilience in Indian forests (Kumar et al. 2016).

- *Soil Profile Analysis:* Soil profile analyses in Indian forest soils reveal that both soil pH and calcium carbonate ($CaCO_3$) content are significantly higher directly beneath oxalogenic tree species such as Terminalia alata compared to soil sampled away from these trees. This increased alkalinity and carbonate concentration indicate an active oxalate–carbonate pathway (OCP) in the rhizosphere, where oxalate produced and deposited by plant roots is transformed by oxalotrophic bacteria, driving local pH elevation and biomineralization of stable carbonates (Kumar et al. 2016). These microbial processes strengthen soil carbon sequestration potential by converting a portion of plant-assimilated carbon into long-lived mineral forms. Enhanced soil pH and $CaCO_3$ under oxalogenic trees also reflect robust microbial enzymatic activity, healthy mycorrhizal networks, and improved nutrient cycling, all vital for ecosystem functioning and resilience (Kumar et al. 2016). Such plant-microbe-driven alterations in soil chemistry and mineral deposition provide a key mechanism for natural climate mitigation and reinforce the ecological value of oxalogenic species in sustaining soil fertility and terrestrial carbon sinks.
- *Microbial Diversity:* Indian forest soils in Uttarakhand exhibit remarkable oxalotrophic bacterial diversity, with genera such as Rhizobium, Stenotrophomonas, Ochrobactrum, Pseudomonas, Mesorhizobium, Arthrobacter, Methylobacterium, Agromyces, and Cellulosimicrobium identified from the rhizospheres of oxalogenic trees like Terminalia alata (Kumar et al. 2016). Arbuscular mycorrhizal fungi (AMF) show notable spore diversity, further enriching the rhizospheric microbiome. Both classical culturing and modern molecular techniques 16S rDNA gene sequencing, denaturing gradient gel electrophoresis, and functional gene markers (e.g., frc gene for oxalotrophic activity)—have confirmed that these microbes actively mediate the oxalate–carbonate pathway (OCP) and promote biomineralization (Kumar et al. 2016). The resulting soil calcium carbonate acts as a long-lived carbon sink, stabilizing atmospheric carbon. These diverse microbial communities provide functional redundancy, enhance nutrient cycling, and strengthen soil aggregation, reinforcing ecosystem productivity and resilience under environmental change. Their synergistic roles underscore the fundamental contribution of plant-microbe partnerships to carbon stabilization and ecosystem functioning in Himalayan forest soils (Kumar et al. 2016).

11.6 Management and Future Directions

11.6.1 Harnessing Plant-Microbe Interactions for Climate Mitigation

- *Agroforestry and Reforestation:* Integrating oxalogenic tree species and their associated microbial partners into agroforestry and reforestation strategies offers significant climate mitigation and ecosystem resilience benefits. Oxalogenic trees, such as Terminalia alata, produce and deposit calcium oxalate into soils, which oxalotrophic bacteria metabolize via the oxalate–carbonate pathway (OCP), resulting in stable calcium carbonate ($CaCO_3$) precipitation with residence times up to 100,000 years, far exceeding organic soil carbon (Kumar et al. 2016; Syed et al. 2020). This inorganic carbon sink traps atmospheric CO_2, enhances soil structure, and boosts fertility, particularly in nutrient-poor and arid regions. Mycorrhizal fungi complement these effects by increasing root biomass, nutrient uptake, and organic carbon stabilization in soil aggregates (Pan et al. 2025; Barman et al. 2025). Agroforestry systems globally demonstrate superior soil organic carbon sequestration, with carbon accumulation rates reaching 0.99 Mg C ha^{-1} yr^{-1}, supported by enhanced microbial activity and diversified carbon inputs (Pan et al. 2025; Barman et al. 2025). The synergy among oxalogenic plants, oxalotrophic bacteria, and mycorrhizal fungi sustains long-term carbon storage, supports biodiversity, and strengthens soil health. Strategic planting of oxalogenic species in agroforestry and reforestation programs thus represents a natural, cost-effective, and scalable approach to climate change mitigation and sustainable land management (Syed et al. 2020; Kumara et al. 2023; Barman et al. 2025).

- *Biofertilizers and Microbial Inoculants:* The application of plant growth-promoting microbes (PGPMs) as biofertilizers and microbial inoculants offers a sustainable approach to enhancing crop productivity and soil health under climate stress. PGPMs improve nutrient uptake by biological nitrogen fixation, phosphate solubilization, and siderophore production, which increase nutrient bioavailability to plants (Kumar et al. 2016; Pattnaik et al. 2021; Ghimirey et al. 2024). Additionally, PGPMs modulate phytohormone levels, such as auxins and cytokinins, enhancing root growth and water-use efficiency, crucial under drought and temperature stress (Brito-Lopez et al. 2025). These microbes also induce systemic resistance against pathogens, reducing disease incidence and promoting plant resilience (Pattnaik et al. 2021). By improving nutrient cycling and soil structure, biofertilizers reduce reliance on harmful chemical inputs, mitigating environmental risks and enhancing soil fertility (Ghimirey et al. 2024). Microbial inoculants, including nitrogen-fixing bacteria such as Rhizobium and phosphate-solubilizing bacteria, act synergistically with plants, sustaining productivity and ecosystem functioning under adverse climatic conditions (Brito-Lopez et al. 2025). Thus, harnessing PGPMs is a promising strategy for climate-resilient agriculture and sustainable land management (Pattnaik et al. 2021; Brito-Lopez et al. 2025).
- *Restoration of Degraded Lands:* Plant-microbe partnerships offer a promising approach for rehabilitating degraded soils and restoring ecosystem functions under climate change. These interactions employ multiple mechanisms to enhance soil recovery: beneficial microbes, including arbuscular mycorrhizal fungi (AMF) and plant growth-promoting rhizobacteria (PGPR), improve soil structure, water retention, and aggregate stability, which are critical for combating erosion and enhancing soil fertility (Coban et al. 2022; Tian et al. 2023).

Symbiotic microbes facilitate nutrient cycling, particularly nitrogen fixation and phosphorus solubilization, supporting plant establishment in nutrient-poor degraded soils. Root-associated microorganisms enhance soil organic carbon (SOC) sequestration by promoting root biomass production and stabilizing carbon inputs through microbial necromass and soil aggregation (Wang et al. 2025; Song et al. 2024). Additionally, diverse plant communities supported by robust microbial networks exhibit greater resilience to climate stressors, accelerating ecosystem recovery. Recent studies demonstrate that ecological restoration primarily enhances carbon stocks by reducing soil carbon loss through improved soil structure and reduced erosion, rather than solely through increased plant carbon input (Song et al. 2024). Thus, leveraging plant-microbe partnerships provides a sustainable, nature-based solution for degraded land restoration and climate adaptation (Coban et al. 2022; Tian et al. 2023).

11.6.2 Research Perspectives

- *Biogeographical Patterns:* Further research is urgently needed to map the distribution and activity of oxalogenic systems across diverse ecosystems to better understand their contribution to global carbon cycling and climate mitigation. While oxalogenic plants and oxalotrophic bacteria are known in tropical forests, deserts, and agricultural soils, their biogeographical extent remains underexplored, particularly in marine and extreme environments where recent metagenomic studies have identified novel oxalotrophic lineages (Sonke and Trembath-Reichert 2023; Hervé et al. 2016). Understanding spatial variability in oxalogenic plant distribution, oxalotrophic microbial diversity, and environmental factors (soil pH, calcium availability, moisture) influencing oxalate-carbonate pathway (OCP) activity is critical for quantifying regional and global carbon sequestration potential (Cailleau et al. 2014; Syed et al. 2020). Expanding phylogenetic analyses of key oxalotrophic genes (frc, oxc) and ecological mapping will reveal functional redundancy, environmental adaptations, and ecosystem-specific contributions to carbon stabilization (Hervé et al. 2016; Cowan et al. 2024). These insights can guide targeted restoration and land management strategies, optimizing natural CO_2 sequestration and ecosystem resilience under climate change (Syed et al. 2020; Cowan et al. 2024).
- *Functional Redundancy and Adaptation:* Understanding the adaptability and functional redundancy of microbial communities is essential for maintaining ecosystem services amid climate change. Functional redundancy, where multiple microbial taxa exhibit overlapping ecological roles, provides resilience by buffering key processes such as nutrient cycling and carbon turnover despite species loss (Ramond et al. 2025). Microbial resilience is supported by gene functional overlap and adaptive mechanisms including gene regulation, horizontal gene transfer, and metabolic flexibility, enabling persistence under environmental stress (Puente-Sánchez et al. 2024). These dynamics allow ecosystems to maintain critical plant-microbe interactions necessary for nutrient uptake, stress toler-

ance, and carbon sequestration (Allison and Martiny 2008). Integrating molecular approaches with ecological modeling enhances prediction of ecosystem responses to climate stress and guides sustainable management strategies, prioritizing conservation of functionally important taxa (Ramond et al. 2025; Puente-Sánchez et al. 2024). This understanding is vital for sustaining ecosystem functioning and resilience in the face of ongoing climate variability.

11.7 Conclusions

Plant-microbe interactions are central to the functioning and resilience of terrestrial ecosystems, especially as climate change accelerates. These interactions regulate nutrient cycling, soil fertility, plant health, and carbon sequestration, underpinning ecosystem stability and productivity in the face of rising temperatures, altered precipitation, and increased abiotic stress (Cailleau et al. 2004; Kumar et al. 2016; Syed et al. 2020). The rhizosphere, a dynamic interface between plant roots and soil, hosts diverse microbial communities—including mycorrhizal fungi and plant growth-promoting rhizobacteria (PGPR)—that facilitate nutrient acquisition, enhance stress tolerance, and suppress pathogens, thereby supporting plant adaptation to environmental fluctuations (Chen et al. 2024; Bhattacharyya et al. 2021).

A compelling example of tightly coupled plant-microbe processes is the oxalate–carbonate pathway (OCP). In this pathway, oxalogenic plants such as Terminalia alata deposit oxalate in soils, which is then metabolized by oxalotrophic bacteria, resulting in the precipitation of stable calcium carbonate ($CaCO_3$). This process creates a long-lived inorganic carbon sink, with mineral carbon persisting for up to 100,000 years—far exceeding the residence time of soil organic carbon (Cailleau et al. 2004). Field studies in Indian forests demonstrate that soils beneath oxalogenic trees exhibit higher pH and $CaCO_3$ content, reflecting active OCP and enhanced carbon sequestration potential (Kumar et al. 2016). The synergy among oxalogenic plants, saprophytic fungi, and oxalotrophic bacteria not only stabilizes carbon but also improves soil structure and fertility, supporting ecosystem services and agricultural productivity.

Harnessing these plant-microbe partnerships through informed management practices—such as integrating oxalogenic species into agroforestry and reforestation, and applying microbial inoculants—offers scalable, cost-effective strategies for climate change mitigation and sustainable land use (Barman et al. 2025; Pattnaik et al. 2021). Biotechnological interventions, including the development of tailored biofertilizers and the conservation of functionally important microbial taxa, can further enhance soil health, crop resilience, and carbon storage (Ghimirey et al. 2024; Brito-Lopez et al. 2025). Restoration of degraded lands using plant-microbe consortia accelerates soil recovery, improves water retention, and increases ecosystem resilience to climate extremes (Coban et al. 2022; Tian et al. 2023).

Maintaining microbial diversity and functional redundancy is crucial for buffering ecosystems against climate-induced disturbances. Diverse microbial communities provide resilience by ensuring that key processes—such as decomposition, nutrient cycling, and disease suppression—persist even when some taxa are lost or suppressed (Ramond et al. 2025; Puente-Sánchez et al. 2024). Advances in molecular ecology and ecological modeling are enhancing our ability to map oxalogenic systems, understand microbial adaptation, and predict ecosystem responses to climate stress, guiding targeted conservation and land management strategies (Syed et al. 2020; Cowan et al. 2024).

In summary, plant microbe interactions exemplified by the OCP are a vital, underutilized tool for climate change mitigation and ecosystem sustainability. Leveraging these relationships through research, management, and biotechnology can unlock new strategies for enhancing carbon sequestration, restoring soil health, and sustaining ecosystem services in a rapidly changing world.

References

Ali S, Tyagi A, Mir RA, Rather IA, Anwar Y, Mahmoudi H (2023) Plant beneficial microbiome a boon for improving multiple stress tolerance in plants. Front Plant Sci 14:1266182

Allison SD, Martiny JB (2008) Resistance, resilience, and redundancy in microbial communities. Proc Natl Acad Sci 105(supplement_1):11512–11519

Backer R, Rokem JS, Ilangumaran G, Lamont J, Praslickova D, Ricci E, Subramanian S, Smith DL (2018) Plant growth-promoting rhizobacteria: context, mechanisms of action, and roadmap to commercialization of biostimulants for sustainable agriculture. Front Plant Sci 9:1473

Bardgett RD (2011) Plant-soil interactions in a changing world. F1000 Biol Rep 3:16

Barman S, Bhattacharyya R, Singh C, Rathore AC, Singhal V, Biswas DR, Ahmed N, Das S, Kumar S, Jat SL, Das TK (2025) Long-term agroforestry enhances soil organic carbon pools and deep soil carbon sequestration in the Indian Himalayas. Front Environ Sci 13:1568564

Begum N, Qin C, Ahanger MA, Raza S, Khan MI, Ashraf M, Ahmed N, Zhang L (2019) Role of arbuscular mycorrhizal fungi in plant growth regulation: implications in abiotic stress tolerance. Front Plant Sci 10:1068

Bennett AE, Groten K (2022) The costs and benefits of plant–arbuscular mycorrhizal fungal interactions. Annu Rev Plant Biol 73(1):649–672

Beugnon R, Eisenhauer N, Lochner A, Blechinger MJ, Buhr PE, Cesarz S, Farfan MA, Ferlian O, Rompeltien Howard AJ, Huang Y, Kuhlmann BS (2025) Sustainable land use enhances soil microbial respiration responses to experimental heat stress. Glob Chang Biol 31(4):e70214

Bhattacharyya A, Pablo CH, Mavrodi OV, Weller DM, Thomashow LS, Mavrodi DV (2021) Rhizosphere plant-microbe interactions under water stress. In: Advances in applied microbiology, vol 115. Academic Press, pp 65–113

Boyle JA, Murphy BK, Ensminger I, Stinchcombe JR, Frederickson ME (2024) Resistance and resilience of soil microbiomes under climate change. Ecosphere 15(12):e70077

Braissant O, Verrecchia EP, Aragno M (2002) Is the contribution of bacteria to terrestrial carbon budget greatly underestimated. Naturwissenschaften 89(8):366–370

Braissant O, Cailleau G, Aragno M, Verrecchia EP (2004) Biologically induced mineralization in the tree *Milicia excelsa* (Moraceae): its causes and consequences to the environment. Geobiology 2(1):59–66

Bravo D, Braissant O, Cailleau G, Verrecchia E, Junier P (2015) Isolation and characterization of oxalotrophic bacteria from tropical soils. Arch Microbiol 197(1):65–77

Brito-Lopez C, van der Wielen N, Barbosa M, Karlova R (2025) Plant growth–promoting microbes and microalgae-based biostimulants: sustainable strategy for agriculture and abiotic stress resilience. Philos Trans R Soc Lond B Biol Sci 380(1927):20240251

Cailleau G, Braissant O, Verrecchia EP (2004) Biomineralization in plants as a long-term carbon sink. Naturwissenschaften 91(4):191–194

Cailleau G, Braissant O, Dupraz C, Aragno M, Verrecchia EP (2005) Biologically induced accumulations of CaCO3 in orthox soils of Biga, Ivory Coast. Catena 59(1):1–17

Cailleau G, Braissant O, Verrecchia EP (2011) Turning sunlight into stone: the oxalate-carbonate pathway in a tropical tree ecosystem. Biogeosci Discuss 8:1755–1767. https://doi.org/10.5194/bg-8-1755-2011

Cailleau G, Mota M, Bindschedler S, Junier P, Verrecchia EP (2014) Detection of active-oxalate-carbonate pathway ecosystems in the Amazon Basin: global implications of a natural potential C sink. Catena 116:132–141. https://doi.org/10.1016/j.catena.2013.12.017

Chen M, Arato M, Borghi L, Nouri E, Reinhardt D (2018) Beneficial services of arbuscular mycorrhizal fungi–from ecology to application. Front Plant Sci 9:1270

Chen Q, Song Y, An Y, Lu Y, Zhong G (2024) Mechanisms and impact of rhizosphere microbial metabolites on crop health, traits, functional components: a comprehensive review. Molecules 29(24):5922

Coban O, De Deyn GB, van der Ploeg M (2022) Soil microbiota as game-changers in restoration of degraded lands. Science 375(6584):abe0725

Conradi T, Eggli U, Kreft H, Schweiger AH, Weigelt P, Higgins SI (2024) Reassessment of the risks of climate change for terrestrial ecosystems. Nat Ecol Evol 8(5):888–900

Cowan DA, Babenko D, Bird R, Botha A, Breecker DO, Clarke CE, Francis ML, Gallagher T, Lebre PH, Nel T, Potts AJ (2024) Oxalate and oxalotrophy: an environmental perspective. Sustain Microbiol 1(1):qvad004

Crutzen PJ (2016) Geology of mankind. In: Paul J. Crutzen: a pioneer on atmospheric chemistry and climate change in the Anthropocene. Springer International Publishing, Cham, pp 211–215

Das N, Mandal S (2022) Microbial populations regulate greenhouse gas emissions in Sundarban mangrove ecosystem, India. Acta Ecol Sin 42(6):641–652

De Vries FT, Shade A (2013) Controls on soil microbial community stability under climate change. Front Microbiol 4:265

Dumas B, Freyssinet G, Pallett KE (1995) Tissue-specific expression of germin-like oxalate oxidase during development and fungal infection of barley seedlings. Plant Physiol 107(4):1091–1096

Durand N, Monger HC, Canti MG, Verrecchia EP (2018) Calcium carbonate features. In: Stoops G, Marcelino V, Mees F (eds) Interpretation of micromorphological features of soils and regoliths. Elsevier, Amsterdam, pp 149–194. https://doi.org/10.1016/B978-0-444-53156-8.00009-X

Duret M, Wallner A, Buée M, Aziz A (2024) Rhizosphere microbiome assembly, drivers and functions in perennial ligneous plant health. Microbiol Res 287:127860

Dutton MV, Evans CS (1996) Oxalate production by fungi: its role in pathogenicity and ecology in the soil environment. Can J Microbiol 42(9):881–895

Fox TR, Comerford NB (1990) Low-molecular-weight organic acids in selected forest soils of the southeastern USA. Soil Sci Soc Am J 54(4):1139–1144

Franceschi VR, Nakata PA (2005) Calcium oxalate in plants: formation and function. Annu Rev Plant Biol 56:41–71

Fu F, Li J, Li Y, Chen W, Ding H, Xiao S (2023) Simulating the effect of climate change on soil microbial community in an Abies georgei var. smithii forest. Front Microbiol 14:1189859

Gadd GM (1999) Fungal production of citric and oxalic acid: importance in metal speciation, physiology and biogeochemical processes. Adv Microb Physiol 41:47–92

GeeksforGeeks (2025) Effects of climate change on terrestrial ecosystems. GeeksforGeeks

Ghimirey V, Chaurasia J, Acharya N, Dhungana R, Chaurasiya S (2024) Biofertilizers: a sustainable strategy for enhancing physical, chemical, and biological properties of soil. Innov Agric 7:1–11

Glick BR (2012) Plant growth-promoting bacteria: mechanisms and applications. Scientifica 2012(1):963401

Hao Y, Mao J, Bachmann CM, Hoffman FM, Koren G, Chen H, Tian H, Liu J, Tao J, Tang J, Li L (2025) Soil moisture controls over carbon sequestration and greenhouse gas emissions: a review. npj Clim Atmos Sci 8(1):16

Hervé V, Junier T, Bindschedler S, Verrecchia E, Junier P (2016) Diversity and ecology of oxalotrophic bacteria. World J Microbiol Biotechnol 32(2):28

Jayaramaiah RH, Martins CS, Egidi E, Macdonald CA, Wang JT, Liu H, Reich PB, Delgado-Baquerizo M, Singh BK (2025) Soil function-microbial diversity relationship is impacted by plant functional groups under climate change. Soil Biol Biochem 200:109623

Khawula S, Daniel AI, Nyawo N, Ndlazi K, Sibiya S, Ntshalintshali S, Nzuza G, Gokul A, Keyster M, Klein A, Niekerk LA (2025) Optimizing plant resilience with growth-promoting Rhizobacteria under abiotic and biotic stress conditions. Plant Stress 17:100949

Kumar V, Khulbe K, Tamta S, Srivastava R, Sharma AK (2016) Biochemical and molecular identification of oxalate-oxidizing bacteria isolated from rhizosphere of Biomenralizing Tree, Terminalia Alata from Kumaun Himalaya, India. Int J Curr Res (Ijcr) 8(11):42179–42182

Kumara K, Pal S, Chand P, Kandpal A (2023) Carbon sequestration potential of agroforestry systems in Indian agricultural landscape: a meta-analysis. Ecosyst Serv 62:101537

Lahlali R, Taoussi M, Laasli SE, Gachara G, Ezzouggari R, Belabess Z, Aberkani K, Assouguem A, Meddich A, El Jarroudi M, Ait Barka E (2024) Effects of climate change on plant pathogens and host-pathogen interactions. Crop Environ 3(3):159–170

Libert B, Franceschi VR (1987) Oxalate in crop plants. J Agric Food Chem 35(6):926–938

Ling N, Wang T, Kuzyakov Y (2022) Rhizosphere bacteriome structure and functions. Nat Commun 13(1):836

Luo C, He Y, Chen Y (2025) Rhizosphere microbiome regulation: unlocking the potential for plant growth. Curr Res Microb Sci 8:100322

Makela M, Galkin S, Hatakka A, Lundell T (2002) Production of organic acids and oxalate decarboxylase in lignin-degrading white rot fungi. Enzym Microb Technol 30:542–549

Mason ARG, Salomon MJ, Lowe AJ, Cavagnaro TR (2023) Microbial solutions to soil carbon sequestration. J Clean Prod 417:137993

Meza C, Valenzuela F, Yáñez M, Cabeza RA, Ramos P, Plaza A, Carrasco B, Flores-Castañón N, Mesquita-Neto J, Arce-Johnson P, Banerjee A (2025) Auxin producing plant growth promoting bacteria enhance temperature stress tolerance in Chilean common bean landraces. Sci Rep 15(1):30359

Muhammad I, Lv JZ, Wang J, Ahmad S, Farooq S, Ali S, Zhou XB (2022) Regulation of soil microbial community structure and biomass to mitigate soil greenhouse gas emission. Front Microbiol 13:868862

Nega A (2025) Climate change impacts on agriculture: a review of plant diseases and insect pests in Ethiopia and East Africa, with adaptation and mitigation strategies. Adv Agric 2025(1):5606701

Osono T (2020) Functional diversity of ligninolytic fungi associated with leaf litter decomposition. Ecol Res 35(1):30–43

Pan J, Chen S, He D, Zhou H, Ning K, Ma N, Li K, Liao D, Mi W, Wu Q, Zhang C (2025) Agroforestry increases soil carbon sequestration, especially in arid areas: a global meta-analysis. Catena 249:108667

Pattnaik S, Mohapatra B, Gupta A (2021) Plant growth-promoting microbe mediated uptake of essential nutrients (Fe, P, K) for crop stress management: microbe–soil–plant continuum. Front Agron 3:689972

Prasanna AN, Parween S, Froehlich K, Schmidt T, Shekhawat K, Prabhu DD, Saad MM, Hirt H (2025) Oxalotrophic bacteria in desert and drylands: enzymatic pathways and carbon sequestration. bioRxiv:2025–2002

Puente-Sánchez F, Pascual-García A, Bastolla U, Pedrós-Alió C, Tamames J (2024) Cross-biome microbial networks reveal functional redundancy and suggest genome reduction through functional complementarity. Commun Biol 7(1):1046

Qiao J, Zheng J, Li S, Zhang F, Zhang B, Zhao M (2024) Impact of climate warming on soil microbial communities during the restoration of the inner Mongolian desert steppe. Front Microbiol 15:1458777

Ramond P, Galand PE, Logares R (2025) Microbial functional diversity and redundancy: moving forward. FEMS Microbiol Rev 49:fuae031

Rastogi G, Sani RK (2011) Molecular techniques to assess microbial community structure, function, and dynamics in the environment. In: Microbes and microbial technology: agricultural and environmental applications. Springer New York, New York, pp 29–57

Rowley MC, Estrada-Medina H, Tzec-Gamboa M, Rozin A, Cailleau G, Verrecchia EP et al (2017) Moving carbon between spheres, the potential oxalate carbonate pathway of Brosimum alicastrum Sw.; Moraceae. Plant Soil 412:465–479
Sahin N (2003) Oxalotrophic bacteria. Res Microbiol 154:399–407
Saini R, Sharma S (2019) Climate resilient microbes in sustainable crop production. In: Contaminants in agriculture and environment: health risks and remediation, vol 1, p 264
Sharma B, Singh BN, Dwivedi P, Rajawat MVS (2022) Interference of climate change on plant-microbe interaction: present and future prospects. Front Agron 3:725804
Singh BK, Delgado-Baquerizo M, Egidi E, Guirado E, Leach JE, Liu H, Trivedi P (2023) Climate change impacts on plant pathogens, food security and paths forward. Nat Rev Microbiol 21(10):640–656
Song J, Wan S, Zhang K, Hong S, Xia J, Piao S, Wang YP, Chen J, Hui D, Luo Y, Niu S (2024) Ecological restoration enhances dryland carbon stock by reducing surface soil carbon loss due to wind erosion. Proc Natl Acad Sci U S A 121(46):e2416281121
Sonke A, Trembath-Reichert E (2023) Expanding the taxonomic and environmental extent of an underexplored carbon metabolism—oxalotrophy. Front Microbiol 14:1161937
Sun Q, Li J, Finlay RD, Lian B (2019) Oxalotrophic bacterial assemblages in the ectomycorrhizosphere of forest trees and their effects on oxalate degradation and carbon fixation potential. Chem Geol 514:54–64
Syed S, Buddolla V, Lian B (2020) Oxalate carbonate pathway—conversion and fixation of soil carbon—a potential scenario for sustainability. Front Plant Sci 11:591297
Tamer AA, Aragno M (1980) Isolement, caractérisation et essaid identifcation de bactéries capab lesd'utiliserl'oxalatecommeseule source de carbone et d'énergie. Bull Soc Neuchâtel Sci Nat 103:91–104
Tian Q, Zhang X, Yi H, Li Y, Xu X, He J, He L (2023) Plant diversity drives soil carbon sequestration: evidence from 150 years of vegetation restoration in the temperate zone. Front Plant Sci 14:1191704
Tripathi S, Dabrast S, Kundu S, Saini DK, Jamal H, Meena RK, Somayanda I, Varma A, Bahuguna RN, Jagadish SK (2025) Harnessing the plant-associated microbiome: a sustainable solution for enhancing crop resilience to abiotic stresses and problematic soils. Plant Stress 101033
Tuason MMS, Arocena JM (2009) Calcium oxalate biomineralization by *Piloderma fallax* in response to various levels of calcium and phosphorus. Appl Environ Microbiol 75:7079–7085
U.S. Environmental Protection Agency (2025) Climate change impacts on ecosystems. EPA
Vejan P, Abdullah R, Khadiran T, Ismail S, Nasrulhaq Boyce A (2016) Role of plant growth promoting rhizobacteria in agricultural sustainability—a review. Molecules 21(5):573
Verrecchia EP, Braissant O, Cailleau C (2006) The oxalate-carbonate pathway in soil carbon storage: the role of fungi and oxalotrophic bacteria. In: Fungi Biogeochem Cycles, pp 289–310
Wang Y, Li D, Yang Q, Lu JL, Jiang Y, Su T, Milinga A, Shi Q, Liu W, Yang H, Fang M (2025) Effects of plant carbon inputs and soil microbe on soil organic carbon accumulation in different tropical vegetation restoration strategies. Ecol Eng 215:107616

CRISPR and Gene Editing for Climate-Resilient Crops

12

Jyoti Singh, Shikha Thakur, Akanksha Singh, Abhishek Kumar, Palak Chauhan, Anamika Pal, and Kanika Dulta

Abstract

Abiotic stresses like drought, heat, flooding, waterlogging, salinity, and soil degradation are all made worse by climate change, posing an increasing threat to plant-based agriculture and endangering world crop output and food security. Extreme weather events brought on by rising CO_2 levels and global temperatures have harmed livelihoods, agriculture, and biodiversity. The growing demand for food production frequently results in increased land requirements, which exacerbates the loss of biodiversity. Conventional breeding has increased crop stress tolerance, but it is ineffective at addressing quickly changing climatic circumstances due to its slow speed and lack of precision. On the other hand, contemporary genome editing technologies—particularly CRISPR-Cas9, in addition to ZFNs and TALENs—provide very specific, effective, and heritable alterations for creating crop types that are climate-tolerant. By targeting regulatory genes, editing stress-responsive transcription factors like DREB, NAC, and WRKY, and modifying hormonal and ROS signaling pathways, CRISPR technology has been used to improve important traits like heat and flood resistance, drought and salinity tolerance, and nutrient-use efficiency. Additionally, it permits conditional gene expression through the editing of stress-inducible promoters. In order to increase yields, shelf life, sweetness, aroma, and resilience, CRISPR has been

J. Singh · S. Thakur · A. Kumar · P. Chauhan · A. Pal
Department of Biotechnology, School of Applied and Life Sciences, Uttaranchal University, Dehradun, Uttarakhand, India

A. Singh
Department of Food Technology, School of Applied and Life Sciences, Uttaranchal University, Dehradun, Uttarakhand, India

K. Dulta (✉)
Center of Advanced Innovation Technologies, VŠB-Technical University of Ostrava, Ostrava-Poruba, Czech Republic

H. C. Joshi et al. (eds.), *Climate-Responsive Plant Biology*,
https://doi.org/10.1007/978-981-95-7354-7_12

used in crops such as rice, wheat, maize, and tomatoes. Crop biofortification, which increases the amount of protein, carbohydrates, vitamins, and oleic acid, has also been made possible by CRISPR-Cas9, increasing crop yield and nutritional value. CRISPR has the potential to enable smallholder farmers to adjust to climate change by streamlining editing operations. Innovations in guide RNA design, base editors, and delivery techniques like protoplast transformation and biolistics keep increasing its accuracy and usefulness. Future developments in multiplex editing, integrating synthetic biology, and editing complex polygenic characteristics will strengthen CRISPR's position in climate-smart, sustainable agriculture.

Keywords

Abiotic stresses · Climate change · Crop productivity · Food security · CRISPR-Cas9 · Genome editing · TALENs · ROS signaling

12.1 Introduction

The world is experiencing severe climate change, as evidenced by rising carbon dioxide (CO_2) levels and high global temperatures. Extreme occurrences have resulted, negatively impacting agriculture, biodiversity, and the human community, among other aspects of the Earth. The major primary source of greenhouse gas emissions is human activities, which are the main cause of climate change, raising the temperature by 1.5 °C since 1850 globally (Karavolias et al. 2021; Nunez et al. 2019). The most significant environmental issue that has a negative impact on agricultural productivity worldwide is climate change (CC) (Ennet and Amusa 2010).

The primary source of this climate change is the buildup of greenhouse gases (GHGs) in the atmosphere, which intensifies the greenhouse effect. Global processes like climate change, which affects agriculture practices, are interconnected, and their interdependence is more significant as the gap between global population growth and food production widens. Reduction in crop yield is the result of the changes in temperature, rainfall, and precipitation, and severe weather events are predicted to lower crop yields in many developing world regions, especially sub-Saharan Africa and portions of Asia (Gornall et al. 2010).

The way that agriculture is practiced in industrialized and developing nations differs greatly, which causes the agricultural sector to contribute differently to climate change. Large numbers of livestock, poor waste management, inappropriate use of agrochemicals, and poor land management all contribute to the agriculture sector's higher GHG emissions in developing nations. As a result of their reliance on agriculture, emerging nations are more severely affected by CC (Yohannes 2016).

For sustaining human life in order to produce food and fiber, weather and climate works in coordination. Agriculture is an economic activity that is expected to be susceptible to change in climate variability, which involves natural processes that require nutrients, temperature, precipitation, and other conditions in a fixed proportion (Vuren et al. 2009). With the enhancement of agricultural sequestration and reduction in GHG emissions, the agriculture sector plays a potent role in mitigating climate.

Human health, food security, and agriculture are all affected by climate change. It impacts a number of crops and results in sharp yield losses, which seriously disrupt the environment, society, and the global economy. Food insecurity will increase globally as a result of the effects on vital staple crops including rice, maize, and wheat. Although there are many different ways to improve crop traits, a new path for functional genomics and agricultural growth has been made possible by CRISPR-associated protein 9 (CRISPR/Cas9)-mediated genome modification and clustered regularly interspaced short palindromic repeats (Kaur et al. 2025a, b).

A new path for functional genomics and agricultural enhancement has been made possible by genome alteration mediated by CRISPR-associated protein 9 (CRISPR/Cas9). Because CRISPR/Cas endonucleases and the genetic engineering tools they produce are highly accurate, adaptable, more specific, and simple to build, they can produce climate-smart or resilient crops that can withstand severe environments and fight food insecurity. Several crops have had their genomes edited using the CRISPR/Cas9 method to increase their climate resilience. Bacteria and archaea have adapted CRISPR/Cas systems, which are varied. The fundamental CRISPR/Cas9 system gave rise to a variety of tools used today, including CRISPRi, CRISPRa, base editors, gene knock-in, targeted protein tagging, and viral-mediated editing (Anzalone et al. 2020; Kampmann 2018).

Wheat is one of the crops that can be enhanced with CRISPR/Cas9 since rising temperatures can drastically alter wheat grain quality by reducing grain-filling time, which will alter wheat's gluten composition. High temperatures can alter the gliadin-to-glutenin ratio, which reduces wheat flour's ability to form dough. Nitrogen levels can also be lowered by prolonged heat waves, and finally the grain's protein content (Liu and Wang 2021). It was challenging to create climate-tolerant wheat using traditional breeding methods because of the hexaploidy of the DNA. Stable hereditary alterations in wheat were successfully obtained using CRISPR/Cas9 technology. CRISPR constructions have been introduced into immature wheat embryos via agrobacterium-mediated transformation (Howells et al. 2018; Zhang et al. 2019).

Biolistic and protoplast transfection techniques have also been used to create transgene-free wheat plants resulting from CRISPR/Cas9 genome editing (Liang et al. 2017; Zhang et al. 2016). A staple crop, rice (Oryza sativa L.), is consumed at a rate of roughly 21% worldwide and 76% in Asian nations. In 2022, 776.5 million tons of rice were produced globally, with Asia accounting for almost 90.5% of the average production, and with China and India being the two largest producers. Because of its flavor and adaptability to many different ethnic cuisines, rice is in high demand (Castanho et al. 2023). Rice also contains moderate amounts of vitamin B, phosphorus, iron, calcium, and protein, as well as other nutrients, primarily complex carbohydrates.

The rice bran contains the majority of nutrients, including proteins, vitamins, minerals, and antioxidants. Rice is an excellent source of a balanced diet because it is free of fat, cholesterol, and sodium and includes all of the essential amino acids except lysine (Sasaki and Burr 2000). However, rice harvests have suffered in a number of ways from farms to consumers as a result of the current climate change scenarios (Zhao et al. 2016).

The primary stressors that can have a direct impact on several aspects of rice, including grain size, quality, yield, nutritional composition, and appearance, are rising temperatures, carbon dioxide, salinity, rainfall, pests, drought, and diseases. Therefore, in order to create climate-smart rice crops, researchers have been experimenting with many approaches, one of which is the reliable and adaptable gene editing method known as the CRISPR/Cas9 system. RNPs, particle bombardment, and Agrobacterium-mediated transformation have all been used to optimize CRISPR/Cas9-mediated gene editing in rice (Miao et al. 2013; Shan et al. 2013). Additionally, a multiplexing strategy was effectively developed for rice. Researchers recently used combined biolistic and Agrobacterium-mediated methods to achieve ultra-multiplexing targeting 49 genes in rice (Wu et al. 2024a, b, c).

Disease-free production of plants is considered to be one of the most beneficial uses of biotechnology, which can be attained by techniques such as micropropagation. Through the process of micropropagation, disease-free banana plantlets can be created from plant tissues. It has all the potential advantages of being a cutting-edge method that is both reasonably priced and simple to use (Jain et al. 2021). Fortified crops, which are enriched with nutrients to help malnourished children grow, turn out to be a great food source, for example, "Protato." This genetically modified crop (GMP) potato has between one-third and half as much protein as a regular potato. All the essential amino acids, including methionine and lysine, are present in considerable amount. This "Protato" has the potential to be a very useful food source in nations where potatoes are a key staple (Rajasree and Pugalendhi 2021).

Golden rice is another example of such a crop. The amount of beta-carotene in these genetically engineered rice grains is greater (Qamar et al. 2020). One of the most prevalent issues with many crops worldwide is pest attacks; these crops can be used as a fodder crops or other crops grown for food. BT cotton is one example of such a crop. To enable the production of a specific protein in cotton, the genes of the common Bacillus thuringiensis (Bt) organism are introduced into the crop. For many insects, the protein is extremely poisonous. With the help of biotechnology, the created BT cotton reduces insect attacks, which eventually results in a considerable increase in production (Qaim 2020).

In genetic engineering, targeted and shortgun methods are two distinct but related strategies. Utilizing these methods, transgenic plants with the capacity to confer drought tolerance are produced (Khan and Khan 2010). Future prosperity primarily depends on the availability of affordable, sustainable, safe, and equitable energy. One of the newer developments in recent years is the production of biofuel, for instance, using *Chlorella vulgaris* as a feedstock.

Higher temperatures cause tomatoes to ripen more quickly, but low temperatures might ruin their flavor. To address that issue, a California business called Calgene genetically modified tomatoes. To solve the problem, they created the Flavr Savr variety of tomatoes. To guarantee ripening, an enzyme called polygalacturonase breaks down pectin. Tomatoes were genetically altered by scientists to produce less enzyme. For that particular objective, antisense RNA is employed (Adenle 2011).

Among the GM foods with improved flavor are tomatoes, eggplant, pepper, cherries, and seedless watermelon. When the seeds are removed from certain fruits, the fruits have superior results, such as having more sugar that is soluble and making them sweeter (Falk et al. 2002).

12.2 Overview of Gene Editing Technologies

Traditional plant breeding, which leverages the innate genetic diversity within or between closely related species through selection and hybridization, has long served as the foundation for crop improvement. Species compatibility and long breeding cycles that can take several generations to stabilize desirable features often limit this strategy. Moreover, recombination may unintentionally transfer unwanted genes during meiosis along with advantageous ones, decreasing overall efficiency. Genetic engineering provides a precise method for introducing, silencing, or changing specific genes to achieve desired phenotypic effects. It transcends species boundaries and makes it possible to integrate traits like insect resistance, drought tolerance, and nutritional improvement. Commercial agriculture has been transformed by plant transformation, which has also made it possible to gain basic insights into plant biology. Even after more than 30 years of technical advancements, transformation and regeneration are still difficult for the majority of crops. Although it depends on genetic transformation and plant regeneration, which are process bottlenecks, genome editing offers new ways to increase agricultural output. While floral dip transformation is an attractive solution inasmuch as it eliminates the need for tissue culture, it is only robustly reproducible in Arabidopsis and its relative *Camelina sativa* (Lu and Kang 2008). Making crop transformation so simple that it may be performed in almost any laboratory is a key research goal for plant biologists. Enhancing plant transformation's ability and effectiveness is essential to maximizing the use of crop genomics information to feed the globe.

By introducing new genes, fixing defective ones, and investigating gene functions, targeted gene modification has made it easier to change genomes. Targeted genome editing is therefore a potent genetics technique. Random mutagenesis is a natural occurrence that causes genomic alterations in organisms. From the use of mutagens, which are physical or chemical agents that may change DNA sequences, to carefully designed biological instruments that can produce gene knock-in and knock-out, genome editing has advanced. These biological tools include mega nucleases, zinc finger nucleases (ZFNs), transcription activator-like effector nucleases (TALENs), and clustered regularly interspaced short palindromic repeats/Cas9 (CRISPR/Cas9). Researchers can quickly and affordably incorporate changes thanks to these versatile genome editing technologies. Numerous animals and cell types are capable of reproducing these sequence-specific changes to the DNA. Significant scientific breakthroughs have been seen in domains including agriculture, human gene therapy, and synthetic biology due to its wide range of applications. In particular, these methods have helped find novel characteristics in plants and create plants that are more resistant to biotic and abiotic challenges and produce higher agricultural yields (Bhagtaney et al. 2023).

12.3 CRISPR Applications in Developing Climate-Resilient Traits

Physical factors such as heat, cold, salt, drought, and heavy metals significantly lower agricultural output worldwide (Table 12.1). In order to cope with such abiotic stresses, transgenic and conventional breeding methods have been extensively employed to reduce the dangers of these environmental stressors (Fig. 12.1).

Drought is a significant issue for crops because it reduces the amount of water and nutrients available, both of which are essential for plant survival and growth. This leads to decreased crop production, growth retardation, and even plant death, based on the plant's developmental stage, the drought's intensity and length, and the genetic lineage of the plant. One of the most difficult traits to research, categorize, and understand is the capacity to tolerate drought, since it is a very complicated trait regulated by several genes.. CRISPR technology, which stands for Clustered Regularly Interspaced Short Palindromic Repeats, has transformed plant molecular breeding, opening a new avenue for crop improvement (Rai et al. 2023).

In certain nations, crop yield sustainability during drought appears to be a major challenge. Plants have evolved complex mechanisms with a variety of physiological and morphological ways to withstand stress since the severity of drought changes over time and area (Ansari et al. 2019). Drought affects crops through varying levels of evasion, avoidance, and adaptation (Sourour et al. 2017).

In plant management, it is essential to take advantage of genetic traits that increase drought resistance while preserving high production. Recombinant and traditional breeding techniques may be used to increase the drought resilience of maize, rice, soybean, and wheat. Historically, the most successful way to cultivate plants in water-stressed environments was through traditional breeding. However, those same tactics appear to be expensive, time-consuming, and labor-intensive. Molecular markers have already been essential in attempting to depict plant genetic heterogeneity under environmental stressors (Queiroz et al. 2019). Numerous quantitative trait loci (QTLs) linked to increasing drought resilience have already been found in a variety of crops; however, the accuracy and consistency of QTL identification remain a problem (Khan et al. 2018). In light of this, genome editing appears to have been very effective in increasing crop resistance to abiotic and biotic stresses (Shinwari et al. 2020). CRISPR-Cas gene-editing systems have received widespread acclaim for their adaptability and user-friendliness. The entire strategic approach makes use of a guide RNA and an intricate endonuclease (Cas endonuclease) that modifies DNA strands to create double-stranded DNA breaks, which are subsequently repaired by endogenic cellular repair mechanisms, resulting in the creation of novel genetic variations (Brokowski and Adli 2019).

Table 12.1 Illustrates how the key crops were genetically modified using CRISPR/Cas9 to withstand drought stress

S.No.	Crops	Gene of interest	Function of the Gene	Characteristics	Reference
1.	Tomato	SlNPR1	Transcription coactivator	Decrease in the mutants' ability to withstand drought	Li et al. (2019)
2.	Tomato	SlLBD40	Transcription Factor	Tolerance towards drought in the knockout lines	Liu et al. (2020)
3.	Tomato	SAPK2	Signal Transduction	Mutants deemed more susceptible to drought stress	Lou et al. (2017)
4.	Wheat	TaERF3	Transcription Factor	Enhanced drought Tolerance	Kim et al. (2018)
5.	Rice	OsDSL	Transcription Factor	Moderate resistance towards osmotic stress and high tolerance to NaCl at stage of seedlings	Kumar et al. (2020)
6.	Maize	ARGOS8	Transcription factor	Increased grain yield in the field (under stress conditions)	Shi et al. (2017)
7.	Soybean	GmHdz4	Transcription factor	Higher drought tolerance	Zhong et al. (2022)

The CRISPR-Cas platform has been used skillfully to develop resistance to a wide range of abiotic stressors, which include salinity, dryness, submersion, and the toxicity of heavy metal ions (Raza et al. 2020).

Across the world, salinity is becoming a bigger threat to agriculture. This is especially crucial for growing crops that are sensitive to salt, like wheat and rice. About 10% of the planet's arable land is used to grow rice, the second-largest crop in the world and the main source of food for 50% of the world's population (FAO, 2015). The large range of salinity susceptibility among rice cultivars indicates a great deal of genetic variability that can be used to pinpoint the genes and proteins that are crucial for rice salt tolerance. Understanding the underlying physiological, molecular, and biochemical processes, as well as identifying relevant genes and gene networks, is crucial for creating crops that can withstand salinity (Foyer and Dhankher 2018). Crops can be made more salt-tolerant by using proteomics and genomics.

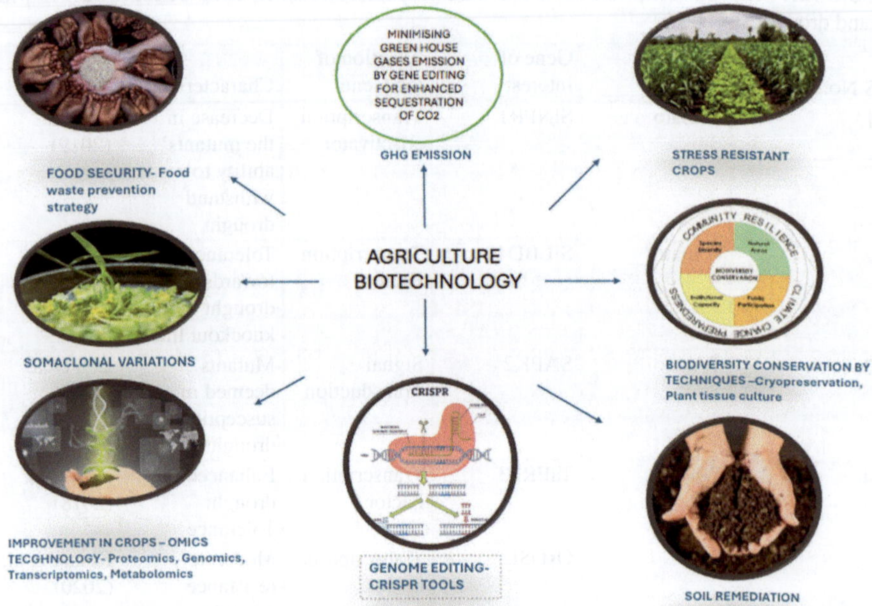

Fig. 12.1 The many aspects of the contribution of biotechnology towards crop improvement for resilient food systems while also contributing to environmental protection

Crop yields and plant productivity decrease when exposed to extreme temperatures (freezing, chilling, or HT). The world's semiarid regions are especially susceptible to the unpredictable weather brought on by climate change. According to Djanaguiraman et al. (2017), HT stress has detrimental effects on the fertility of both pollen and pistil tissues, with the pistil being more vulnerable than pollen.

12.4 Target Genes and Pathways for Climate Resilience

12.4.1 Regulatory Genes for Abiotic Stress—Transcription Factors

Abiotic and biotic stressors affect every stage of a plant's phenology. They have a negative impact on plant growth and productivity, and they are the main factor limiting agricultural yield. Pathogenic bacterial, fungal, viral, and oomycete attacks are examples of biotic stressors. Abiotic stressors include oxidative stress, heat, cold, radiation, heavy metals, drought, and soil salinization. Plants must be able to adapt to these pressures and respond to a variety of environmental stresses in order to survive and pass on to the next generation. A number of transcription factors (TFs) that can operate as switches to regulate the expression of putative stress-related genes have been discovered in recent years. In plant biological systems, these transcription factors are essential for controlling transcriptional reprogramming during development. They can also help plants become more resilient to biotic and abiotic stressors. Therefore, the creation of transgenic plants with increased stress tolerance depends on the identification and characterization of key genes involved in plant stress responses. Numerous TF families, including WRKY, AP2 (APETALA2)/ERF (ethylene-responsive factor), and NAC (NAM, ATAF1/2, CUC1/2), are particular to plants and have significant and distinct functions in regulating plant-specific processes. TFs are typically described as proteins that attach to particular DNA sequences and have the ability to either activate or inhibit the transcription of particular target genes. They frequently take part in regulating a number of crucial cellular functions, including the reaction to messages from other cells (Jiang et al. 2012).

12.4.2 Hormonal and ROS Signaling Pathways

Plants are constantly subjected to a variety of environmental challenges because they are sessile organisms (Suzuki 2016). Hormonal and Reactive oxygen species (ROS) are important factors that control how plants develop and react to stress. Signalling networks of ROS and plant growth regulators interact during organisms' life cycles to provide the proper environmental and developmental response. Stress is typically linked to both enhanced and inadequate ROS production in the photosynthetic (such as leaves) and non-photosynthetic (such as seeds) tissues of plants. In severe situations, ROS production can be fatal to individual cells, an organ, or even an organism. ROS play a significant role in controlling the growth and development of plants because of their capacity to interact with all cellular components. In plant cells, reactive oxygen species (ROS) are adaptable molecules that mediate a range of physiological responses, such as hormone signaling, development, programmed cell death (PCD), and gravitropism. Numerous other signaling pathways, including nitric oxide and the stress hormones ethylene, salicylic acid, and jasmonic acid, are either directly or indirectly impacted by ROS signals. Whether a cell survives or dies depends on how these pathways interact and balance (Oracz 2016).

12.4.3 Genome Editing of Stress-Inducible Promoters

Ironically, the fundamental characteristics that determine a plant's agronomic and financial worth are also some of the most susceptible to increasing environmental stressors. The cellular, physiological, and biochemical equilibrium is upset by a variety of abiotic and biotic stressors, including heat, drought, salinity, infections, and herbivores, which ultimately degrade fruit quality and result in large output losses. Mechanistic understanding of the molecular foundations of stress resilience is necessary to address such issues. As master regulators of gene expression, transcription factors (TFs) are essential for coordinating how plants react to external stimuli. These TF activities can now be precisely manipulated thanks to recent developments in CRISPR–Cas genome editing, providing a potent way to reorganize stress-responsive networks with previously unheard-of specificity (Singha et al. 2022). To maintain homeostasis and sustain productivity in the face of environmental stress, plants use a variety of physiologic and biochemical strategies, such as decreased leaf expansion, wilting, leaf abscission, increased root growth, and changes in relative water content (RWC) and membrane stability. These physiologic responses are frequently accompanied by elevated production of reactive oxygen species (ROS), reactive nitrogen species (RNS), and free radicals, which are essential for signaling but can also cause lipid peroxidation, membrane disruption, enzyme inactivation, and cellular damage if not strictly controlled (Rahman et al. 2024). Abscisic acid (ABA) is a key mediator of abiotic stress adaptation, controlling stomatal closure to reduce transpirational water loss and preserve water-use efficiency (WUE) in challenging circumstances (Trejo-Paniagua et al. 2024). Transcriptional regulatory networks, which incorporate environmental and hormonal cues into precise gene expression programs, coordinate these reactions.

12.5 Crop-Specific Case Studies

12.5.1 Rice

Rice *(Oryza sativa L.)*, a grain consumed by almost half of the world's population and playing a major role in ensuring global food security, has been notably impacted by climate change in terms of agricultural production. Drought, heat, salinity, heavy metals, rice blast, and bacterial blight are just a few of the biotic and abiotic factors that can severely reduce rice yields and jeopardize global food security. In this context, a number of plant breeding and biotechnological methods have been employed to produce rice cultivars capable of adapting to climate change. Gene editing (GE) technology has transformed crop development in the modern era. Crop improvement has been transformed in recent years by gene editing (GE) technologies. Plant researchers, particularly rice breeders and geneticists, are increasingly using the CRISPR/Cas (Clustered Regularly Interspaced Short Palindromic Repeats/CRISPR-associated protein) system because it is one of the most practical, reliable, economical, and labor-efficient GE technologies. Since 2013, when the CRISPR/Cas-based GE techniques have been used to produce a number of trait-specific climate-resilient rice lines (Fig. 12.2). Several earlier reports have been released attesting to the effective use of GE technologies for rice development (Shaheen et al. 2023).

By changing its genetic code, genetic alteration enables the possibility of steady phenotypic performance. Genes from alien species are often introduced, gene expression levels are altered, and gene function is totally disrupted. As a result, several gene editing methods, including TALENs, ZFNs, CRISPR/Cas, Prime Editing, and Base Editing, have been created and used to accomplish these objectives. The development of these technologies has had a significant influence on current rice breeding initiatives by revolutionizing the exact manipulation of specific nucleic acids. Modification of the coding regions of functional genes was the main

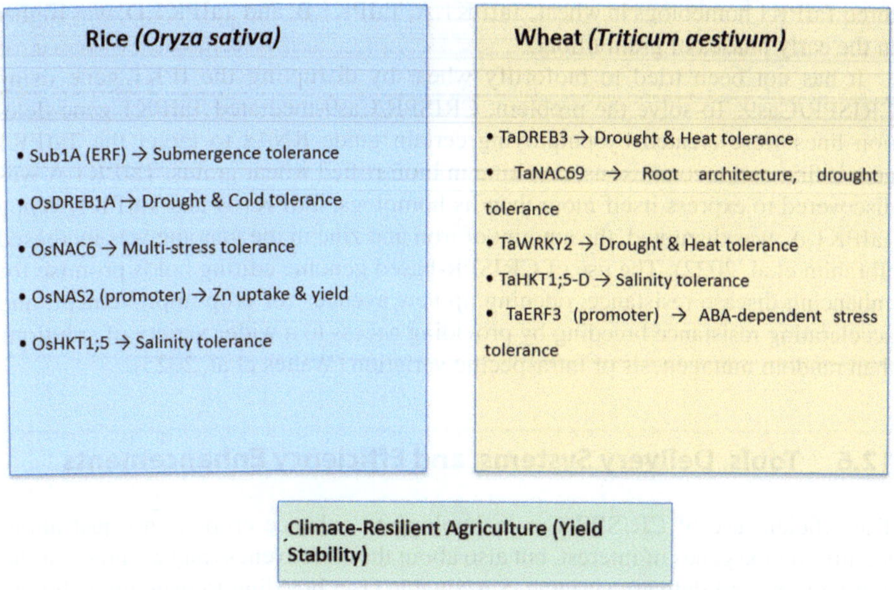

Fig. 12.2 Comparative overview of stress-responsive pathways in rice and wheat

focus of rice gene editing for a long time. Although this strategy has produced some encouraging outcomes, it has also shown certain shortcomings. For instance, resulting in undesirable aberrant traits, including decreased yield, heightened susceptibility to illnesses, and blooming stoppage (Wu et al. 2024a, b, c).

12.5.2 Wheat

Wheat *(Triticum aestivum),* the most extensively produced crop, is a staple meal for 35% of the world's population and accounts for 20% of daily calories and proteins, making it a crucial component of global food security (Fig. 12.2). Wheat output must also rise at an extraordinary rate of 1.7% annually to fulfil the demands of the world's expanding population (Gupta et al. 2023). Salinity, drought, high temperatures, and heavy metals are examples of abiotic stressors that have a significant impact on plant growth and development and can reduce crop production by 50%. In these situations, creating crop varieties with increased flexibility to flourish in a range of environmental conditions is crucial. The disadvantage of conventional breeding is that it reduces genetic variability and fitness, even while it greatly improves productivity (Liu et al. 2020; Wolter et al. 2019). Although there are many different ways to enhance agricultural traits, CRISPR-associated protein 9 (CRISPR/Cas9)-mediated genome modification and clustered regularly interspaced short palindromic repeats have created a new path for functional genomics and crop development (Kaur et al. 2025a, b). It was discovered that the expression abundance of the three TaIPK1 homeologs in wheat, TaIPK1.A, TaIPK1.B, and TaIPK1.D, was higher in the early phases of grain filling.

It has not been tried to biofortify wheat by disrupting the IPK1 gene using CRISPR/Cas9. To solve the problem, CRISPR/Cas9-mediated TaIPK1 gene deletion lines were created by employing certain guide RNAs to target the TaIPK1 gene's first and second exons, resulting in biofortified wheat grains. TaIPK1.A was discovered to express itself more than its homologs, TaIPK1.B and TaIPK1. When TaIPK1.A, was disrupted, the amount of iron and zinc in the genome was increased (Ibrahim et al. 2022). The use of CRISPR-based genome editing holds promise for enhancing disease resistance, opening up new avenues for crop improvement, and accelerating resistance breeding by providing access to a wider variety of solutions than random mutagenesis or intraspecific variation (Waites et al. 2025).

12.6 Tools, Delivery Systems, and Efficiency Enhancements

The efficient use of CRISPR-Cas technology to enhance crops is not just about identifying the genes of interest, but also about the effectiveness and accuracy of the editing tools and delivery systems. Sustainable crop breeding to overcome climate change requires high editing accuracy, consistent integration or transient expression of editing machinery, and minimal off-target alterations to create sustainable results. In this section, the recent developments in CRISPR-Cas variants, optimization of guide RNA, and delivery methods that lead to improved efficiency of gene editing in plants are highlighted.

12.6.1 CRISPR-Cas Variants and Base Editors

The canonical CRISPR-Cas9 system has been extensively used in plant gene editing, but more recent variants are more specific, apt to broader targeting ranges, and have various editing capabilities. For example, one of the protospacer adjacent motif (PAM) requirements offered by Cas12a (Cpf1) is T-rich, allowing genome-wide editing in areas of the genome that are inaccessible to Cas9. Similarly, Cas13 is capable of binding RNA instead of DNA, which opens the prospects for temporal regulation of transcripts that respond to stress.

Base editors, that is fusion of a catalytically inactive form of Cas protein with an enzyme, e.g., cytidine deaminase or adenine deaminase, permit substitutions of single nucleotides, but do not cause a double-strand break. This methodology comes in handy especially when point mutations that are related to drought or heat tolerance have to be introduced. More importantly, prime editing systems have added to the toolkit, allowing more accurate insertions, deletions, and base conversions with lower off-target activity. These innovations offer potent opportunities to develop crops to resist abiotic stresses such as salinity, drought, and extreme temperatures (Anzalone et al. 2019; Li et al. 2018; Mishra et al. 2020).

12.6.2 Guide RNA Design and Off-Target Minimization

The efficiency of CRISPR editing mostly relies on the construction of guide RNAs (gRNAs), which bind Cas proteins to particular genomic loci. Online tools (e.g., CRISPR-P, CHOPCHOP, CRISPOR) and computational algorithms (e.g., CRISPR-P) help to determine gRNAs that have high on-target activity and low off-target activity. Off-target mutation may affect the stability of the trait and be a problem for biosafety. Some of the strategies to reduce them are:

- High-fidelity Cas9 variants (e.g., eSpCas9, SpCas9-HF1) have less non-specific binding.
- The use of shorter or chemically modified gRNAs, which increase binding specificity.
- Utilization of deep sequencing and off-target prediction tools across the genome on a global scale to validate results.

In the case of crops that are environmentally friendly and require polygenic properties, it is possible to use gRNA multiplexing carefully to enable the simultaneous editing of multiple loci, which increases breeding efficiency and preserves the genotype (Doench et al. 2016; Zhang et al. 2019; Tang et al. 2019).

12.6.3 Gene Delivery Methods (Agrobacterium, Biolistics, Protoplasts)

One of the most challenging processes in plant genome editing is the introduction of the elements of CRISPR into the plant cell. Some of these approaches have been simplified, each possessing its own advantages:

Agrobacterium-Mediated Transformation This is a technique that is largely effective, affordable, and most economical, especially in dicotyledonous crops. It allows stable integration but is usually restricted in monocots.

Biolastic (Gene Gun) Delivery This is capable of direct delivery of DNA-coated particles into plant tissues and is thus applicable to a wide range of crops, including recalcitrant monocots. It can, however, result in multiple insertions or damage to tissues.

Protoplast Transfection This process consists of the introduction of CRISPR-Cas components into plant cells with no cell wall by PEG-mediated uptake or electroporation. Although it provides temporary expression and does not integrate, protoplast plant regeneration may be difficult.

New options are nanoparticle-based delivery systems and virus-mediated vectors, which have a higher efficiency, independence on species, and low genome damage (Altpeter et al. 2016; Woo et al. 2015; Demirer et al. 2019).

12.7 Regulatory, Ethical, and Biosafety Consideration

The swift implementation of CRISPR-based technologies to improve crops has been met with serious questions on governance, ethics, and environmental safety. Although the impact of climate-resilient agriculture could be highly positive, the regulatory environment in the global community is still disjointed, and the concept of research adoption into practice is still influenced by the social perceptions of the population.

12.7.1 Regulatory Frameworks Across Countries

Countries have different policies regarding biotechnology, and therefore regulation of CRISPR-edited crops varies. For example, the United States Department of Agriculture (USDA) does not apply rigorous GMO regulations to some CRISPR-edited crops when they do not carry the foreign DNA (Waltz 2016). Contrastingly, the European Union (EU) implements strict GMO regulations for the use of gene-edited crops, as legally determined by the European Court of Justice in 2018 (Callaway 2018). Other countries, including Argentina, Brazil, and Japan, have taken a more laissez-faire approach, and genome-edited crops are not viewed as transgenic organisms when no foreign genetic material is introduced (Eckerstorfer et al. 2019a, b). India and China are working on the development of better guidelines, with a view to reconciling innovation, biosafety, and trade. This regulatory difference presents a problem for the international harmonization and trade of gene-edited crops.

12.7.2 Public Perception and Ethical Concerns

Gene-edited crops are largely determined by the public perception of the success. Polls indicate that customers tend to perceive CRISPR more as accurate and less dangerous than conventional GMOs, and there is still a certain lack of trust because of the scandals about biotechnology in the past (Shew et al. 2018). Ethical issues include:

Equity and Access Fears that smallholder farmers will be left out in the event that technology is monopolized by bigger companies.

Food Sovereignty There is an argument as to whether local people will have control over seed systems.

Scope of Application It will raise questions of whether editing should be restricted to stress resistance or extended to other characteristics such as shelf life, yield or even aesthetic improvements.

The development of trust should be through programs of open communication, involvement of the farmer, and governance structures that will not be based only on commercial benefits but also on the interests of the people (Metcalfe 2021). There are also issues of fair benefit sharing, which are emphasized by ethical debates, particularly on crops that are vital in developing regions.

12.7.3 Risk Assessment and Environmental Safety

The CRISPR-edited crops are evaluated regarding unintended off-target mutations, effects on the ecology, and gene flow to wild relatives. Even though research findings indicate that off-target effects in plants are relatively rare compared to animals, they still need to be examined by using rigorous genome-wide analyses (Zhang et al. 2019). Potential risks to the environment are that there could be disruption of the ecosystems, non-target organism unintended effects, and management of gene-edited traits released into the environment. Global standards, including those of the Cartagena Protocol on Biosafety consider the adoption of precautionary measures and post-release surveillance (Eckerstorfer et al. 2019a, b). Finally, a science-based, transparent, and context-specific risk assessment construct is needed to strike the balance between innovation in CRISPR, ecological sustainability, and food security.

12.8 Future Prospects and Challenges

A revolutionary method for crop development, multiplex genome editing allows for the simultaneous change of several genes to improve complicated features. Thanks to recent developments, ultra-multiplex CRISPR/Cas9 systems have been created that can edit multiple loci in a single transformation event (Wu et al. 2024a, b, c). Optimizing delivery strategies and guaranteeing accurate editing without off-target impacts are still difficult tasks, nevertheless. By enhancing guide RNA design and anticipating possible off-target locations, the combination of genome editing with artificial intelligence (AI) shows promise in overcoming these obstacles. Multiplex genome editing is a groundbreaking technique for crop growth that enables the simultaneous modification of numerous genes to enhance complex traits. An ultra-multiplex CRISPR/Cas9 system that may modify multiple loci in a single transformation event has been developed as a result of recent advancements. However, it is still challenging to ensure correct editing without off-target effects and optimize distribution techniques. The combination of genome editing and artificial intelligence (AI) holds promise for solving these challenges by improving guide RNA design and predicting potential off-target areas (Lim et al. 2024).

12.9 Conclusion

Due to rising temperatures, unpredictable rainfall, and an increase in abiotic stressors including heat, salt, and drought, climate change poses a danger to global agricultural output and food security. CRISPR/Cas9, base editors, prime editors, and other gene editing technologies provide precise tools for improving crop resilience, explaining the workings of these technologies and how they can be used to create crops that can withstand climate change in order to meet future challenges. The base editors permit direct base conversion without causing double-strand breaks, the prime editors permit accurate insertions, deletions, and substitutions, and CRISPR permits targeted modifications of plant DNA. Crop tolerance to heat, salt, and drought stress can be improved by comprehending and adjusting important regulator genes involved in stress responses. Gene editing can enhance characteristics linked to oxidative stress response, photosynthesis, membrane integrity, ion homeostasis, osmotic adjustment, stress response pathways, heat shock response, root growth, and water-use efficiency. Promising developments include phenomics, artificial intelligence (AI)/machine learning (ML), gene editing technologies, and integration with genomics. Off-target effects, delivery strategies, and regulatory obstacles are some of the difficulties that need to be addressed. In summary, CRISPR/Cas9 and other genome editing technologies offer strong instruments for modifying positive regulator genes to improve agricultural climate resilience. Researchers can create crop varieties more resilient to the complex issues brought on by climate change by focussing on important regulatory genes and signalling pathways involved in stress tolerance mechanisms. To fully realize the potential of these technologies in guaranteeing global food security in the face of rapidly shifting environmental conditions, a comprehensive strategy that incorporates epigenetic regulation, small RNA-mediated processes, and plant microbiome interactions is essential.

References

Adenle AA (2011) Response to issues on GM agriculture in Africa: are transgenic crops safe? BMC Res Notes 4:2–7. https://doi.org/10.1186/1756-0500-4-388

Altpeter F, Springer NM, Bartley LE, Blechl AE, Brutnell TP, Citovsky V et al (2016) Advancing crop transformation in the era of genome editing. Plant Cell 28(7):1510–1520

Ansari WA, Atri N, Pandey M, Singh AK, Singh B, Pandey S (2019) Influence of drought stress on morphological, physiological and biochemical attributes of plants: A review. Biosci Biotechnol Res Asia 16:697–709

Anzalone AV, Randolph PB, Davis JR, Sousa AA, Koblan LW, Levy JM et al (2019) Search-and-replace genome editing without double-strand breaks or donor DNA. Nature 576(7785):149–157

Anzalone AV, Koblan LW, Liu DR (2020) Genome editing with CRISPRCas nucleases, base editors, transposases and prime editors. Nat Biotechnol 38:824–844. https://doi.org/10.1038/s41587-020-0561-9

Bhagtaney L, Sundarrajan P (2023) An overview of tools for genome editing: ZFNs, mega nucleases, and TALENs. CRISPR/Cas-mediated genome editing in plants:37–64

Brokowski C, Adli M (2019) CRISPR ethics: moral considerations for applications of a powerful tool. J Mol Biol 431:88–101

Callaway E (2018) CRISPR plants now subject to tough GM laws in European Union. Nature 560(7716):16–17

Castanho A, Guerra M, Brites C, Oliveira JC, Cunha LM (2023) Design thinking for food: remote association as a creative tool in the context of the ideation of new rice-based meals. Int J Gastron Food Sci 31:100664. https://doi.org/10.1016/j.ijgfs.2023

Demirer GS, Zhang H, Goh NS, González-Grandío E, Landry MP (2019) Carbon nanotube–mediated DNA delivery without transgene integration in intact plants. Nat Protoc 14(10):2954–2971

Dhankher OP, Foyer CH (2018) Climate resilent crops for improving global food security and safety. Wiley, Plant, Cell and Environment. https://doi.org/10.1111/pce.13207

Djanaguiraman M, Perumal R, Jagadish SVK, Ciampitti IA, Welti R, Prasad PVV (2017) Sensitivity of sorghum pollen and pistil to high-temperature stress. Plant Cell Environ 41:1065–1082

Doench JG, Fusi N, Sullender M, Hegde M, Vaimberg EW, Donovan KF et al (2016) Optimized sgRNA design to maximize activity and minimize off-target effects of CRISPR-Cas9. Nat Biotechnol 34(2):184–191

Eckerstorfer MF, Dolezel M, Heissenberger A, Miklau M, Reichenbecher W, Steinbrecher RA, Waßmann F (2019a) An EU perspective on biosafety considerations for plants developed by genome editing and other new genetic modification techniques (nGMs). Front Bioeng Biotechnol 7:31

Eckerstorfer MF, Engelhard M, Heissenberger A, Simon S, Teichmann H (2019b) Plants developed by new genetic modification techniques—comparison of existing regulatory frameworks in the EU and non-EU countries. Front Bioeng Biotechnol 7:26

Enete AA, Amusa AT (2010) Challenges of agricultural adaptation to climate change in Nigeria: a synthesis from the literature. Field Actions. Science Reports

Falk MC, Chassy BM, Harlander SK et al (2002) Food biotechnology: benefits and concerns. J Nutr 132:1384–1390. https://doi.org/10.1093/jn/132.6.1384

Food and Agriculture Organization of the United Nations (2015). FAOSTAT. Rome: Food and Agriculture Organization of the United Nations

Gornall J, Betts R, Burke E (2010) Implications of climate change for agricultural productivity in the early twenty-first century. Philos Trans R Soc Lond Ser B Biol Sci 365:2973–2989

Gupta A, Hua L, Zhang Z, Yang B, Li W (2023) CRISPR-induced miRNA156-recognition element mutations in TaSPL13 improve multiple agronomic traits in wheat. Plant Biotechnol J 21(3):536–548

Howells RM, Craze M, Bowden S, Wallington EJ (2018) Efficient generation of stable, heritable gene edits in wheat using CRISPR/Cas9. BMC Plant Biol 18(1):215. https://doi.org/10.1186/s12870-018-1433-z.6

Ibrahim S, Saleem B, Rehman N, Zafar SA, Naeem MK, Khan MR (2022) CRISPR/Cas9 mediated disruption of inositol Pentakisphosphate 2-kinase 1 (TaIPK1) reduces phytic acid and improves iron and zinc accumulation in wheat grains. J Adv Res 37:33–41

Jain A, Singh N, Kumari S, Khan S (2021) Bioentrepreneurship in agricultural biotechnology. In: Agarwal S, Kumari S, Khan S (eds) Bioentrepreneurship and transferring technology into product development. IGI Global, Hershey, pp 183–200

Jiang Y, Zeng B, Zhao H, Zhang M, Xie S, Lai J (2012) Genome-wide transcription factor gene prediction and their expressional tissue-specificities in maize F. J Integr Plant Biol 54(9):616–630

Kampmann M (2018) CRISPRi and CRISPRa screens in mammalian cells forprecision biology and medicine. A.C.S. Chem Biol 13:406–416. https://doi.org/10.1021/acschembio.7b00657

Karavolias NG, Horner W, Abugu MN, Evanega SN (2021) Application of gene editing for climate change in agriculture. Front Sustain Food Syst 5:685801. https://doi.org/10.3389/fsufs.2021.685801

Kaur N, Qadir M, Francis DV, Alok A, Tiwari S, Ahmed ZF (2025a) CRISPR/Cas9: a sustainable technology to enhance climate resilience in major staple crops. Frontiers in Genome Editing 7:1533197

Kaur N, Qadir M, Francis DV, Anshu A, Tiwari S, Ahmed Z (2025b) CRISPR/Cas9: A sustainable technology to enhance climate resilence in major staple crops. Frontiers in Genome Editing. https://doi.org/10.3389/fgeed.2025.1533197

Khan S, Khan J (2010) Drought tolerant wheat cultivar (raj) for rainfed areas of Kpk, Pakistan. Pakistan J Agric Sci 47:355–359

Khan A, Pan X, Najeeb U, Tan DKY, Fahad S, Zahoor R, Luo H (2018) Coping with drought: stress and adaptive mechanisms, and management through cultural and molecular alternatives in cotton as vital constituents for plant stress resilience and fitness. Biol Res 51

Kim D, Alptekin B, Budak H (2018) CRISPR/Cas9 genome editing in wheat. Functional Integ Genom 18:31–41

Li Chao LC, Zong Yuan ZY, Wang Yan Peng WY, Jin Shuai JS, Zhang Ding Bo ZD, Song Qian Na SQ, … Gao Cai Xia GC. (2018). Expanded base editing in rice and wheat using a Cas9-adenosine deaminase fusion

Li R, Liu C, Zhao R, Wang L, Chen L, Yu W, Zhang S, Sheng J, Shen L (2019) CRISPR/Cas9-mediated SlNPR1 mutagenesis reduces tomato plant drought tolerance. BMC Plant Biol 19:38

Liang Z, Chen K, Li T, Zhang Y, Wang Y, Zhao Q et al (2017) Efficient DNAfree genome editing of bread wheat using CRISPR/Cas9 ribonucleoprotein complexes. Nat Commun 8:14261. https://doi.org/10.1038/ncomms14261

Lim SR, Lee SJ (2024) Multiplex CRISPR-Cas genome editing: next-generation microbial strain engineering. J Agric Food Chem 72(21):11871–11884

Liu L, Zhang J, Xu J, Li Y, Guo L, Wang Z, Zhang X, Zhao B, Guo YD, Zhang N (2020) CRISPR/Cas9 targeted mutagenesis of SlLBD40, a lateral organ boundaries domain transcription factor, enhances drought tolerance in tomato. Plant Sci 301:110683

Lou D, Wang H, Liang G, Yu D (2017) OsSAPK2 confers abscisic acid sensitivity and tolerance to drought stress in rice. Front Plant Sci 8:993

Lu C, Kang J (2008) Generation of transgenic plants of a potential oilseed crop Camelina sativa by agrobacterium-mediated transformation. Plant Cell Rep 27(2):273–278

Metcalfe D (2021) Ethics of the inheritable genetic modification of human embryos via CRISPR technology: Pauli Murray and the institutionalized control of Black women's reproductive lives. The University of Iowa

Miao J, Guo D, Zhang J, Huang Q, Qin G, Zhang X et al (2013) Targeted mutagenesis in rice using CRISPR-Cas system. Cell Res 23:1233–1236. https://doi.org/10.1038/cr.2013.123

Mishra R, Joshi RK, Zhao K (2020) Base editing in crops: current advances, limitations and future implications. Plant Biotechnol J 18(1):20–31

Nunez S, Arets E, Alkemade R, Verwer C, Leemans R (2019) Assessing the impacts of climate change on biodiversity: is below 2° C enough? Clim Chang 154:351–365. https://doi.org/10.1007/s10584-019-02420-x

Oracz K, Karpiński S (2016) Phytohormones signaling pathways and ROS involvement in seed germination. Front Plant Sci 7:864

Qaim M (2020) Bt cotton, yields and farmers' benefits. Nat Plants 6:1318–1319. https://doi.org/10.1038/s41477-020-00788-8

Qamar S, Tantray AY, Bashir SS et al (2020) Golden rice: genetic engineering, promises, present status and future prospects. In: A. R (ed) Rice research for quality improvement: genomics and genetic engineering. Springer, Singapore, pp 581–604

Queiroz MS, Oliveira CE, Steiner F, Zuffo AM, Zoz T, Vendruscolo EP, Silva MV, Mello BFFR, Cabra RC, Menis FT (2019) Drought stresses on seed germination and early growth of maize and sorghum. J Agric Sci 11:310–318

Rahman MM, Ghosh PK, Akter M, Al Noor MM, Rahman MA, Keya SS et al (2024) Green vanguards: harnessing the power of plant antioxidants, signal catalysts, and genetic engineering to combat reactive oxygen species under multiple abiotic stresses. Plant Stress 13:100547

Rai GK, Khanday DM, Kumar P, Magotra I, Choudhary SM, Kosser R, Kalunke R, Giordano M, Corrado G, Rouphael Y, Pandey S (2023) Enhancing crop resilience to drought stress through CRISPR-Cas9 genome editing. Plants 12:2306. https://doi.org/10.3390/plants12122306

Rajasree V, Pugalendhi L (2021) Breeding vegetables for nutritional security. In: Jovandaric LPE-MZ (ed) veganism–a fashion trend or food as a medicine. Intech open, Rijeka, p Ch.7

Raza, A.; Charagh, S.; Razzaq, A.; Javed, R.; Khan RSA, Hasanuzzaman, M. (2020) Brassicaceae plants response and tolerance to drought stress: Physiological and molecular interventions. In The Plant Family Brassicaceae: Biology and Physiological pp. 229–261

Santosh Kumar VV, Verma RK, Yadav SK, Yadav P, Watts A, Rao MV, Chinnusamy V (2020) CRISPR-Cas9 mediated genome editing of drought and salt tolerance (OsDST) gene in indica mega rice cultivar MTU1010. Physiol Mol Biol Plants 26:1099–1110

Sasaki T, Burr B (2000) International Rice genome sequencing project: the effort to completely sequence the rice genome. Curr Opin Plant Biol 3:138–141. https://doi.org/10.1016/s1369-5266(99)00047-3

Shaheen N, Ahmad S, Alghamdi SS, Rehman HM, Javed MA, Tabassum J, Shao G (2023) CRISPR-Cas system, a possible "savior" of rice threatened by climate change: an updated review. Rice 16(1):39

Shan Q, Wang Y, Li J, Zhang Y, Chen K, Liang Z et al (2013) Targeted genome modification of crop plants using a CRISPR-Cas system. Nat Biotechnol 31:686–688. https://doi.org/10.1038/nbt.2650

Shew AM, Nalley LL, Snell HA, Nayga RM Jr, Dixon BL (2018) CRISPR versus GMOs: public acceptance and valuation. Glob Food Sec 19:71–80

Shi J, Gao H, Wang H, Lafitte HR, Archibald RL, Yang M, Hakimi SM, Mo H, Habben JE (2017) ARGOS 8 variants generated by CRISPR-Cas9 improve maize grain yield under field drought stress conditions. Plant Biotechnol J 15:207–216

Shinwari ZK, Jan SA, Nakashima K, Yamaguchi-Shinozaki K (2020) Genetic engineering approaches to understanding drought tolerance in plants. Plant Biotechnol Rep 14:151–162

Singha DL, Das D, Sarki YN, Chowdhury N, Sharma M, Maharana J, Chikkaputtaiah C (2022) Harnessing tissue-specific genome editing in plants through CRISPR/Cas system: current state and future prospects. Planta 255(1):28

Sourour A, Afef O, Mounir R, Mongi BY (2017) A review: morphological, physiological, biochemical and molecular plant responses to water deficit stress. Int J Eng Sci 6:1–4

Suzuki N (2016) Hormone signaling pathways under stress combinations. Plant Signal Behav 11(11):e1247139

Tang X, Ren Q, Yang L, Bao Y, Zhong Z, He Y et al (2019) Single transcript unit CRISPR 2.0 systems for robust Cas9 and Cas12a mediated plant genome editing. Plant Biotechnol J 17(7):1431–1445

Trejo-Paniagua BO, Caamal-Chan MG, Cruz-Rodríguez RI, Lam-Gutiérrez A, Ruiz-Lau N (2024) Physiological, biochemical, and molecular response in siete caldos chili pepper plants (Capsicum frutescens) exposed to water deficit. Horticulturae 10(6):558

Vuren DPV, Ochola WO, Riha S, Gampietro M, Ginze H, et al. (2009). Outlook on agricultural change and its drivers. In: McIntyre BD, Herren, HR, Wakhungu J, Watson RT (ed.). Agriculture at a crossroads, Island Press, Washington, DC

Waites J, Achary VMM, Syombua ED, Hearne SJ, Bandyopadhyay A (2025) CRISPR-mediated genome editing of wheat for enhancing disease resistance. Front Genome Editing 7:1542487

Waltz E (2016) Gene-edited CRISPR mushroom escapes US regulation. Nature 532(7599):293

Wang X, Liu F (2021) Effects of elevated CO_2 and heat on wheat grain quality. Plants 10:1027. https://doi.org/10.3390/plants10051027

Wolter F, Schindele P, Puchta H (2019) Plant breeding at the speed of light: the power of CRISPR/Cas to generate directed genetic diversity at multiple sites. BMC Plant Biol 19(1):176

Woo JW, Kim J, Kwon SI, Corvalán C, Cho SW, Kim H et al (2015) DNA-free genome editing in plants with preassembled CRISPR-Cas9 ribonucleoproteins. Nat Biotechnol 33(11):1162–1164

Wu B, Luo H, Chen Z, Amin B, Yang M, Li Z et al (2024a) Rice promoter editing: an efficient genetic improvement strategy. Rice 17(1):55

Wu S, Kyaw H, Tong Z, Yang Y, Wang Z, Zhang L et al (2024b) A simple and efficient CRISPR/Cas9 system permits ultra-multiplex genome editing in plants. The Crop J 12(2):569–582

Wu S, Kyaw H, Tong Z, Yang Y, Wang Z, Zhang L et al (2024c) A simple and efficient CRISPR/Cas9 system permits ultra-multiplex genome editing in plants. Crop J.12:569–582. https://doi.org/10.1016/j.cj.2024.01.010

Yohannes H (2016) A review on relationship between climate change and agriculture. Earth Science and Climate Change

Zhang Y, Liang Z, Zong Y, Wang Y, Liu J, Chen K et al (2016) Efficient and transgene-free genome editing in wheat through transient expression of CRISPR/Cas9 DNA or RNA. Nat Commun 7:12617. https://doi.org/10.1038/ncomms12617

Zhang Y, Malzahn AA, Sretenovic S, Qi Y (2019) The emerging and uncultivated potential of CRISPR technology in plant science. Nat Plants 5(8):778–794

Zhao C, Piao S, Wang X, Huang Y, Ciais P, Elliott J et al (2016) Plausible riceyield losses under future climate warming. Nat Plants 3:16202. https://doi.org/10.1038/nplants.2016.202

Zhong X, Hong W, Shu Y, Li J, Liu L, Chen X, Islam F, Zhou W, Tang G (2022) CRISPR/Cas9 mediated gene-editing of GmHdz4 transcription factor enhances drought tolerance in soybean (Glycine max [L.] Merr.). Front Plant Sci:13

Socioeconomic Impacts of Climate Change on Agriculture and Plant Systems

13

Rishika Ajay, Sonakshi Beri, Sanju, and Vivek Kumar

Abstract

The alteration in climatic patterns and its commensurate impact pose a serious threat to the contemporary scenario due to the multifaceted challenges it creates for the agriculture and plant systems. The undesirable increase in earth's temperature, alteration in precipitation patterns, extremity of weather fluctuations have negatively impacted crop production and food security and, as a corollary, the nutritional deficiencies, poor health of livestock, and soaring food prices. Climate change and the consequential increase in CO_2 level have also adversely affected the pedosphere and hydrology of the planet, antagonistically impacting the microhabitats and the associated microbiota. Studies conducted across the globe, including Asia, Africa, and Europe, show that the impact of climate change goes beyond the economic aspect of destabilizing income among the vulnerable sections to an exacerbation of migration and social unrest. The worst sufferers are the poorer nations wherein a significant proportion of the population relies on climate-sensitive agriculture for sustenance and livelihood, but the economic restraints limit their capacity to adapt according to the changing climate. International cooperation and a multidisciplinary approach can help alleviate these socioeconomic challenges. Strategies like Climate Smart Agriculture and India's strategy of Panchamrit to tackle climate change hold great potential in this regard. Further, region-specific mitigation strategies and inclusive policy formulation, along with the acknowledgment of traditional knowledge, can help to adapt efficiently in the long run and realize the dream of a sustainable future.

R. Ajay · S. Beri · Sanju · V. Kumar (✉)
Department of Botany, IPGGPGCC Haldwani, Kumaun University, Nainital, Uttarakhand, India

H. C. Joshi et al. (eds.), *Climate-Responsive Plant Biology*,
https://doi.org/10.1007/978-981-95-7354-7_13

Keywords

Climate change · Sustainable agriculture · Climate smart agriculture · Socioeconomic challenges

13.1 Introduction

Among the most serious problems of today is climate change, an international concern mainly caused by anthropogenic activities that has numerous socioeconomic impacts on various aspects of social life, such as water supply, agriculture, and economic development. Driven by long-term accumulation of greenhouse gases (GHGs), it has resulted in increased levels of CO_2, higher temperatures, and changed precipitation patterns, including a spike in the frequency of extreme weather events (Rocha et al. 2022). These changes disrupt soil properties, biogeochemical cycles, water availability, and plant development at the same time, harming agricultural productivity, biodiversity, and ecological equilibrium. While CO_2 enrichment promotes photosynthesis, the response is tempered by nutrient scarcity and phenological inconsistencies (Oishy et al. 2025).

Agriculture, a backbone of global food security, being largely reliant on climate, is growing progressively more prone to the intricate consequences of climate change. Predictions suggested that the west coast and southern regions will experience high-temperature regimes with detrimental impacts on agriculture. Variation in rainfall and temperature directly affects crop growth and maturity, increasing vulnerability to various biotic and abiotic stresses that cause 30–50% losses in global output (Rajput et al. 2021; Chaudhry and Sidhu 2022). Additionally, a significant increase in the variety of pests and pathogens due to climate change is a threat to crop protection as it raises the occurrence and extremity of plant diseases (Harvey et al. 2023). These climate-induced disruptions are severely affecting crop physiology and overall yield, which eventually creates a major challenge to global food security and agroecosystems.

Climate change has critically impacted major aspects of food security by altering crop yields, affecting crop physiology (Bibi and Rahman 2023), soil health, water resources, and livestock productivity. These climatic disruptions extend beyond local contexts, as interconnected global food systems carry regional shocks throughout international supply, resulting in higher food prices and decreased nutritional diversity. To effective these challenges, it is necessary to incorporate adaptation and mitigation strategies. Adaptation involves modifying farming practices in response to shifting climatic patterns, like changing planting time, implementing water management techniques, and fostering the use of improved resilient crop genotypes (Rao et al. 2024). Globally, agriculture is a contributor as well as a sufferer of the climate change, as practices like fertilizer application, enteric fermentation, rice farming, and land use changes increase greenhouse gas emissions. On the other side, farming becomes less productive and more vulnerable due to fluctuating climatic conditions, especially in developing countries. The two-fold role highlights

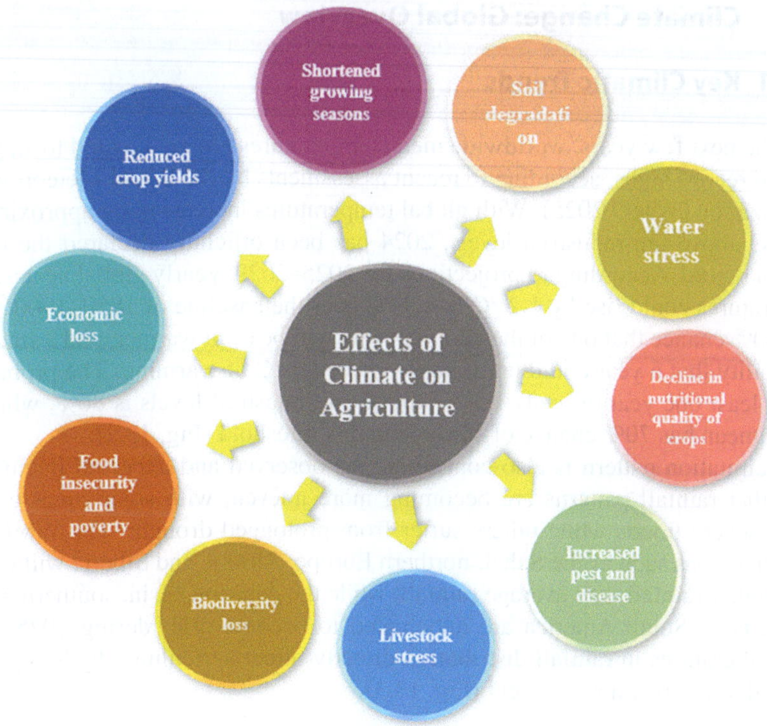

Fig. 13.1 Effects of climate change on agriculture

the urgent demand for long-term solutions that balance between environmental preservation and food production (Fig. 13.1).

Climate change is a crucial topic at present because the catastrophic consequences are occurring more quickly and intensely than expected, forcing the planet closer to irreversible thresholds. The effects contribute to a "threat multiplier" that heightens existing global challenges that involve poverty, inequality, and human security, while also offering new structural hazards for the economy, social stability, and health. There is mismatch between growing needs and decreasing output as a result of population growth, urbanization, and changing consumption patterns all contributing to increased food demand. The bulk of agricultural producers in many areas are smallholder farmers, who are among the most vulnerable due to their lack of resources, financial limitations, and institutional assistance.

The main objective of this chapter is to explore how the changing climate affects plant and agricultural ecosystems, specifically how it influences crop productivity, development, and the quality of the soil. Apart from these biophysical impacts, the chapter emphasizes the socioeconomic consequences, particularly for small-scale farmers, rural livelihoods, and food safety on a national and international level. It also covers adaptation and mitigation practices like adaptive technologies, climate-smart agriculture, and efficient source use. The objective is to offer strategies for creating economically viable, sustainable, and agricultural sectors.

13.2 Climate Change: Global Overview

13.2.1 Key Climatic Trends

Over the next few years, worldwide mean temperatures are anticipated to stay at or close to record highs, according to recent assessments by the World Meteorological Organization (WMO 2025). With global temperatures increasing by approximately 1.55 °C above pre-industrial levels, 2024 has been officially declared the hottest year on record. According to projections for 2025–2029, yearly global near-surface temperatures could rise by 1.2 °C to 1.9 °C over the baseline of 1850–1900. There is an 80% chance that one of the coming years will be even warmer, and a small 1% probability that a year could temporarily exceed 2 °C of warming. The probability that at least one year exceeds 1.5 °C above pre-industrial levels is 86%, while the 5-year mean has 70% chance of surpassing this threshold (Fig. 13.2).

Precipitation pattern is also concerning, as observed and expected fluctuations show that rainfall patterns are becoming more uneven, with some areas coming across severe floods while others suffer from prolonged drought. The WMO predicts that areas such as the Sahel, northern Europe, Alaska, and Siberia will experience higher rainfall than average rainfall, while the Amazon basin, southern Africa, and parts of South America are anticipated to become drier during 2025–2029. These alterations in rainfall distribution directly affect agriculture, hydrology, and global disaster risk management (Fig. 13.3).

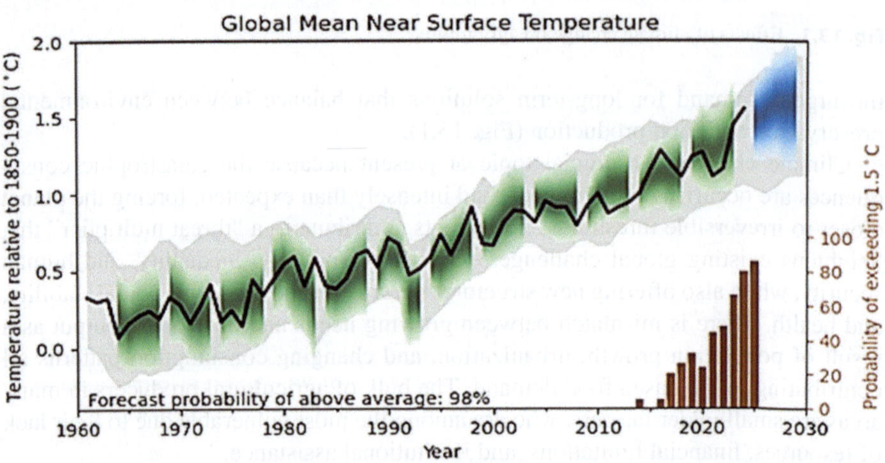

Fig. 13.2 Multi-annual predictions of global mean near-surface temperature relative to 1850–1900. Annual global mean observations are shown in black, hindcasts in green, and forecasts in blue. Shading indicates 90% confidence intervals, with intensity reflecting likelihood. Brown bars (right axis) show the probability of exceeding 1.5 °C in at least one of the following 5 years. The forecast probability of above-average temperature (relative to 1991–2020) is 98%. (Source: World meteorological organization, WMO (2025))

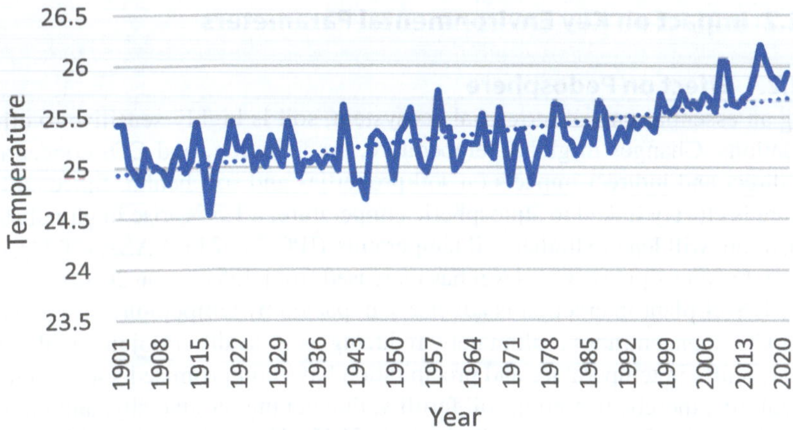

Fig. 13.3 Annual mean temperature in India (1901–2021), showing an overall increasing trend. The solid blue line represents observed annual mean temperatures, while the dotted line indicates the long-term trend. (Source: India Meteorological Department (2022))

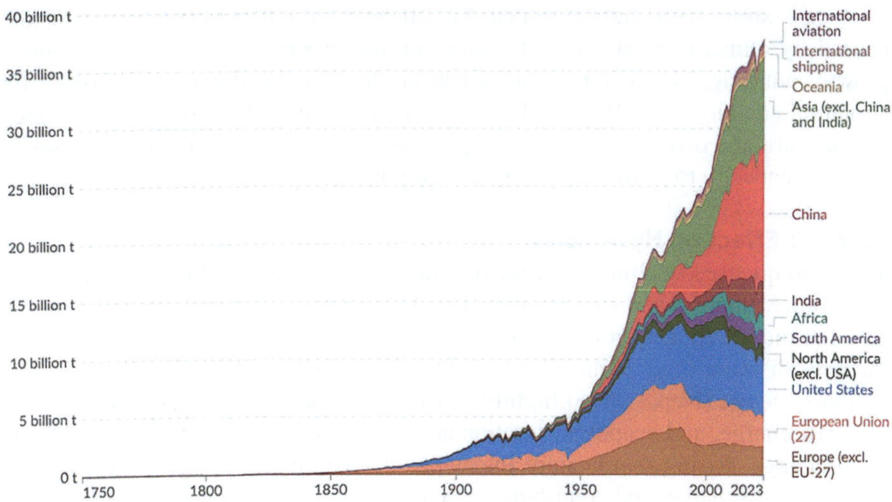

Fig. 13.4 The annual CO_2 emissions from fossil fuels and industry, broken down by world regions and countries from 1750 to 2023. (Source: Global Carbon Budget (2024))

Human activities in recent years have been responsible for emitting around 35 billion tons of carbon dioxide annually (IEA 2024). Since the pre-industrial period, atmospheric carbon dioxide levels have risen by more than 40%, rising from nearly 280 ppm in the eighteenth century to 419 ppm in 2023 (IPCC 2013; Lan et al. 2023) (Fig. 13.4).

13.2.2 Impact on Key Environmental Parameters

13.2.2.1 Effect on Pedosphere

Being an essential part of terrestrial ecosystem, soil is highly sensitive to climatic fluctuations. Changes in global temperature, precipitation, and CO_2 concentration have direct and indirect impacts on soil properties and soil health. Since soil temperature is closely linked to atmospheric temperature, a 1.5 °C rise in global surface temperature will lead to higher soil temperature (IPCC 2021). NASA (2024), documented that atmospheric CO_2 level has increased from 365 ppm in 2002 to 427 ppm in 2024. Soil-plant interaction is adversely impacted by temperature stress and elevated CO_2 concentration, ultimately reducing soil health (Gojon et al. 2023). Climate shifts in temperature and precipitation have compromised soil organic carbon stability, thereby depleting soil fertility, threatening biodiversity, and intensifying soil salinization processes (Lal et al. 2021; Hassani et al. 2021). Climate change-driven drought induces a wide spectrum of ecological consequences, such as decline in biodiversity of plants, reduced photosynthesis, hampered root growth, and variations in soil microbial population, by influencing plant-soil physiochemical attributes (Furtak and Wolińska 2023).

A case study from Italy revealed that changes in soil quality and soil health caused by climatic events strongly influence the processes that support biomass growth, with significant impact on global production (Bonfante et al. 2019). By structural equation modelling studies, Sünnemann et al. (2023) found that elevated temperature and drought conditions deplete soil multifunctionality, thereby weakening its capacity to promote agricultural output.

13.2.2.2 Effect on Hydrology

The consequences of climate change on water resources are multifaceted, affecting availability, quality, and distribution of water resources. Altered rainfall patterns result in more frequent droughts and intense rainfall events, which in turn heighten the risks of water scarcity and flooding. Declined water availability influences irrigation systems. (Nepal 2016) highlighted that snowfall pattern, melting of snow, water discharge, and evapotranspiration are highly vulnerable to the effects of climate change. Rising temperature intensifies glacier retreat, triggers flash floods, alters river discharge, and contributes to sea level rise, which together have adverse effects on freshwater quality (Abbas et al. 2022). Borrelli et al. (2020) suggested that across all global scenarios, climate change trends point toward a more active hydrological cycle, raising global water erosion by 30–66%.

13.2.2.3 Effect on Atmospheric CO_2

Rising temperature and CO_2 levels influence plant growth and long-term agricultural productivity and also affect the nutritional quality of crops. Studies confirm that higher CO_2 concentrations lower zinc, iron, and protein levels in staple crops (Myers et al. 2014). Elevated CO_2 can be beneficial for certain C3 crops like barley, rice, wheat, cotton, and soybean, with increment in production up to 30% under doubled CO_2 conditions. Other crops like maize show minimal response (<10%). Importantly, these benefits are conditional—temperature rise, water scarcity, or nutrient limitations can decrease or even reverse positive effect of CO_2. (Varanasi et al. 2016). Increased CO_2 additionally alters composition of biomass through fostering the formation of sugar, which raises carbon content but lowers protein levels, particularly in cereal grains, thus lowering the wheat's baking quality (Högy et al. 2013). Yet, increased CO_2 will boost weed growth (Ziska 2001).

13.3 Direct and Indirect Impacts on Agriculture and Plant Systems

13.3.1 Crop Yield and Production

13.3.1.1 Crop Growth and Growing Season

Elevated global temperatures affect crop growth by accelerating crop maturity, shortening grain-filling period, and lowering the production of major food crops such as wheat, rice, and maize (Lobell et al. 2011). The variations in climate also influence the timing and duration of growing seasons, and many areas have reported shorter crop cycles (Xiao et al. 2021) along with phenological changes like earlier flowering (Gordo and Sanz 2010). In the Mediterranean, crop growing seasons are projected to increase by 27–47 days under different climatic scenarios (Christiansen et al. 2011). Studies predicted a noticeable shortening in growing season of wheat and rice, leading to altered development stages (Ahmad et al. 2023). Alterations in daylight duration affect crop growth cycles (Ozkaynak 2013).

13.3.1.2 Crop Yield and Variability

Severe climate shifts projected considerable risk to crop productivity, with yield losses estimated up to 7–23% without adaptation (Rezaei et al. 2023). Persistent climate changes and greenhouse gas emissions decrease the output of wheat by 5–50%, rice by 20–30%, and maize by 20–45% by 2100 (Hassan et al. 2022). Kumar and Parikh (2001) projected huge changes in the climate would lead to substantial drops in the crop yields, consequently affecting the agricultural production. In addition, climate variability accounts for nearly 60% of crop yield fluctuations (Ray et al. 2015; Aryal et al. 2020). Yield reduction of about 6–18% has been reported in arid and semi-arid zones of the world due to shortenings of growing season, inadequate water supply, and high temperature stress at reproductive phases of crops. (Ullah et al. 2019). Similarly, Habib-ur-Rahman et al. (2022) reported rice and wheat yield reductions of 15–17% and 12–14%, respectively, under mid-century scenarios using DSSAT and APSIM models. Projections revealed that in major agricultural model inter-comparison studies, yield will fluctuate in multiple agricultural zones across low and high latitudes; however, production will be severely affected in low-latitudes areas.

13.3.2 Soil Health and Microbial Ecology

Changes in rainfall patterns, alterations in temperature, and frequent extreme weather events directly influence soil moisture, soil structure, soil erosion, and compaction, ultimately affecting soil fertility. Soil compaction decreases pore space, which increases bulk density, and it has been reported that soil bulk density higher than 1.2 Mg/m^3 decreases maize yield by 30% (Shah et al. 2017). Soil compaction further inhibits microbial activity and carbon release, resulting in reduced uptake of nutrients by plants (Shoaib et al. 2022). Increasing temperature also reduces total soil organic carbon by 10.6% and limits soil ability to hold moisture, sustain crop development, and resist erosion (Ma et al. 2024).

Rising temperature and CO_2 levels increase the release of organic acids like nitric acid and oxalic acid into the soil, which accelerates microbial decomposition of organic matter (Ferdush et al. 2023) leads to reduction in soil pH, and induce soil acidification (Zhang et al. 2023). Frequent altered rainfall patterns affect concurrent soil acidification and soil salinity (Dal Molin et al. 2020). High soil acidity constrains the accessibility of vital nutrients such as phosphorus and molybdenum while accelerating toxic element aluminum to the extent that hampers plant growth (Rahman et al. 2024). Globally, soil salinity impacted around 20% of irrigated land, especially in arid regions where scarce irrigation induces salt accumulation (Shrivastava and Kumar 2015).

Fluctuation in soil moisture as a result of irregular rainfall and drought can interrupt microbial communities and nutrient cycling, changing the carbon, nitrogen, and phosphorus balance, leading to decline in soil fertility and yield (Furtak and Wolińska 2023). Extreme precipitation gives rise to waterlogging, diminishing soil aeration, suppressing microbial processes, and affecting root respiration ultimately affecting nutrient cycling (Topali et al. 2024). Oxygen-deprived conditions, the result of waterlogging, inhibit the functioning of nitrifying bacteria and therefore favor denitrification, leading to release of nitrous oxide (N_2O) that promotes global warming (Hui et al. 2024).

Climate change affects soil erosion through fluctuations in rainfall intensity and plant cover. Greater runoff results in erosion of nutrient-rich topsoil, severely decreasing soil fertility. Decreasing enzyme potential also reduces soil potential, hampers biogeochemical cycling, and degrades ecosystem resistance and recovery (Luo et al. 2017). Mild increase in temperature may improve microbial activity, but increased temperature can hamper their activity and further harm soil biodiversity and ecosystem health.

13.3.3 Livestock and Food Security

Food availability is directly affected by reductions in crop yields and livestock productivity brought on by climate change. Reducing crop yields converts to decreased food supplies, while heat stress and decreasing forage quality adversely affect milk, meat, and egg production. Decline in food accessibility can lead to price hikes and increased resource competition, especially in vulnerable areas (Schmidhuber and Tubiello 2007).

Livestock might get affected by changing climate in two ways: either directly through warmer temperatures or indirectly through shifts in both quality and quantity of grassland forage. Research on the consequences of climate change on livestock is limited. Hot summer temperatures have been reported to reduce livestock appetite, leading to decrease in weight gain and also observed that livestock productivity in the U.S. drop by 10% for cow/calf and dairy farms in Appalachia, the Southeast, the Delta States, the Southern Plains, and Texas if temperature jumped by 5.0 °C (Adams et al. 1998).

The changing climate impacts the frequency of plant diseases, along with their chronological and geographical pattern, and also has an influence on pathogen invasion, spore formation rate, sporulation rate, strength, replication rate, and spore dispersal direction and distance (Ahanger et al. 2013). Fluctuating climatic events affect pathogen biology and infectivity and also alter plant microclimate and its susceptibility to getting infected. These shifting patterns may affect soil microbial communities and canopy pathosystems, potentially affecting their beneficial roles. (Elad and Pertot 2014).

Increased CO_2 concentration has the potential to increase pest and disease issues by altering the nutritional content of the plants and thereby rendering crops more prone to pathogen infestation or pest occurrence (Skendžić et al. 2021). Precision irrigation, drought-tolerant agriculture production, and pest suppression represent a few adaptation techniques currently used in order to reduce these effects (Nikolaou et al. 2020). Due to regional context and technical access, its effectiveness varies greatly by location (Rana 2019).

13.3.4 Microhabitats and Biodiversity

According to the IPCC (2023), in terrestrial, aquatic, cryospheric, and marine ecosystems, climate change has currently resulted in severe damage along with eventually irreversible 75 losses. The ranges of several species have currently expanded towards the poles or higher-elevations, demonstrating that climate change has caused major alterations in geographic distribution of species (Hickling et al. 2006). When species are unable to adjust or scatter rapidly under shifting weather patterns, they may exceed their survival thresholds, resulting in the extinction of species (Schweiger et al. 2008; Zhu et al. 2012). These range shifts trigger species expanding or decreasing in their current and former ranges, respectively. Eventually, these shifts result in changes in the biomes that will ultimately have an effect on global carbon cycle. Species of alpine plants are widely spreading upwards due to global warming (Lenoir et al. 2008), extension of the boreal forest towards northward (Danby and Hik 2007), spreading of shrubs within dryland ecosystem (Schlesinger et al. 1990), and shrub invasion in arctic tundra (Wookey et al. 2009).

13.4 Socioeconomic Impacts

13.4.1 On Susceptible Communities and Nations

As seen from the previous segment, climate change extends beyond just being an environmental concern to a critical economic issue that can jeopardize global equity and stability. The hostile effects are disproportionately borne by the communities and the nations with restrained financial resources, dilapidated infrastructure, heavy adherence to climate-sensitive sectors and limited adaptive capacity i.e., the 'vulnerable'. These communities mostly reside in the developing nations now termed as the global south. These effects can be direct or indirect.

13.4.1.1 Direct Effects

13.4.1.1.1 Agriculture and Nutritional Security

Farming has been the foundation of human sustenance and a major sector of employment for a significant proportion of the populace. But such livelihood is fragile as it is driven by climatic variations and it disproportionately affects the rural poor of the developing nations that rely on agriculture for sustenance. Erratic rainfall and drought pattern, heat, stress, increased intensity of cyclones and typhoons threaten the agriculture and plant systems by decreasing the yield of staple crops. A scarcity of grains and crops cause price surge and make markets volatile. Soaring food price is a significant contributor of food insecurity, also driving farmers to poverty by reducing agricultural income. Since a large share of income is spent on food by the poor households, the poverty crisis deepens. The meagre income of a farmer is insufficient to meet all the basic demands of the household and as a consequence they suffer from malnourishment, lack of education and other basic amenities thus reinforcing the vicious cycle and trapping the vulnerable sections in chronic health and economic crisis. Failure of crops and the consequential monetary loss often force farmers to borrow money from informal sources like moneylenders in the villages. The harassment by moneylenders drive farmers to commit suicide in nations like India. Per +1 °C addition in the planetary average surface temperature, the global crop output would fall by 5.5×10^{14} kcal annually which is 4.4% of recommended daily intake (Hultgren et al. 2025).

13.4.1.1.2 Health Risks and Economic Security

The socioeconomic susceptibilities are exacerbated by health concerns driven by climate change. The increasing temperature translates into an increased prevalence of vector-borne diseases, especially in the tropical regions. For instance, an increase in water-borne diseases has been observed in South Asian countries like Bangladesh due to frequent and intense floods driven by climate change. For the last few years, many lives have been claimed in India as a consequence of heat waves. Studies have claimed that such heatwaves and dry spells decrease labor productivity, particularly among outdoor workers like construction laborers and farmers. Failure of crops due to altering climate causes food scarcity, especially among the children of poor households, thus imperiling social capital of the nations and their future.

13.4.1.1.3 Deprivation of Shelter and Resources

Frequent rainfall, cyclones, flashfloods, landslides, and rising sea levels destroy shelters and other hard infrastructure. Typhoons have claimed infrastructure in the USA for years. People lose property annually to cyclones and river erosion in countries like Bangladesh in South Asia and China in Southeast Asia. Small island states like Tuvalu are experiencing land loss as a result of climate change, thus risking their sovereignty.

13.4.1.1.4 Displacement and Migration

Natural disasters have become more frequent due to climate change and, as a corollary, the displacement of millions into overcrowded urban slums. Overcrowding leads to an increase in competition for resources, unsanitary living conditions leading to increased proliferation of diseases, and increased criminal activities as diverse sections of people reside together, and cultural heterogeneity might fuel social tension. Thus, the sink region faces challenges of accommodation, employment, and social integration. Apart from such difficulties, many island nations like the Maldives and Kiribati are facing existential crises due to rising sea levels, underscoring the widespread adversity climate change can present to the nations.

13.4.1.2 Indirect Effects (Unfold Over Time)

13.4.1.2.1 Entrenched Poverty and Reduced Income

Failure of agriculture and the consequential reduction in income entraps the household into poverty. Also, recurrent weather extremities deplete livestock, household savings, and assets. Efforts to endure such crises are sought by selling land, livestock, and other possession. This reduces their capacity to recover and perpetuates poverty.

13.4.1.2.2 Educational Loss

Loss of infrastructure, displacement of families, and depletion of assets and resources force children to drop out from schools in order to support household incomes. This halts the development of human capital and turns the demographic dividend.

13.4.1.2.3 Gender Disparity

Women and girls have for ages been responsible for household chores like collection of water and food and caregiving. Climate change disproportionately impacts them by increasing their workload and limiting their educational and economic opportunities.

13.4.1.2.4 Psychosocial Stress

Climatic vulnerability of agriculture and plant systems and its associated consequences, like loss of livelihood, displacement, health risks, etc., cause mental agony and decrease the productivity of communities.

13.4.1.2.5 Economic Disparity Between Nations and Social Tension

The vulnerable communities and nations have an increased risk of getting affected by adverse climatic patterns, as they have limited resources to adapt to climate change. This leads to widening of global inequalities, as the most vulnerable communities reside in the developing world and small island nations.

13.4.1.2.6 Compromised Human Capital Development

Long-term development of human capital is hindered by reduced productivity and its consequences like malnutrition, entrenched poverty, health insecurity, and education loss (Table 13.1).

Table 13.1 Disintegration of culture and tradition with the loss of livelihood and climate-driven displacement leads to diminishing of social cohesion

Consequences	Elucidation	References
Reduced agricultural output	Temperature and precipitation alteration reduce yield of major crops like rice, wheat, maize, etc.	Pereira (2017)
Health insecurity	Rising temperature leads to heat stress for both human and livestock and depletes their productivity. Food scarcity leads to malnutrition	Parkes et al. (2022)
Economic insecurity	Reduction in yield translates to reduction in agricultural income and entrenched poverty	Adesete et al. (2023)
Migration, displacement and unrest	Climate-induced migration due to loss of livelihood increases the competition for resources	Adesete et al. (2023)
Increase in pest outbreaks	Warming increases incidence, severity and range of insect and pest that damage crops	Kaushik et al. (2023)
Shift in cultivation zones	Local farming gets affected due to shifting of zones induced by climate change	Malhi et al. (2021)
Infrastructure damage	Weather extremities damage transportation and storage infrastructure of crops	Yiadom et al. (2023)
Change in land use	Population explosion and increased demand for food and climate induced displacement, urbanization, etc. have led to clearing of forested lands and biodiversity loss	Okeleye et al. (2023)

13.4.1.2.7 Fading Away of Culture

13.5 Case Studies

13.5.1 Africa

Africa is an abode to a diversity of the world's vulnerable tribes and diverse climate patterns ranging from humid equatorial to dry tropical and subtropical Mediterranean (Couvreur et al. 2021). According to scientific assessments by IPCC's general circulation modelling (GCM), sub-Saharan Africa is expected to warm faster than the global average (Shepard 2019). Climate change-induced rises in maximum and minimum temperatures, both diurnal and monthly, can be noticed in Ethiopia, Kenya, and Tanzania (Gebrechorkos et al. 2019). And this has reduced agricultural productivity of the region. Further, yields of six major crops—rice, wheat, maize, sorghum, cassava, and soybean—are projected to decrease by 39.4% by 2098 (Hultgren et al. 2025). In Ethiopia, higher temperatures have increased the incidence, and severity of coffee berry disease caused by the fungus *Colletotrichum kahawae* and have resulted in its spread to higher altitude regions of East Africa (Ghini et al. 2011). Also, the outbreaks of desert locust and *Spodoptera frugiperda* (fall armyworm) have accelerated (Ashagidigbi et al. 2022; Paudel Timilsena et al. 2022). The emergence of novel plant diseases and transboundary outbreaks are also consequences of climate change (Meyer et al. 2023). These concerns have exacerbated agricultural and economic losses. Increasing temperatures and changes in rainfall patterns increase the risk of infectious diseases in humans and drive the already economically fragile communities to deeper poverty and inequality. Severe hunger risk would threaten more than 40 million people in sub-Saharan Africa by 2050 (Wiebe et al. 2019). Variability of precipitation is observed, with some regions facing intense dry spells, like Sahel and South Africa, while heavier rainfall is observed in Eastern Africa. Economies like Ghana and Central and West African countries are the worst sufferers, as they rely heavily on rainfall and lack irrigation infrastructure (Sarku 2023). These regions also lack the capacity to adapt to climate change due to limited economic and institutional means (Sultan and Gaetani 2016). El Nino events of 2015–2016 decreased cereal output by 24% (SADC 2022). Estimates state that by 2050, the yield of major cereals would plummet, leading to price surge of 25% (Schilling et al. 2020). Since food scarcity demands food imports, inflation would rise and impact exchange rates of currencies, thus exacerbating poverty levels. All this would drive the already impoverished population and vulnerable subgroups like women, aged, crippled children, and substance abusers to extreme socio-economic deprivation (Benevolenza and DeRigne 2019). This would eventually increase the social unrest of the region, which is already entrapped in political instability.

13.5.2 Asia

Asia is acknowledged as the most populous subcontinent on Earth, and the majority of the continent's agriculture is dominated by the vulnerable groups like small farmers, women laborers, and poor households. Developing countries of Asia are more prone to decreased output as they rely on old technologies (Lybbert and Sumner 2012). Since 60% of the world's population is concentrated in Asia, a decrease in output and increased hunger risk could push millions to malnutrition. Reshaping of Asian agriculture, with a reduction in rice and wheat productivity, is driven by climate change. Rice and wheat, Asia's staple crops, are highly heat sensitive. Projections state that in Pakistan by 2050, climate change-emanated reduction in rice and wheat productivity would reduce the country's real GDP by 19.5 billion USD (Khan et al. 2020). Similarly, in Nepal climate change would lead to 10% decrease in real GDP in the highest impact scenario (Chalise et al. 2017). Yield changes across major crops are expected to decrease by 29.1% by 2098 (Hultgren et al. 2025). Water stress in transboundary river basins would also increase geopolitical tension in the already sensitive and hostile geography of Asia. Coastal countries in South Asia like Bangladesh are already facing amplified occurrence and severity of cyclones, river erosion, and other disasters that undermine food production and distribution, livelihood, access to potable water, and increasing slum habitation and unsanitary conditions (Mahapatra et al. 2021; Gregory et al. 2005). The natural disasters limit freshwater sources and arable lands. Saltwater intrusion leads to salinization of soil and disrupts cultivation. The International Centre for Climate Change and Development projected that by this century, coastal Bangladesh would be submerged with a sea level rise between 12% and 18% (Huq et al. 2024). Most exposed are the poor communities, as they have the least adaptive capacity. The scenario in India is even more concerning, as a significant proportion of poor households practice climate-sensitive farming and monocropping (Ramachandran 2014). Being a low-altitudinal country, the yields would be negatively impacted with an increase of temperature by 1–2 °C (Bhat and Pandit 2020). As observed in Himachal Pradesh, India, the apple productivity has declined due to soaring temperatures, and the apple-growing belt has shifted to higher elevations (Rana et al. 2013; Singh et al. 2016). Climate stress has resulted in a shift from major cereal crops to legumes and non-food crops in states like Odisha, India (Birthal et al. 2021). Studies have shown that yields of rice, wheat, and maize would be lost owing to alteration in rainfall regimes by 10–30% and 2–12% respectively (Panda et al. 2019; Senapati 2022; Guntukula 2020; Saravanakumar et al. 2022). Climate change-induced natural disasters like floods and drought have caused crop losses, nutritional insecurity, price rise, supply chain disruptions, as observed in Tamil Nadu, Bihar, Odisha, and Kerala (Vishnu and Sridharan 2016; Kumar et al. 2016; Sam et al. 2021). India suffered an estimated economic loss of 159 USD in 2021 alone (World Economic Forum 2024). Studies conducted between 2013 and 2014 in the villages of Jalna district of Maharashtra, India, observed an adverse impact of 60% on local crop production and farmer income due to drought, and this decreases the well-being of their children (Vedeld et al. 2014). This contributes to the high incidence of stunting, wasting, and undernourishment among Indian children (Bankole 2020). Thus, climate change has a multifaceted impact on Indian society through its influence on agriculture, supply chains, and natural disaster-led losses.

13.5.3 Europe

Europe faces a contrast in the impacts of climate change between the northwestern and southeastern regions. Studies have shown that with a longer growing season accompanied by higher CO_2 concentrations and milder winters, North and West Europe would gain in the short term due to a 10–30% increase in the yields of wheat, maize, and spring barley. Favorable temperature would lead to a northward shift in crop zones for sunflower, soy, maize. Whereas heat stress, drought, and water scarcity would reduce the yield of maize and wheat by 25% and also reduce food security and income in Southeast Europe, causing rural unemployment. This would increase the North-South economic divide. The worst impacted would be Mediterranean agriculture as high temperature and decreased rainfall adversely affect the vineyards of France, Italy, and Spain, causing a more northern relocation. A rise in sea level and saltwater intrusion would leave the fertile Dutch lowlands unsuitable for cultivation. By 2098, the staple crop output is projected to reduce by 21.5% (Hultgren et al. 2025). Overall, the economic losses suffered as a result of climate change is projected to be between €1.5 and €4 billion by 2050. Thus, in the longer run climate change would risk food and economic security of Europe (Table 13.2).

13.6 Adopted, Adaptive, and Mitigation Strategies

The greatest concerns in the contemporary times are climate change-driven food and nutritional insecurity, and environmental degradation, alongside the spillover socioeconomic implications. To adapt and counteract these adverse effects, sustainable practices are the need of the hour. Some of the initiatives are discussed below:

13.6.1 At International Level

13.6.1.1 Resource Support

Global organizations like the World Bank, Global Climate Fund, and Global Economic Facility provide financial assistance to the developing countries to support climate-resilient agricultural practices.

13.6.1.2 Knowledge Sharing and Technology Transfer

Insights on sustainable agriculture practices (CSA) and climate resilience areshared globally by the collaboration of institutions like the Consultative Group on

Table 13.2 Climate change and its adverse effects

Country	Effect of climate change	Impacted crops	Economic loss (USD)	References
Bangladesh	Increased sea level and weather extremities	Major cereals, cash crops, fruit trees, etc.	Projected to be 2.6 billion USD annually by 2050	World Bank (2021)
Brazil	Temperature, precipitation and other alterations in environment	Cassava, cocoa, corn, soybeans, rice and wheat	59.55 and 15.59 billion USD (2013–2023)	Fiorini et al. (2024)
China	Weather extremities	Rice, wheat, corn, soybean	595–868 million USD in last 10 years	Duan et al. (2022)
Ethiopia	Increase in temperature, rainfall and unpredictable weather	Maize, sorghum and teff	Estimated to be 320–800 million USD annually by 2040	World Bank Group (2024)
India	Unpredictable precipitation, drought etc.	Major cereals, cash crops and fruit trees	159 billion USD in 2021	World Economic Forum (2024)
Indonesia	Extreme and variable weather conditions	Rice, copra, coffee, palm oil, rubber, cocoa	About 4.6 billion USD per annum by 2050	(World Bank (2021)
Kenya	Low rainfall, increased drought	Maize, sweet potato, beans	About 2.7 billion USD by 2050 annually	World Bank (2024)
Pakistan	Global warming and increased natural disasters	Wheat, rice, cotton, sugarcane, maize	19.5 billion USD by 2050 (wheat and rice output)	Syed et al. (2022)
USA	Climate extremities, wildfires	Major crops like corn, soybean, wheat, cotton, sorghum, rice	Billions of USD (projected)	Usgcrp (2018)

International Agricultural Research (CGIAR). The World Bank is an important investor in CSA.

13.6.1.3 Implementation Global Policies

Global framework to reverse climate change is established through treaties like the Paris Agreement, UNFCC, and UNCCD that foster international cohesion.

13.6.1.4 Strengthening Capabilities

Focus on nature-based solutions and local resilience building to ensure food and water security, disaster management, etc.

Key initiatives at international level include:

13.6.1.5 Climate Smart Agriculture (CSA)

According to World Bank one-third of the global emissions are contributed by agri-food system. It is the leading source of methane emission and also consumes about 70% of the freshwater. Developed by Food and Agriculture Organization (FAO), CSA has emerged as a sustainable practice that addresses these issues and increases productivity and resilience and decreases greenhouse gas (GHG) emissions. This is achieved by acknowledging traditional practices and also adopting innovative technologies. For instance, practices like drip irrigation, rainwater harvesting, composting, precision farming, agroforestry, livestock management, integrated pest management, nutrient management, etc. With the help of the International Atomic Energy Agency (IAEA) and FAO and adoption of nuclear-derived CSA practices, farmers in Bangladesh, Lao People's Democratic Republic, Myanmar, Nepal, Pakistan, and Vietnam witnessed increased rice production from 1 to 2.5 tons per hectare.

13.6.1.6 Paris Agreement

Adopted in COP 21 in Paris (2015), it intends to prevent global temperature increase beyond 2 °C and pursue 1.5 °C to achieve the stated sustainable development goals (SDG_s).

13.6.2 Regional Policies: India

The Government of India has formulated and initiated numerous policies to alleviate the adverse impacts of climate change on agriculture and plant systems. Some key initiatives are discussed below:

13.6.2.1 National Innovation on Climate Resilient Agriculture (NICRA)

It is a flagship initiative by the Indian Council of Agriculture Research (ICAR) that strives to innovate climate-resilient agricultural technologies and capacity building in 151 climatically vulnerable districts of India.

13.6.2.2 National Mission on Sustainable Agriculture (NMSA)

This mission is a component of the National Action Plan on Climate Change (NAPCC) and encompasses various submissions like Soil Health Management, Rainfed Area Development, organic farming, etc.

13.6.2.3 Pradhan Mantri Fasal Bima Yojana (PMFY)

It is a financial scheme to support farmers by providing crop insurance against crop losses due to a weather adversity and other natural causes.

13.6.2.4 India's Five-Pronged Pledge: 'PANCHAMRIT'

At COP 26 in Glasgow (UK) in 2021, India unveiled a pledge to mitigate climate change and ensure a resilient future. It includes the following five nectar elements (Panchamrit):

- Expanding renewable energy capacity to 500GW by 2030 to lower dependence on fossil fuels.
- Enhancing the contribution of renewable energy to 50% by 2030.
- Reducing carbon emission to GDP by 45% by 2030 (from 2005 levels).
- One billion tons of carbon emission reduction by 2030.
- Attaining neutral carbon footprint by 2030.

13.6.3 Role of Traditional Knowledge

Climate variability of agriculture can be efficiently addressed through traditional farming practices since they are deeply ingrained in community-based ecological wisdom. The risk of climate extremities like droughts and floods can be countered with practices like mixed cropping, crop rotation, and agroforestry. Similarly, to address water scarcity and soil erosion, traditional practices include terrace farming, mulching, the use of farmyard and green manure, rainwater harvesting structures, tank irrigation, etc. In India a variety of structures like these have been used for years to harvest rainwater. The practice of using and protecting local seed varieties, which are well adapted to local environment maintains biodiversity. Soil carbon sequestration is achieved through composting and organic farming. Such traditional practices are not only ecofriendly but also economical. Such knowledge is transferred across generations and increases community participation and sustainability.

13.7 Way Forward

Regardless of multiple studies, there still exist various unanswered questions of how climate change drives socioeconomic changes by impacting agriculture and plant systems. Most assessments use global and regional models and quantify only major crop losses. There is a gap in predicting climate change shocks at the district or community scale. Most of the studies ignore focus on loss of micronutrients and long-term malnutrition. Little is known about secondary metabolites, crop–pest–disease interactions, and plant–microbe associations under multiple stresses. Socioeconomic impacts on landless laborers, women, and indigenous groups remain untapped. Studies predict reduction in losses suffered by 23–34% in adaptation scenario (Hultgren et al. 2025) but lack clarity on barriers like credit access, knowledge gaps, and incurred cost. The integration of climate-induced displacement and commensurate increase in poverty and political instability is faint. Moreover, there is no holistic valuation of traditional farming practices and indigenous knowledge as climate resilience measures. To address these gaps, the future research should focus on region-specific, high-resolution models that integrate biophysical and socioeconomic variables to predict agricultural outcomes under future climate scenarios. Investments in climate smart agriculture practices and green jobs should be increased. Formulation of gender-sensitive and inclusive adaptation programs, planned relocation strategies to target the vulnerable groups. Promotion of climate literacy, vocational training and 'Global South' knowledge sharing would ensure a climate secure future.

13.8 Conclusion

Global agriculture, food security, and livelihoods are all severely jeopardized by fluctuating climate. Increased CO_2 levels, elevated temperatures and irregular precipitation, all interrupt with crop growth, crop cycle, crop productivity and also deteriorate soil and water availability. Small-scale farmers and developing nations are disproportionately affected by these alterations, which additionally alter cultivation zones, elevate the risk of pests and diseases, destabilize the ecological balance. Socioeconomic impacts are severe, with crop failures, food insecurity, inflation, migration, and health risks, which promote inequality, poverty, and starvation. Vulnerable communities, particularly women and minority groups, face decreased adaptability and limited access to resources, further harming social cohesiveness and social growth. The capacity of resilience can be improved by adaptation and mitigation techniques, such as climate-smart agriculture, traditional agriculture practices, insurance plans, and global agreements like Paris Agreement. Future initiatives should focus on integrated, broad, region-specific approaches combining technological innovations, policy support, community participation, and knowledge sharing. Such concerted action is essential to ensure food security, sustainable agriculture, and equitable resilience in quickly evolving climate.

Acknowledgement The authors gratefully acknowledge the financial support provided by the Uttarakhand State Council for Science and Technology (UCOST), Dehradun, Uttarakhand, India, under Sanctioned Project, Order No. UCS&T/R&D-13/23-24/24525, which made this study possible.

References

Abbas M, Zhao L, Wang Y (2022) Perspective impact on water environment and hydrological regime owing to climate change: a review. Hydrology 9(11):203

Adams RM, Hurd BH, Lenhart S, Leary N (1998) Effects of global climate change on agriculture: an interpretative review. Clim Res 11(1):19–30

Adesete AA, Olanubi OE, Dauda RO (2023) Climate change and food security in selected Sub-Saharan African Countries: AA Adesete et al. Environ Dev Sustain 25(12):14623–14641

Ahanger RA, Bhat HA, Bhat TA, Ganie SA, Lone AA, Wani IA, Ganai SA, Haq S, Khan OA, Junaid JM, Bhat TA (2013) Impact of climate change on plant diseases. Int J Modern Plant Animal Sci 1(3):105–115

Ahmad QUA, Moors E, Biemans H, Shaheen N, Masih I, ur Rahman Hashmi MZ (2023) Climate-induced shifts in irrigation water demand and supply during sensitive crop growth phases in South Asia. Clim Change 176(11):150

Aryal JP, Sapkota TB, Rahut DB, Krupnik TJ, Shahrin S, Jat ML, Stirling CM (2020) Major climate risks and adaptation strategies of smallholder farmers in coastal Bangladesh. Environ Manag 66(1):105–120

Ashagidigbi WM, Orilua OO, Olagunju KA, Omotayo AO (2022) Gender, empowerment and food security status of households in Nigeria. Agriculture 12(7):956

Bankole OM (2020) The impact of climate change on food security: a case study of Karnataka, India. Int J Multidiscipl Educ Res 5(4):26–32

Benevolenza MA, DeRigne L (2019) The impact of climate change and natural disasters on vulnerable populations: a systematic review of literature. J Hum Behav Soc Environ 29(2):266–281

Bhat SU, Pandit AK (2020) Water quality assessment and monitoring of Kashmir Himalayan freshwater springs-a case study. Aquat Ecosyst Health Manage 23(3):274–287

Bibi F, Rahman A (2023) An overview of climate change impacts on agriculture and their mitigation strategies. Agriculture 13(8):1508

Birthal PS, Hazrana J, Negi DS, Bhan SC (2021) Climate change and land-use in Indian agriculture. Land Use Policy 109:105652

Bonfante A, Terribile F, Bouma J (2019) Refining physical aspects of soil quality and soil health when exploring the effects of soil degradation and climate change on biomass production: an Italian case study. Soil 5(1):1–14

Borrelli P, Robinson DA, Panagos P, Lugato E, Yang JE, Alewell C, Wuepper D, Montanarella L, Ballabio C (2020) Land use and climate change impacts on global soil erosion by water (2015-2070). Proc Natl Acad Sci 117(36):21994–22001

Chalise S, Naranpanawa A, Bandara JS, Sarker T (2017) A general equilibrium assessment of climate change–induced loss of agricultural productivity in Nepal. Econ Model 62:43–50

Chaudhry S, Sidhu GPS (2022) Climate change regulated abiotic stress mechanisms in plants: a comprehensive review. Plant Cell Rep 41(1):1–31

Christiansen DE, Markstrom SL, Hay LE (2011) Impacts of climate change on the growing season in the United States. Earth Interact 15(33):1–17

Couvreur TL, Dauby G, Blach-Overgaard A, Deblauwe V, Dessein S, Droissart V, Hardy OJ, Harris DJ, Janssens SB, Ley AC, Mackinder BA (2021) Tectonics, climate and the diversification of the tropical African terrestrial flora and fauna. Biol Rev 96(1):16–51

Dal Molin SJ, Ernani PR, Gerber JM (2020) Soil acidification and nitrogen release following application of nitrogen fertilizers. Commun Soil Sci Plant Anal 51(20):2551–2558

Danby RK, Hik DS (2007) Responses of white spruce (Picea glauca) to experimental warming at a subarctic alpine treeline. Glob Chang Biol 13(2):437–451

Duan H, Yuan D, Cai Z, Wang S (2022) Valuing the impact of climate change on China's economic growth. Econ Anal Policy 74:155–174

Elad Y, Pertot I (2014) Climate change impacts on plant pathogens and plant diseases. J Crop Improv 28(1):99–139

Ferdush J, Paul V, Varco J, Jones K, Sasidharan SM (2023) Consequences of elevated CO2 on soil acidification, cation depletion, and inorganic carbon: a column-based experimental investigation. Soil Tillage Res 234:105839

Fiorini ACO, Angelkorte G, Bakman T, Baptista LB, Cruz T, A Diuana F, Morais TN, Rathmann R, da Silva FT, Tagomori I, Zanon-Zotin M (2024) How climate change is impacting the Brazilian agricultural sector: evidence from a systematic literature review. Environ Res Lett 19(8):083001

Furtak K, Wolińska A (2023) The impact of extreme weather events as a consequence of climate change on the soil moisture and on the quality of the soil environment and agriculture–a review. Catena 231:107378

Gebrechorkos SH, Hülsmann S, Bernhofer C (2019) Changes in temperature and precipitation extremes in Ethiopia, Kenya, and Tanzania. Int J Climatol 39(1):18–30

Ghini R, Hamada E, Figueiredo JS (2011) Impacts of climate change on agriculture: potential effects on the coffee crop. Plant Dis 95(6):770–776

Gojon A, Cassan O, Bach L, Lejay L, Martin A (2023) The decline of plant mineral nutrition under rising CO2: physiological and molecular aspects of a bad deal. Trends Plant Sci 28(2):185–198

Gordo O, Sanz JJ (2010) Impact of climate change on plant phenology in Mediterranean ecosystems. Glob Chang Biol 16(3):1082–1106

Gregory PJ, Ingram JS, Brklacich M (2005) Climate change and food security. Philos Trans R Soc B, Biol Sci 360(1463):2139–2148

Guntukula R (2020) Assessing the impact of climate change on Indian agriculture: evidence from major crop yields. J Public Aff 20(1):e2040

Habib-ur-Rahman M, Ahmad A, Raza A, Hasnain MU, Alharby HF, Alzahrani YM, Bamagoos AA, Hakeem KR, Ahmad S, Nasim W, Ali S (2022) Impact of climate change on agricultural production; issues, challenges, and opportunities in Asia. Front Plant Sci 13:925548

Harvey JA, Tougeron K, Gols R, Heinen R, Abarca M, Abram PK, Basset Y, Berg M, Boggs C, Brodeur J, Cardoso P (2023) Scientists' warning on climate change and insects. Ecol Monogr 93(1):e1553

Hassan MU, Aamer M, Mahmood A, Awan MI, Barbanti L, Seleiman MF, Bakhsh G, Alkharabsheh HM, Babur E, Shao J, Rasheed A (2022) Management strategies to mitigate N2O emissions in agriculture. Life 12(3):439

Hassani A, Azapagic A, Shokri N (2021) Global predictions of primary soil salinization under changing climate in the 21st century. Nat Commun 12(1):6663

Hickling R, Roy DB, Hill JK, Fox R, Thomas CD (2006) The distributions of a wide range of taxonomic groups are expanding polewards. Glob Chang Biol 12(3):450–455

Högy P, Poll C, Marhan S, Kandeler E, Fangmeier A (2013) Impacts of temperature increase and change in precipitation pattern on crop yield and yield quality of barley. Food Chem 136(3–4):1470–1477

Hui D, Ray A, Kasrija L, Christian J (2024) Impacts of climate change and agricultural practices on nitrogen processes, genes, and soil nitrous oxide emissions: a quantitative review of meta-analyses. Agriculture 14(2):240

Hultgren A, Carleton T, Delgado M, Gergel DR, Greenstone M, Houser T, Hsiang S, Jina A, Kopp RE, Malevich SB, McCusker KE (2025) Impacts of climate change on global agriculture accounting for adaptation. Nature 642(8068):644–652

Huq S, Khan M, Islam AS, Mirza AB (2024) Climate change impacts in Bangladesh: what climate change means for a country and its people

IEA (2024) CO2 emissions in 2023. IEA, Paris

India Meteorological Department (IMD) (2022) Climate summary of India 2021: annual and seasonal mean temperature series (1901–2021). Ministry of Earth Sciences, Government of India

IPCC (2013) Climate change 2013: The physical science basis. Working Group I contribution to the fifth assessment report of the Intergovernmental Panel on Climate Change

IPCC (2021) Climate Change 2021 : The Physical Science Basis. Contribution of Working Group I to the Sixth Assessment Report of the Intergovernmental Panel on Climate Change

IPCC (2023) Climate change 2023: synthesis report. Contribution of Working Groups I, II and III to the Sixth Assessment Report of the Intergovernmental Panel on Climate Change

Kaushik B, Singh K, Tiwari DK, Singh UK (2023) Impact of climate change on crop yield due to pests and crop diseases: future projections. Microsc Microanal 29:56

Khan MA, Tahir A, Khurshid N, Husnain MIU, Ahmed M, Boughanmi H (2020) Economic effects of climate change-induced loss of agricultural production by 2050: a case study of Pakistan. Sustainability 12(3):1216

Kumar K, Parikh J (2001) Socio-economic impacts of climate change on Indian agriculture. Int Rev Environ Strateg 2(2)

Kumar V, Cheng SYC, Singh AK (2016) Impact of flood on rural population and strategies for mitigation: a case study of Darbhanga District, Bihar State, India. Contemp Rural Soc Work J 8(1):5

Lal R, Monger C, Nave L, Smith P (2021) The role of soil in regulation of climate. Philos Trans R Soc B 376(1834):20210084

Lan X, Trans P, Thoning K (2023) Trends in globally-averaged CO2 determined from NOAA Global Monitoring Laboratory Measurements

Lenoir J, Gégout JC, Marquet PA, de Ruffray P, Brisse H (2008) A significant upward shift in plant species optimum elevation during the 20th century. Science 320(5884):1768–1771

Lobell DB, Schlenker W, Costa-Roberts J (2011) Climate trends and global crop production since 1980. Science 333(6042):616–620

Luo Z, Feng W, Luo Y, Baldock J, Wang E (2017) Soil organic carbon dynamics jointly controlled by climate, carbon inputs, soil properties and soil carbon fractions. Glob Chang Biol 23(10):4430–4439

Lybbert TJ, Sumner DA (2012) Agricultural technologies for climate change in developing countries: policy options for innovation and technology diffusion. Food Policy 37(1):114–123

Ma S, Cao Y, Lu J, Ren T, Cong R, Lu Z, Zhu J, Li X (2024) Response of soil aggregation and associated organic carbon to organic amendment and its controls: a global meta-analysis. Catena 237:107774

Mahapatra B, Walia M, Rao CAR, Raju BMK, Saggurti N (2021) Vulnerability of agriculture to climate change increases the risk of child malnutrition: evidence from a large-scale observational study in India. PLoS One 16(6):e0253637

Malhi GS, Kaur M, Kaushik P (2021) Impact of climate change on agriculture and its mitigation strategies: a review. Sustainability 13(3):1318

Meyer P, Mwangi W, Kuzmina V (2023) Transboundary plant disease outbreaks and climate change: strategies for prevention. Fungal Biol Rev 42(4):243–257

Myers SS, Zanobetti A, Kloog I, Huybers P, Leakey AD, Bloom AJ, Carlisle E, Dietterich LH, Fitzgerald G, Hasegawa T, Holbrook NM (2014) Increasing CO2 threatens human nutrition. Nature 510(7503):139–142

NASA (2024) Global climate change: vital signs of the planet. National Aeronautics and Space Administration

Nepal S (2016) Impacts of climate change on the hydrological regime of the Koshi river basin in the Himalayan region. J Hydro Environ Res 10:76–89

Nikolaou G, Neocleous D, Christou A, Kitta E, Katsoulas N (2020) Implementing sustainable irrigation in water-scarce regions under the impact of climate change. Agronomy 10(8):1120

Oishy MN, Shemonty NA, Fatema SI, Mahbub S, Mim EL, Raisa MBH, Anik AH (2025) Unravelling the effects of climate change on the soil-plant-atmosphere interactions: a critical review. Soil Environ Health 3:100130

Okeleye SO, Okhimamhe AA, Sanfo S, Fürst C (2023) Impacts of land use and land cover changes on migration and food security of north central region, Nigeria. Land 12(5):1012

Ozkaynak E (2013) Effects of air temperature and hours of sunlight on the length of the vegetation period and the yield of some field crops. Ekoloji 22(87):58–63

Panda A, Sahu N, Behera S, Sayama T, Sahu L, Avtar R, Singh RB, Yamada M (2019) Impact of climate variability on crop yield in Kalahandi, Bolangir, and Koraput districts of Odisha, India. Climate 7(11):126

Parkes B, Buzan JR, Huber M (2022) Heat stress in Africa under high intensity climate change. Int J Biometeorol 66(8):1531–1545

Paudel Timilsena B, Niassy S, Kimathi E, Abdel-Rahman EM, Seidl-Adams I, Wamalwa M, Tonnang HE, Ekesi S, Hughes DP, Rajotte EG, Subramanian S (2022) Potential distribution of fall armyworm in Africa and beyond, considering climate change and irrigation patterns. Sci Rep 12(1):539

Pereira L (2017) Climate change impacts on agriculture across Africa. In: Oxford research encyclopedia of environmental science

Rahman SU, Han JC, Ahmad M, Ashraf MN, Khaliq MA, Yousaf M, Wang Y, Yasin G, Nawaz MF, Khan KA, Du Z (2024) Aluminum phytotoxicity in acidic environments: a comprehensive review of plant tolerance and adaptation strategies. Ecotoxicol Environ Saf 269:115791

Rajput VD, Minkina T, Kumari A, Harish, Singh VK, Verma KK, Mandzhieva S, Sushkova S, Srivastava S, Keswani C (2021) Coping with the challenges of abiotic stress in plants: new dimensions in the field application of nanoparticles. Plants 10(6):1221

Ramachandran N (2014) Factoring climate change into the nutrition scenario. In: Persisting undernutrition in India: causes, consequences and possible solutions. Springer India, New Delhi, pp 165–178

Rana AW (2019) Comprehensive institutional review for climate resilient agriculture. Intl Food Policy Res Inst

Rana RS, Bhagat RM, Kalia V, Lal H, Sen V (2013) Indigenous perceptions of climate change vis-a-vis Mountain Agricultural activities in Himachal Pradesh, India. Indian J Tradit Knowl 12(4):596–604

Rao CS, Baral K, Chanadana VM, Jagadesh M, Karthik R (2024) Climate change adaptation and mitigation in Indian agriculture. J Agrometeorol 26(2):137–148

Ray DK, Gerber JS, MacDonald GK, West PC (2015) Climate variation explains a third of global crop yield variability. Nat Commun 6(1):5989

Rezaei EE, Webber H, Asseng S, Boote K, Durand JL, Ewert F, Martre P, MacCarthy DS (2023) Climate change impacts on crop yields. Nat Rev Earth Environ 4(12):831–846

Rocha J, Oliveira S, Viana CM, Ribeiro AI (2022) Climate change and its impacts on health, environment and economy. In: One health. Academic Press, pp 253–279

Sam AS, Abbas A, Padmaja SS, Sathyan AR, Vijayan D, Kächele H, Kumar R, Müller K (2021) Flood vulnerability and food security in eastern India: a threat to the achievement of the sustainable development goals. Int J Disaster Risk Reduct 66:102589

Saravanakumar V, Lohano HD, Balasubramanian R (2022) A district-level analysis for measuring the effects of climate change on production of rice: evidence from Southern India. Theor Appl Climatol 150(3):941–953

Sarku R (2023) Farmers' perspectives on water availability in the lower Volta Delta region in Ghana. Reg Environ Chang 23(4):163

Schilling J, Hertig E, Tramblay Y, Scheffran J (2020) Climate change vulnerability, water resources and social implications in North Africa. Reg Environ Chang 20(1):15

Schlesinger WH, Reynolds JF, Cunningham GL, Huenneke LF, Jarrell WM, Virginia RA, Whitford WG (1990) Biological feedbacks in global desertification. Science 247(4946):1043–1048

Schmidhuber J, Tubiello FN (2007) Global food security under climate change. Proc Natl Acad Sci 104(50):19703–19708

Schweiger O, Settele J, Kudrna O, Klotz S, Kühn I (2008) Climate change can cause spatial mismatch of trophically interacting species. Ecology 89(12):3472–3479

Senapati AK (2022) Weather effects and their long-term impact on agricultural yields in Odisha, East India: agricultural policy implications using NARDL approach. J Public Aff 22(3):e2498

Shah AN, Tanveer M, Shahzad B, Yang G, Fahad S, Ali S, Bukhari MA, Tung SA, Hafeez A, Souliyanonh B (2017) Soil compaction effects on soil health and cropproductivity: an overview. Environ Sci Pollut Res 24(11):10056–10067

Shepard D (2019) Global warming: severe consequences for Africa: new report projects greater temperature increases. Africa Renewal 32(3):34–34

Shoaib M, Banerjee BP, Hayden M, Kant S (2022) Roots' drought adaptive traits in crop improvement. Plants 11(17):2256

Shrivastava P, Kumar R (2015) Soil salinity: a serious environmental issue and plant growth promoting bacteria as one of the tools for its alleviation. Saudi J Biol Sci 22(2):123–131

Singh N, Sharma DP, Chand H (2016) Impact of climate change on apple production in India: a review. Curr World Environ 11(1):251

Skendžić S, Zovko M, Živković IP, Lešić V, Lemić D (2021) The impact of climate change on agricultural insect pests. Insects 12(5):440

Sultan B, Gaetani M (2016) Agriculture in West Africa in the twenty-first century: climate change and impacts scenarios, and potential for adaptation. Front Plant Sci 7:1262

Sünnemann M, Beugnon R, Breitkreuz C, Buscot F, Cesarz S, Jones A, Lehmann A, Lochner A, Orgiazzi A, Reitz T, Rillig MC (2023) Climate change and cropland management compromise soil integrity and multifunctionality. Commun Earth Environ 4(1):394

Syed A, Raza T, Bhatti TT, Eash NS (2022) Climate impacts on the agricultural sector of Pakistan: risks and solutions. Environ Challenges 6:100433

Topali C, Antonopoulou C, Chatzissavvidis C (2024) Effect of waterlogging on growth and productivity of fruit crops. Horticulturae 10(6):623

Ullah A, Ahmad I, Ahmad A, Khaliq T, Saeed U, Habib-ur-Rahman M, Hussain J, Ullah S, Hoogenboom G (2019) Assessing climate change impacts on pearl millet under arid and semi-arid environments using CSM-CERES-Millet model. Environ Sci Pollut Res 26(7):6745–6757

Varanasi A, Prasad PV, Jugulam M (2016) Impact of climate change factors on weeds and herbicide efficacy. Adv Agron 135:107–146

Vedeld T, Salunke SG, Aandahl G, Lanjekar P (2014) Governing extreme climate events in Maharashtra, India. Final report on WP3. 2: extreme risks, vulnerabilities and community-based adaptation in India (EVA): a pilot study. CIENS-TERI

Vishnu CR, Sridharan R (2016) A case study on impact of Chennai floods: supply chain perspective. Ind Eng J 9(8):12–16

Wiebe K, Robinson S, Cattaneo A (2019) Climate change, agriculture and food security: impacts and the potential for adaptation and mitigation. In: Sustainable food and agriculture, pp 55–74

Wookey PA, Aerts R, Bardgett RD, Baptist F, Bråthen KA, Cornelissen JH, Gough L, Hartley IP, Hopkins DW, Lavorel S, Shaver GR (2009) Ecosystem feedbacks and cascade processes: understanding their role in the responses of Arctic and alpine ecosystems to environmental change. Glob Chang Biol 15(5):1153–1172

World Bank (2021) Indonesia marine economy: unlocking the potential of the blue economy. World Bank. Retrieved from. https://www.worldbank.org/en/news/press-release/2021/03/25/sustainable-ocean-economy-key-for-indonesia-prosperity

World Bank (2024) Kenya economic update: transforming agricultural productivity to achieve food security for all. World Bank. https://www.worldbank.org/en/country/kenya/publication/kenya-economic-update-transforming-agricultural-productivity-to-achieve-food-security-for-all

World Economic Forum (2024) Helping's agriculture sector build climate resilience. World Economic Forum. https://www.weforum.org/agenda/2024/08/income-insurance-india-climate-resilience/

World Meteorological Organization (2025) WMO Global Annual to Decadal Climate Update (2025–2029). World Meteorological Organization, Geneva

Xiao D, Zhang Y, Bai H, Tang J (2021) Trends and climate response in the phenology of crops in Northeast China. Front Earth Sci 9:811621

Yiadom EB, Dziwornu RK, Mawutor JK, Amankwah RF (2023) Exploring the relationship between extreme weather events, urbanization, and food insecurity: institutional quality perspective. Environ Challenges 13:100775

Zhang S, Zhu Q, de Vries W, Ros GH, Chen X, Muneer MA, Zhang F, Wu L (2023) Effects of soil amendments on soil acidity and crop yields in acidic soils: a world-wide meta-analysis. J Environ Manag 345:118531

Zhu K, Woodall CW, Clark JS (2012) Failure to migrate: lack of tree range expansion in response to climate change. Glob Chang Biol 18(3):1042–1052

Ziska LH (2001) Changes in competitive ability between a C4 crop and a C3 weed with elevated carbon dioxide. Weed Sci 49(5):622–627

Green Nanotechnology for Enhancing Plant Resilience

14

Pal Neha, Kausar Arisha, and Shahzad Anwar

Abstract

Climate change has intensified abiotic stresses such as drought, salinity, and extreme temperatures, posing severe threats to global agricultural productivity. Green nanotechnology offers innovative, sustainable solutions to enhance plant resilience against these challenges. This chapter explores eco-friendly nanoparticle synthesis methods, focusing on plant- and microbe-mediated approaches that reduce environmental footprints. It further elucidates how nanomaterials improve stress tolerance through enhanced nutrient uptake, modulation of antioxidant defense systems, and regulation of stress-responsive gene expression. Real-world applications are evaluated, including nano-fertilizers, nano-pesticides, and nano-sensors, demonstrating their potential to strengthen fragile agricultural systems. Additionally, the chapter addresses critical concerns regarding nanoparticle toxicity, environmental safety, scalability, and their role in supporting a circular economy. By linking research innovations to practical strategies, this work provides a roadmap for integrating nanotechnology into climate-resilient and sustainable farming practices. Future perspectives highlight the importance of interdisciplinary research and responsible deployment for the next generation of green nano-agriculture.

Keywords

Green nanotechnology · Abiotic stress · Plant resilience · Nano-fertilizers · Sustainable agriculture · Antioxidant defense

P. Neha (✉) · K. Arisha · S. Anwar
Plant Biotechnology Section, Department of Botany, Aligarh Muslim University, Aligarh, India

H. C. Joshi et al. (eds.), *Climate-Responsive Plant Biology*,
https://doi.org/10.1007/978-981-95-7354-7_14

14.1 Introduction

Nanobiotechnology has played a major role in advancing the production of metal oxide nanoparticles, providing biomedical and environmental applications compared to conventional physical and chemical techniques (Samuel et al. 2022). Abiotic stresses such as drought, salinity, pest infestations, and extreme temperature variations remain major threats to crop growth and agricultural productivity (Jahan et al. 2021; Alharbi et al. 2021; Hasan et al. 2020a, b). In recent years, several agricultural innovations such as nano-pesticides, nano-fertilizers, and nano-sensors have been developed. These nano technologies help farming work better and last longer by using fewer resources and creating less waste than usual farming methods (Servin et al. 2015; Liu and Lal 2015). Current advances in nanotechnology have introduced two primary mechanisms for pest management: the direct use of nanostructures as toxic agents and the application of nanocarriers to improve the efficacy of conventional pesticide compounds (Kah et al. 2021). Alongside pest control, several metallic nanoparticles—including silver (Ag NPs), gold (Au NPs), iron (Fe NPs), iron sulphide (FeS2 NPs), titanium dioxide (TiO2 NPs), copper oxide (CuO NPs), and zinc-based nanoparticles (Zn NPs, ZnO NPs)—have shown significant potential as nano-fertilizers. These nanoparticles have been shown to positively influence seed germination, promote plant growth, and improve stress tolerance in a variety of crops (Taran et al. 2017; Latef et al. 2017a). Silver nanoparticles have been extensively studied for their ability to improve seed germination (Ibrahim 2016; Singh et al. 2016b), support plant growth and development (Kim et al. 2018), and enhance physiological processes like gas exchange under stressful environmental conditions (Wang et al. 2020). However, with the growing stress on natural resources and the intensifying need to reduce post-harvest losses, there is an urgent demand for sustainable approaches that do not harm the environment. In this regard, integrating biofertilizers and biopesticides presents a sustainable solution, providing eco-friendly alternatives for maintaining agricultural productivity (Kumar and Singh 2015).

Biofertilizers play an important role in sustainable agriculture by improving nutrient availability and improving plant resilience under stress. Certain microorganisms play a crucial role by solubilizing key nutrients such as phosphorus and zinc, fixing atmospheric nitrogen, and enhancing the availability of other macro- and micronutrients for plants. This process supports plant growth even when facing biotic stress by ensuring sufficient nutrient availability (Singh et al. 2022). Additionally, these microbes enhance plant stress tolerance by inducing genes responsible for phytohormone production and stress-related metabolites (Moustafa-Farag et al. 2019). Under drought conditions, when water is scarce, biofertilizers can stimulate the production of important plant hormones such as cytokinins, gibberellins, abscisic acid, and indole-3-acetic acid. Together, these hormones promote root growth, increase root surface area, and encourage the development of lateral roots and root hairs, thereby improving the plant's capacity to absorb water from dry soil (Kenneth et al. 2019; Raza et al. 2019). Biopesticides, derived from natural organisms and substances such as plants and inorganic compounds, offer eco-friendly alternatives to chemical pesticides. They function through diverse modes of action, excluding neurotoxic mechanisms, making them safer for the environment and non-target organisms (Tijjani et al. 2016; Marrone 2019). Biopesticides are generally divided into three main categories: microbial biopesticides, biochemical biopesticides, and plant-incorporated protectants (Kumar 2012; Ibrahim and Shawer 2014; Leahy et al. 2014). Despite their environmental benefits, the broader adoption of biopesticides has been hindered by several limitations, including lower efficacy, short persistence, variable performance in field conditions, and relatively high production costs.

Nanotechnology can help make biopesticides work better and be safer for the environment. It allows these natural pesticides to be more effective at targeting pests while reducing harm to non-target plants, animals, and people. This means we can protect crops better with less environmental damage. By improving the stability of active ingredients and reducing toxicity, nanotechnology-based biopesticides offer targeted pest control with minimal ecological impact (Khot et al. 2012; Agrawal and Rathore 2014; Koul 2019). The use of biosynthesized nanoparticles in agricultural systems supports the goals of sustainable development, offering dual benefits in crop protection and productivity enhancement. Nanomaterials serve as efficient carriers of agrochemicals, enabling the precise delivery of nutrients to plants. This targeted approach not only promotes better crop growth and yield but also minimizes input waste. Additionally, nanotechnology supports the nano-biosensors developments, which is important for early detection and management of crop diseases and pests (Singh et al. 2015).

14.2 Improved Nutrient Uptake with Green Nanotechnology

Green-synthesized nanoparticles greatly enhance nutrient acquisition by plants. Unlike conventional fertilizers, nanoparticles such as zinc oxide (ZnO NPs), iron oxide (Fe_3O_4 sNPs), and silica-based nanomaterials exhibit high surface reactivity, improved solubility, and targeted delivery, allowing for more efficient absorption and transport of essential macro- and micronutrients (Rastogi et al. 2017; Naderi and Shahraki 2013). These nanomaterials promote root absorption due to their nanoscale size (around 1 to 100 nm) and expanded surface area, which enhances nutrient utilization, especially under stressful environmental conditions such as drought and salinity, which commonly impede standard fertilization strategies (Dimkpa and Bindraban 2018). For example, nano-formulations of iron and zinc have been found to improve chlorophyll biosynthesis and nitrogen assimilation, directly translating to improved plant vigor and yield.

14.3 Activation of Antioxidant Defense Systems

Plants exposed to abiotic stresses frequently accumulate reactive oxygen species (ROS), which cause cellular oxidative damage and impair growth. Green nanotechnology enhances the plant's natural antioxidant defense system. Specifically, applying nanoparticles like selenium (SeNPs), cerium oxide (CeO_2NPs), and zinc oxide (ZnONPs) has been linked to increased activity of essential antioxidant enzymes such as superoxide dismutase (SOD), catalase (CAT), and peroxidases (POD) (Iqbal et al. 2019; Zhang et al. 2022; Latef et al. 2017a, b). These enhancements help mitigate oxidative stress, stabilize biological membranes, and preserve metabolic homeostasis during critical periods of environmental stress, thereby improving plant stress tolerance.

14.4 Regulation of Stress-Responsive Genes

One of the most transformative impacts of nanotechnology on plant resilience is the modulation of stress-responsive gene expression. Treating plants with nanoparticles can trigger increased expression of genes that produce protective proteins and metabolites, including heat shock proteins, transcription factors like DREB and WRKY, and Osmoprotectants (Gohari et al. 2020). These genes play central roles in the perception and transduction of stress signals, fortifying plants against abiotic insults like drought, salinity, and temperature extremes. Furthermore, studies have shown that nanomaterials prime plants for faster and stronger defense responses, increasing their ability to survive and recover from environmental challenges.

The adoption of green nanotechnology in agriculture offers a promising approach to enhance plant resilience against various abiotic stresses. By boosting nutrient absorption, reinforcing antioxidant defenses, and modulating stress-responsive gene pathways, nano-based solutions are shaping the future of climate-resilient and sustainable farming.

Recent advancements in plant science have placed significant focus on nanobiotechnology, particularly its role in improving agricultural practices and enhancing plant productivity. The green synthesis of nanoparticles, also known as bio-reduction, can take place through either in vitro or in vivo methods. During this process, biological molecules such as enzymes, proteins, sugars, and various phytochemicals—including flavonoids, phenolics, terpenoids, and cofactors—act as natural agents that reduce and stabilize metal ions, facilitating nanoparticle formation (Prasad 2014).

14.5 Green Nanoparticle Synthesis Using Plant Extracts

Green synthesis using plant-based materials has become an attractive alternative to conventional chemical and microbial methods for nanoparticle production. This eco-friendly approach offers several advantages, including simplicity, rapid synthesis, low cost, non-toxicity, and environmental compatibility. Notably, plant-mediated synthesis tends to be more efficient and faster compared to microbial methods involving bacteria or fungi (Alabdallah and Hasan 2021). A diverse array of plant parts—including leaves, stems, flowers, seeds, bark, and shoots—have been effectively used in nanoparticle biosynthesis due to their abundant phytochemicals that aid in reducing and stabilizing metal ions (Kharissova et al. 2013). The general procedure involves collecting the appropriate plant material, followed by thorough washing with distilled water to eliminate surface contaminants (Roy and Das 2015). The cleaned plant parts are then shade-dried for 10 to 15 days to preserve active constituents, after which they are ground into a fine powder using a blender. To prepare the plant extract, 10 grams of the dried powder are boiled with 100 mL of distilled water, resulting in a bioactive broth. This extract is subsequently filtered to remove insoluble residues. For silver nanoparticle synthesis, a 1 mM solution of silver nitrate ($AgNO_3$) is added to the plant extract, initiating a reduction reaction. The reaction mixture is briefly incubated under agitation, typically in a shaking incubator, where a visible color change signifies the formation of nanoparticles via the reduction of Ag^+ ions to metallic Ag^0 (Fig. 14.1).

The synthesis process is tracked using UV-Visible (UV-Vis) spectroscopy, which detects the unique surface plasmon resonance bands of the nanoparticles. Subsequent characterization is performed using a range of analytical techniques to determine nanoparticle size, shape, distribution, and crystallinity (Roy and Das 2015). This eco-friendly synthesis method not only supports sustainability principles but also offers great promise for producing biocompatible nanoparticles on a large scale for various industrial uses.

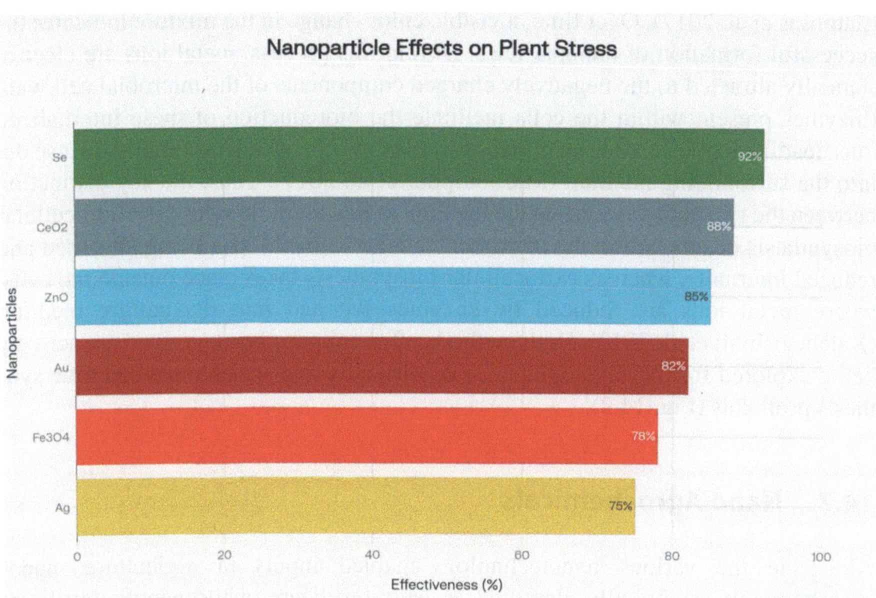

Fig. 14.1 Nanoparticle effects on plant stress

14.6 Microbial Synthesis of Nanoparticles

Microorganisms play a significant role in the green synthesis of nanomaterials through two principal biosynthetic routes: intracellular and extracellular synthesis, both of which depend on the ability of microbes to take up metal ions from their environment and enzymatically convert them into elemental nanoparticles (Li et al. 2011).

In the cell-free synthesis pathway, microbes are initially grown in a suitable nutrient medium. After reaching sufficient growth, the culture undergoes centrifugation to separate the microbial cells from the liquid supernatant. The resulting supernatant, which contains secreted enzymes, particularly reductases, serves as the reaction medium for nanoparticle biosynthesis (Das et al. 2014; Yadav et al. 2015; Al-Dhabi et al. 2018). Upon the addition of metal ions to this enzyme-rich filtrate, the enzymatic reduction of ions leads to the formation of nanoparticles in a cell-free environment (Maroofpour et al. 2019). Characterization of the synthesized nanoparticles is essential to determine their structural and functional attributes. Commonly used techniques such as X-ray diffraction (XRD), scanning electron microscopy (SEM), transmission electron microscopy (TEM), and Fourier transform infrared spectroscopy (FTIR) help in analyzing the morphology, crystallinity, surface chemistry, and distribution of nanoparticles (Jeevan et al. 2012).

In an alternative approach, intracellular synthesis uses the microbial cell machinery for nanoparticle formation. Microbial cultures are maintained in liquid media, after which the cells are harvested and washed thoroughly with sterile distilled water. The biomass pellet is then incubated with an aqueous solution of metal salts under controlled conditions (Castro et al. 2014; Shah et al. 2015; Fernández-Llamosas et al. 2017). Over time, a visible color change in the mixture indicates the successful formation of nanoparticles. During this process, metal ions are electrostatically attracted to the negatively charged components of the microbial cell wall. Enzymes present within the cells facilitate the bioreduction of these internalized ions, leading to the generation of nanoparticle clusters, which eventually diffuse out into the surrounding medium (Maroofpour et al. 2019). Thus, the key distinction between the two pathways lies in the location of nanoparticle synthesis. Intracellular biosynthesis occurs within the microbial cells, with metal ions being absorbed and reduced internally, whereas extracellular biosynthesis takes place outside the cells, where metal ions are reduced by enzymes secreted into the culture medium (Kalabegishvili et al. 2012). Both methods offer unique advantages and are actively being explored for the development of eco-friendly and scalable nanoparticle synthesis protocols (Fig. 14.2).

14.7 Nano Agrochemicals

Alongside the various nanotechnology-enabled inputs in agriculture, nano-agrochemicals are broadly classified as nano-fertilizers, while nanobiofertilizers and nano-pesticides also make up the broad category. These formulations often

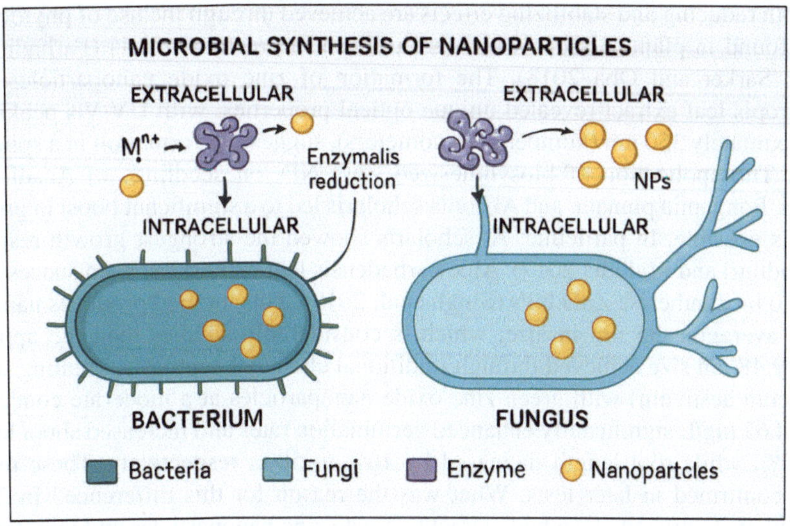

Fig. 14.2 Microbial synthesis of nanoparticles

incorporate nanoparticles as active components through methods such as nanoencapsulation, bioencapsulation, or nanocarrier-based delivery systems. Agricultural applications require improved precision, efficiency, and environmental protection (Mubeen et al. 2023; Tang et al. 2023). Several studies have highlighted the potential of nano-pesticides and nano-fertilizers as promising options for curbing global food insecurity while also supporting sustainable agricultural intensification. By facilitating the targeted delivery of nutrients and active compounds, nano-agrochemicals can enhance crop uptake efficiency and reduce off-target losses, leading to increased crop productivity with lower input use (Ali et al. 2023). Biological entities such as microorganisms, plant-derived extracts (e.g., from leaves, flowers, stems), or agro-industrial biowaste materials are utilized in the production of biogenic or green nano-aromatic compounds, which have high susceptibility to artificial origin. In accordance with circular economy and green chemistry, these nanomaterials that are biologically synthesized are an environmentally friendly and sustainable substitute for conventional synthetic agricultural products (Sharma and Kumar 2024; Sharma et al. 2023).

14.8 Nano Fertilizers

Sustainable agriculture has seen the rise of innovative technologies like nano-fertilizers that deliver specific nutrients. This is an exciting development. Their capacity to deliver nutrients precisely to the intended site in a controlled way is mainly because of their extensive surface area and excellent ability to adsorb substances. The bioactivity and environmental compatibility of zinc oxide nanoparticles (ZnO NPs) synthesized through green methods have led to significant attention.

Both reducing and stabilizing effects are achieved through the use of phytochemicals found in plant extracts, which are utilized in green synthesis (Farhadi et al. 2019; Sarker and Oba 2018). The formation of zinc oxide nanoparticles from Calotropis leaf extract revealed unique optical properties, with UV-Vis peaking at approximately 350 nm (number of nanometers), suggesting formation in a spherical form. The application of biosynthesized ZnO NPs on seedlings of Azadirachta indica, Pongamia pinnata, and Alstonia scholaris led to a significant boost in growth. This is possible. In particular, A. scholaris showed the strongest growth response (Chaudhuri and Malodia 2017). Aloe barbadensis leaf extract has been successfully used to biosynthesise ZnO NPs (Singh et al. 2019). This method produces nanoparticles averaging 35 nm in size, which is considerably smaller than the approximately 48 nm size achieved through traditional chemical synthesis. Treating wheat (Triticum aestivum) with green zinc oxide nanoparticles at a moderate concentration of 62 mg/L significantly enhanced germination rates and increased shoot length by 50%, while root length increased by 10% to 50%, respectively. These results were confirmed in later tests. What was the reason for this difference? In 2022, Ahmed et al. utilized clove bud extract to create zinc nanoparticles and investigated their impact on pea growth patterns and yield. Using concentrations of between 400 and 600 parts per million (ppm) led to the most significant improvements in both growth and crop yield, surpassing traditional zinc sulphate treatments.

Nanoparticles can be produced efficiently in microbial systems, just like in plants. Several microorganisms, particularly those that thrive in metal-rich environments, are highly resistant to metal chelation mediated by intracellular and extracellular proteins. During the biosynthesis of metallic nanoparticles like silver and gold (the most abundant form in alloys), copper (in the rarest range), and zinc, based on their relative abundance in carbon-resistant (Khan et al. 2018). Such processes are similar to natural biomineralization mechanisms and provide an environmentally friendly and affordable alternative to conventional chemical synthesis (Kato and Suzuki 2020; Burketová et al. 2022).

14.9 Nano Pesticides

In recent years, the use of metallic nanoparticles synthesized by biological methods has increased, such as Ag NPs and Au PNs (alloyed aromatic carbon), to control pests or pathogens in agriculture. Several studies have revealed their ability to combat bacterial and fungal infections, making them a valuable substitute for conventional chemical pesticides. The biosynthesis of spherical Au NPs and Ag PNs was accomplished by Costa et al. (2020), who identified the use of the archaeon Haloferax volcanii. Two Gram-negative bacterial strains were identified and successfully treated with nanoparticles, suggesting their potential as broad-spectrum antibacterials. Non-living materials with high levels of antimicrobial properties were created through the extracellular synthesis of Ag nanoparticles from Sporosarcina koreensis strain DC4 (Singh et al. 2016a, b). Biologically synthesized Ag NPs not only express antibacterial effects but also demonstrate significant antifungal activity. The effectiveness of reducing the size of Ag NPs from Prussian (Ibrahim et al.) poae, an important fungal pathogen in cereal grains, was demonstrated by their ability to produce nanometers (Ag NR) sizes between 20 and 50 (Pseudomonas poae CO. 2020). Similarly, Streptomyces sp., an example of the a halogenous bacterium, synthesized large aggregate Ag NPs with a diameter of 16 nm that demonstrated potent fungicidal activity against Fusarium verticillioides in another investigation. Fungal growth was hindered by nanoparticles, which prevented the biosynthesis of ergosterol and compromised the integrity of the membrane (Marathe et al. 2021). Why did this happen? The quick expansion and effective capture of metal ions in microalgae have made them ideal biofactories for nanoparticles. In their research, Rauwel et al. (2015) highlighted the potential of algal systems to produce Ag NP in nanoscale settings, with emphasis on the increased surface area of silver and environmental sustainability. The synthesis of Ag NPs by Chlorella vulgaris, a green microalga, was demonstrated in practice by El-Moslamy et al. (2016). This process is well-known. Nanoparticles demonstrated a potent antifungal activity against Alternaria alternata, which is responsible for leaf spot and plant rot, suggesting their potential in controlling foliar diseases.

A recent study has shown that plant-based nanoformulations are effective for sustainable pest and disease control. This is supported by experimental evidence. By using extracts of Cymbopogon citratus and Prosopis juliflora, Iqbal et al. (2022) were able to control Bemisia tabaci in eggplant by 91.24% through the use of stable neem oil nanoemulsion, which decreased surface tension but improved adhesion to leaves. The effectiveness and bioavailability of nanoagents were emphasized by Wuryantini et al. (2020), who found that tobacco nanoemulsions performed better than crude extracts in controlling citrus phloxи. Antifungal properties have been demonstrated in ZnO nanoparticles that were synthesized in green, particularly for disease control. Through the use of Terminalia bellericum leaf extract, Dhiman et al. (2021) identified ZnO nanoparticles that exhibit potent antioxidant properties against both leaf spot and blight. In a comparison of ZnO-based nanobiohybrids from garlic extract and chemically synthesized ZNO NPs, Arciniegas-Grijalba et al. (2019) found that fungal inhibition rates were 72.4% and 87.1%, respectively. The authors report both types were effective as nanofungicides.

14.10 Nano Sensors

Nano sensors are tiny gadgets that utilize nanotechnology to detect and measure changes in their surroundings. They are particularly suited to monitoring soil conditions (temperature, moisture levels, and nutrient levels). Efficacy and adaptability are the main factors driving interest in Nano biosensors, among others. Essentially, these are small analytical instruments that incorporate a biological recognition element into essentially an analogous system for signal conversion. Through this system, they can recognize specific biological or chemical substances and translate the interaction into a measurable signal. The performance of Nano biosensors varies depending on the context, offering either precise numerical data or semi-quantitative data (Lu et al. 2020; Mokhtarzadeh et al. 2015; Nosrati et al. 2017). Molecular or cellular responses to different types of stress caused by specific stimuli are monitored and quantified through the use of nano biosensors. Nanosensor applications in agriculture have become more prevalent in recent times. Their sensitivity and signal resolution are greatly improved by their specialized nano-interface, which allows them to detect chemical changes at the molecular scale (Kumar et al. 2019). The interface is essential for signal recognition and amplification. According to Ballesteros et al. (2020), advanced electrical components can rapidly convert plant signals into digital data. The way a biosensor transmits signals determines its classification into types such as optical, electrochemical, piezoelectric, pyroelectric, electronic, or gravimetric biosensors, which are suitable for specific sensing purposes (Damborský et al. 2016). The development of a multimodal nanosensor was achieved by Satapathi et al. (2018) through the use of multiple steps of synthetic technology, which included utilizing scleral silica to surround magnetic (Fe2O3) nanoparticles, incorporating an essentially unidirectional spacer arm, and detecting/removing corresponding fluorescent quantum dots. Quantum dots were chosen for their exceptional photophysical and chemical properties, which allowed them to be sensitive to metal ions (Mancini et al. 2008; Yang et al. 2018). A major disadvantage of quantum dots in real-world applications is that it's difficult to separate and recover them, a process that is often overly laborious, time-consuming, and tedious. Pathogen detection involves the use of biosensing elements such as bacterial receptors, antibodies, and lectins. The simplicity of incorporating them into biosensor systems for pathogen analysis and their adaptability make these components widely used (El Ichi et al. 2014; Liu et al. 2014; Li et al. 2014) developed a multichannel sensor that uses gold nanoparticles to identify biofilms and differentiate them based on their physicochemical characteristics. This device is the first of its kind. Its high sensitivity is evident through its ability to distinguish between six distinct biofilm types. Phillips et al. (2008) created a sensor that could identify three different strains of E. coli quickly using hydrophobically modified gold nanoparticles. This is similar in concept to their current application. Silver nanotube-based biosensors have the potential to identify organophosphorus pesticides in agricultural products and surrounding environments (Wu et al. 2011). Disulfide bond regions in the CD4-binding domain of host viruses can be targeted by chitosan-based biosensors, which inhibit both RNA synthesis and the release of extracellular divalent viral agents, consequently, suggesting another means of interfering with viral genetic material. Moreover, there are numerous nanosensors that can identify environmentally toxic pesticides. A pesticide can be identified using the detection of fluorescent nanoparticles, as an illustration. Ytterbium (III) oxide is coated with 3-aminopropyltriethoxysilane and synthesized through a hydrothermal process, as described by Kumar et al. (2020). Electrochemical luminescence sensors can detect glyphosate in environmental samples. They are made by layering graphene, gold nanoparticles, luminol, L-cysteine, and copper (Liu et al. 2020).

14.11 Plants Are Responsible for the Uptake, Transport, and Storage of Nanoparticles

Nanotechnology has quickly become a game-changer in various fields, including agriculture, where engineered NPs are increasingly being investigated for their ability to increase crop productivity, stress tolerance, and nutrient uptake. However, biological interactions between NPs and plants are complex systems dependent on many factors. Key factors include the physical characteristics of nanoparticles, their mode (soil, foliar, or hydroponic), and the concentration employed. However, SO2 is unknown. The primary interface for nanoparticle uptake in higher plants via soil or hydroponic systems is the roots. NPs can be internalized by plants through two main channels (Table 14.1):

1. *Apoplastic Pathway*

This route involves the passive movement of nanoparticles through pores or channels on the root surface. Once NPs enter the root apoplast, they traverse the cell wall and enter the intercellular spaces. Their movement is driven by physical forces such as capillary action, osmotic gradients, and pressure differentials.

2. *Symplastic Pathway*

Unlike the so-called apoplastic route, in the symplastic pathway, nanoparticles are transported from one cell to another either actively or with assistance. Upon

Table 14.1 Summarizing the topic "Uptake, Transport, and Translocation of Nanoparticles in Plants"

Pathway	Description	Mechanism	Factors influencing uptake
Apoplastic pathway	Passive movement of nanoparticles through root surface pores, cell walls, and intercellular spaces	Driven by physical forces such as capillary action, osmotic gradients, and pressure differentials	Particle size, surface charge
Symplastic pathway	Active or facilitated transport from cell to cell through the cytoplasm and plasmodesmata	Transported via intracellular cytoplasm and plasmodesmata to adjacent cells and xylem vessels for long-distance movement	Particle size, surface charge
Foliar uptake	Entry through stomatal openings or the cuticle layer on leaves	Penetration via stomata (>10 nm) or diffusion across the cuticle	Leaf morphology, cuticle thickness, trichomes, leaf exudates
Uptake factors	Size, charge, plant species, and application method influence NP uptake and translocation	Smaller NPs (3–5 nm) penetrate root epidermal cells directly, with mobility influenced by charge	Species-specific responses and NP physicochemical properties

initial entry into the root epidermal cells, NPs are then transported to the cytoplasm and transmitted through plasmodesmata to adjacent cells. This route enables deeper and more targeted translocation within plant tissues. The movement of nanoparticles from root systems to aerial parts is dependent on the presence of xylem vessels (Lv et al. 2019). According to Tripathi et al. (2017), nanoparticles' size is a key factor in their movement through plant tissues. Root epidermal cells can be penetrated by smaller nanoparticles, which are usually within the 3–5 nm range and may benefit from direct contact through osmotic and capillary forces (Lin and Du 2008). Larger particles are less bioavailable and more easily broken down than these ultrafine particles. NP uptake and translocation are significantly impacted by surface charge. Why is this so? Nevertheless, it is often the case that negatively charged particles are more efficiently transported between roots and shoots in plants, indicating that charge has different roles in vascular system mobility patterns.

Direct absorption of nanoparticles onto leaves is a distinct process from their application to roots. Why? By means of stomatal openings or diffusion across the cuticle layer, foliar-applied NPs can penetrate the plant surface. During transpiration, when the stomatal pores are open, nanoparticles with dimensions greater than 10 nm typically enter through a passage in cytoplasmic membranes. Following entry into the body, NPs travel through the apoplastic and other vascular tissues before being transported to the systemic tissues (Ruttkay-Nedecky et al. 2017). Observational characteristics such as leaf shape, cuticular size, and the presence of trichomes or chemical composition of leaf exudates and waxed substances also affect nanoparticle retention and entry (Larue et al. 2014). Foliar NP delivery strategies are more complex due to the significant differences in these factors among different plant species and stages of development. Nanoparticle exposure has been observed to affect specific species like Au NPs (*Oryza sativa*) (Zhu et al. 2012). Furthermore, Au NPs with positive charges are absorbed by the root system, while negatively charged ones are transferred to air. Supplementary research has revealed the utilization and transmission of various engineered nanomaterials in plant operations. It was discovered that Glycine max is capable of absorbing ZnO and Cerium dioxide (CeO2) nanoparticles from the soil. Specifically, CeO2 was transported as intact nanoparticles, and Zn underwent biotransformation into zinc-citrate complexes within plant tissues. It has been shown that CuO NPs move from roots to shoots in *Zea mayus*, and interestingly, they also travel back through the xylem via Phloem (Wang et al. 2012). Essentially, the uptake and mobility of nanoparticles in plants are dependent on several factors, including the application method (such as preparation or scanning for materials), particle size, amount of charge, type of plant material, and the mechanism of internal transport.

14.12 Challenges and Risks

A list of the "Challenges and Risks" of biogenic nano-agrochemicals is presented, with all references arranged organically to match book chapters (Table 14.2).

Table 14.2 synthesizes the challenges, along with corresponding references, that align with published scholarly reports and reviews on biogenic nano-agrochemicals

Challenges and Risks	Details	References
Lower toxicity and eco-friendly synthesis	Biogenic nano-agrochemicals reduce toxicity and are cost-effective due to the use of biological resources	Prokisch et al. (2024)
Broad-spectrum efficacy	Effective against a range of pathogens, including bacteria and fungi	Prokisch et al. (2024)
Ecological risk with excessive use	Overuse can cause ecological imbalance; risk depends on nanoparticle size, shape, and biodegradability	Prokisch et al. (2024), Chaud et al. (2021)
Cellular toxicity	Nanoparticles may disrupt protein transport, membrane permeability, and core cell functions	Chaud et al. (2021)
Unknown long-term ecological fate and mobility	Fate, transport, and toxicological effects in soil and environment remain unclear	Prokisch et al. (2024)
Potential human health risks	Need for comprehensive risk assessment due to possible toxicity and genotoxicity	Chaud et al. (2021)
Knowledge gaps about chronic impacts on soil health, plant systems, and food safety	Long-term effects on agricultural ecosystems and food chain safety are not fully understood	Prokisch et al. (2024), Garg et al. (2024)
Importance of establishing non-toxic thresholds	Defining safe size and concentration levels for nanoparticles to enable safe application	Prokisch et al. (2024)
Need for balanced and judicious use to safeguard environment and human health	Proper selection and dosage are critical to minimize negative impacts	Prokisch et al. (2024), Chaud et al. (2021)
Possibility of nanomaterials adding to environmental pollution if not effectively managed	Improper use can lead to bioaccumulation and contamination of soil and water	Chaud et al. (2021)
Challenges in safety assessment due to the complexity of nanomaterials	Variability in size, charge, and structure makes universal prediction of toxicity difficult	Chaud et al. (2021)

14.13 Conclusion

Biogenic nano-agrochemicals have many potential advantages over conventional methods.... They are very promising. Their toxicity, cost-effectiveness, and ability to combat various pests and diseases make them ideal tools for sustainable farming. They positively impact the environment as they are more environmentally friendly. However, there are still hurdles to be overcome. The characteristics of these nanoparticles may make them hazardous to the environment or living organisms if used excessively or misused. Our comprehension of their long-term impacts on soil health, plants, and food safety is still limited. Careful handling and monitoring of these nanoparticles are essential.

14.14 Future Perspectives

Looking ahead, it's crucial to conduct thorough safety assessments and establish clear guidelines for the safe levels of nanoparticles used in agriculture. Research should focus on understanding how these tiny particles move and change over time in the environment to avoid any unintended harm. Advances in designing nanoparticles with improved shapes, surfaces, and biodegradable properties will help deliver nutrients or pesticides more precisely and minimize side effects. By combining knowledge from different fields, including medicine, we can develop smarter, safer, and more sustainable nano-agrochemicals that are acceptable to farmers and consumers alike. Using natural and sustainable materials, involving the public, and working closely with regulators will be key to responsibly leveraging nanotechnology for feeding the global population while protecting our planet.

References

Agrawal S, Rathore P (2014) Nanotechnology pros and cons to agriculture: a review. Int J Curr Microbiol App Sci 3(3):43–55

Alabdallah NM, Hasan MM (2021) Plant-based green synthesis of silver nanoparticles and their effective role in abiotic stress tolerance in crop plants. Saudi J Biol Sci 28(10):5631–5639

Al-Dhabi NA, Mohammed Ghilan AK, Arasu MV (2018) Characterization of silver nanomaterials derived from marine Streptomyces sp. al-dhabi-87 and its *in vitro* application against multidrug-resistant and extended-spectrum beta-lactamase clinical pathogens. Nanomaterials. 8(5):279

Alharbi BM, Elhakem AH, Alnusairi GS, Soliman MH, Hakeem KR, Hasan MM, Abdelhamid MT (2021) Exogenous application of melatonin alleviates salt stress-induced decline in growth and photosynthesis in *Glycine max* (L.) seedlings by improving mineral uptake, antioxidant, and glyoxalase system. Plant Soil Environ 67(4):208–220

Ali S, Ahmad N, Dar MA, Manan S, Rani A, Alghanem SM, Khan KA, Sethupathy S, Elboughdiri N, Mostafa YS, Alamri SA (2023) Nano-agrochemicals as substitutes for pesticides: prospects and risks. Plants 13(1):109

Arciniegas-Grijalba PA, Patiño-Portela MC, Mosquera-Sánchez LP, Sierra BG, Muñoz-Florez JE, Erazo-Castillo LA, Rodríguez-Páez JE (2019) ZnO-based nanofungicides: synthesis, characterization, and their effect on the coffee fungi *Mycena citricolor* and *Colletotrichum* sp. Mater Sci Eng C 98:808–825

Ballesteros R, Intrigliolo DS, Ortega JF, Ramírez-Cuesta JM, Buesa I, Moreno MA (2020) Vineyard yield estimation by combining remote sensing, computer vision and artificial neural network techniques. Precis Agric 21(6):1242–1262. https://doi.org/10.1007/s11119-020-09717-3

Burketová L, Martinec J, Siegel J, Macůrková A, Maryška L, Valentová O (2022) Noble metal nanoparticles in agriculture: impacts on plants, associated microorganisms, and biotechnological practices. Biotechnol Adv 58:107929

Castro L, Blázquez ML, Munoz JA, González FG, Ballester A (2014) Mechanism and applications of metal nanoparticles prepared by bio-mediated process. Rev Adv Sci Eng 3(3):199–216

Chaudhuri SK, Malodia L (2017) Biosynthesis of zinc oxide nanoparticles using leaf extract of *Calotropis gigantea*: characterization and its evaluation on tree seedling growth in nursery stage. Appl Nanosci 7(8):501–512

Costa MI, Álvarez-Cerimedo MS, Urquiza D, Ayude MA, Hoppe CE, Fasce DP, De Castro RE, Giménez MI (2020) Synthesis, characterization, and kinetic study of silver and gold nanoparticles produced by the archaeon *Haloferax volcanii*. J Appl Microbiol 129(5):1297–1308

Damborský P, Švitel J, Katrlík J (2016) Optical biosensors. Essays Biochem 60(1):91–100. https://doi.org/10.1042/EBC20150010

Das VL, Thomas R, Varghese RT, Soniya EV, Mathew J, Radhakrishnan EK (2014) Extracellular synthesis of silver nanoparticles by the *Bacillus* strain CS 11 isolated from an industrialized area. 3 Biotech. 4(2):121–126

Dhiman S, Singh S, Varma A, Goel A (2021) Phytofabricated zinc oxide nanoparticles as a nanofungicide for management of *Alternaria* blight of *Brassica*. Biometals 34(6):1275–1293

Dimkpa CO, Bindraban PS (2018) Nanofertilizers: new products for the industry? J Agric Food Chem 66(26):6462–6473. https://doi.org/10.1021/acs.jafc.8b02042

El Ichi S, Leon F, Vossier L, Marchandin H, Errachid A, Coste J, Jaffrezic-Renault N, Fournier-Wirth C (2014) Microconductometric immunosensor for label-free and sensitive detection of Gram-negative bacteria. Biosens Bioelectron 54:378–384. https://doi.org/10.1016/j.bios.2013.11.016

El-Moslamy S, Kabeil S, Hafez EJ (2016) Bioprocess development for Chlorella vulgaris cultivation and biosynthesis of anti-phytopathogens silver nanoparticles. J Nanomater Mol Nanotechnol 5: 1. J Nanomater Mol Nanotechnol 9:2

Farhadi F, Khameneh B, Iranshahi M, Iranshahy M (2019) Antibacterial activity of flavonoids and their structure–activity relationship: an update review. Phytother Res 33(1):13–40

Fernández-Llamosas H, Castro L, Blázquez ML, Díaz E, Carmona M (2017) Speeding up bioproduction of selenium nanoparticles by using *Vibrio natriegens* as microbial factory. Sci Rep 7(1):16046

Gohari G, Mohammadi A, Darvishzadeh R, Farsi M (2020) Transcriptional regulation of stress-responsive genes in plants by engineered nanoparticles. Plant Physiol Biochem 156:343–353. https://doi.org/10.1016/j.plaphy.2020.09.005

Hasan MM, Alharby HF, Uddin MN, Ali MA, Anwar Y, Fang XW, Hakeem KR, Alzahrani Y, Hajar AS (2020a) Magnetized water confers drought stress tolerance in *Moringa* biotype via modulation of growth, gas exchange, lipid peroxidation and antioxidant activity. Pol J Environ Stud 1:29

Hasan MM, Ali MA, Soliman MH, Alqarawi AA, Abd_Allah EF, Fang XW (2020b) Insights into 28-homobrassinolide (HBR)-mediated redox homeostasis, AsA–GSH cycle, and methylglyoxal detoxification in soybean under drought-induced oxidative stress. J Plant Interact 15(1):371–385

Ibrahim EA (2016) Seed priming to alleviate salinity stress in germinating seeds. J Plant Physiol 192:38–46

Ibrahim RA, Shawer DM (2014) Transgenic Bt-plants and the future of crop protection (an overview). Int J Agric Food Res 3(1):14–40

Iqbal M, Khan S, Ansari MI (2019) Selenium nanoparticles ameliorate oxidative stress and moderate antioxidant defense and secondary metabolism in wheat under drought stress. Plant Physiol Biochem 146:1–10. https://doi.org/10.1016/j.plaphy.2019.11.010

Iqbal N, Hazra DK, Purkait A, Agrawal A, Kumar J (2022) Bioengineering of neem nano-formulation with adjuvant for better adhesion over applied surface to give long term insect control. Colloids Surf B Biointerfaces 209:112176

Jahan MS, Shu S, Wang Y, Hasan MM, El-Yazied AA, Alabdallah NM, Hajjar D, Altaf MA, Sun J, Guo S (2021) Melatonin pretreatment confers heat tolerance and repression of heat-induced senescence in tomato through the modulation of ABA-and GA-mediated pathways. Front Plant Sci 12:650955

Jeevan P, Ramya K, Rena AE (2012) Extracellular biosynthesis of silver nanoparticles by culture supernatant of *Pseudomonas aeruginosa*. Indian J Biotechnol 11(1):72–76

Kah M, Johnston LJ, Kookana RS, Bruce W, Haase A, Ritz V, Dinglasan J, Doak S, Garelick H, Gubala V (2021) Comprehensive framework for human health risk assessment of nanopesticides. Nat Nanotechnol 16(9):955–964

Kalabegishvili TL, Kirkesali EI, Rcheulishvili AN, Ginturi EN, Murusidze IG, Pataraya DT, Gurielidze MA, Tsertsvadze GI, Gabunia VN, Lomidze LG, Gvarjaladze DN (2012 Feb)

Synthesis of gold nanoparticles by some strains of *Arthrobacter* genera. J Mater Sci Eng A: Struct Mater: Prop Microstruct Process 2(2):164–173

Kato Y, Suzuki M (2020) Synthesis of metal nanoparticles by microorganisms. Crystals 10(7):589

Kenneth OC, Nwadibe EC, Kalu AU, Unah UV (2019) Plant growth-promoting rhizobacteria (PGPR): a novel agent for sustainable food production. Am J Agric Biol Sci 14(35):54

Khan T, Abbas S, Fariq A, Yasmin A (2018) Microbes: Nature's cell factories of nanoparticles synthesis. In: Exploring the realms of nature for nanosynthesis. Springer International Publishing, Cham, pp 25–50

Kharissova OV, Dias HR, Kharisov BI, Pérez BO, Pérez VM (2013) The greener synthesis of nanoparticles. Trends Biotechnol 31(4):240–248

Khot LR, Sankaran S, Maja JM, Ehsani R, Schuster EW (2012) Applications of nanomaterials in agricultural production and crop protection: a review. Crop Prot 35:64–70

Kim DY, Kadam A, Shinde S, Saratale RG, Patra J, Ghodake G (2018) Recent developments in nanotechnology transforming the agricultural sector: a transition replete with opportunities. J Sci Food Agric 98(3):849–864

Koul O (ed) (2019) Nano-biopesticides today and future perspectives. Academic Press

Kumar S (2012) Biopesticides: a need for food and environmental safety. J Biofertil Biopestic 3(4):1–3

Kumar S, Singh A (2015) Biopesticides: present status and the future prospects. J Fertil Pestic 6(2):1–2

Kumar S, Nehra M, Dilbaghi N, Marrazza G, Hassan AA, Kim KH (2019) Nano-based smart pesticide formulations: emerging opportunities for agriculture. J Control Release 294:131–153. https://doi.org/10.1016/j.jconrel.2018.12.012

Kumar S, Sachdeva S, Chaudhary S, Chaudhary GR (2020) Assessing the potential application of bio-compatibly tuned nanosensor of Yb2O3 for selective detection of imazapyr in real samples. Colloids Surf A Physicochem Eng Asp 593:124612. https://doi.org/10.1016/j.colsurfa.2020.124612

Larue C, Castillo-Michel H, Sobanska S, Cécillon L, Bureau S, Barthès V, Ouerdane L, Carrière M, Sarret G (2014) Foliar exposure of the crop *Lactuca sativa* to silver nanoparticles: evidence for internalization and changes in Ag speciation. J Hazard Mater 264:98–106

Latef AA, Alhmad MA, Abdelfattah KE (2017a) The possible roles of priming with ZnO nanoparticles in mitigation of salinity stress in lupine (*Lupinus termis*). Plants 36:60–70. https://doi.org/10.1007/s00344-016-9618-x

Latef AAHA, Alhmad MFA, Qahtan AA (2017b) The interactive effect of zinc oxide nanoparticles and salinity on physiological and molecular responses of wheat (Triticum aestivum L.). Plant Physiol Biochem 123:1–13. https://doi.org/10.1016/j.plaphy.2017.12.004

Leahy J, Mendelsohn M, Kough J, Jones R, Berckes N (2014) Biopesticide oversight and registration at the US Environmental Protection Agency. In: Biopesticides: state of the art and future opportunities, pp 3–18

Li X, Xu H, Chen ZS, Chen G (2011) Biosynthesis of nanoparticles by microorganisms and their applications. J Nanomater 2011(1):270974

Li BB, Clements WR, Yu XC, Shi K, Gong Q, Xiao YF (2014) Single nanoparticle detection using split-mode microcavity Raman lasers. Proc Natl Acad Sci 111(41):14657–14662. https://doi.org/10.1073/pnas.1408453111

Liu R, Lal R (2015) Potentials of engineered nanoparticles as fertilizers for increasing agronomic productions. Sci Total Environ 514:131–139

Liu X, Lei Z, Liu F, Liu D, Wang Z (2014) Fabricating three-dimensional carbohydrate hydrogel microarray for lectin-mediated bacterium capturing. Biosens Bioelectron 58:92–100. https://doi.org/10.1016/j.bios.2014.02.056

Liu H, Chen P, Liu Z, Liu J, Yi J, Xia F, Zhou C (2020) Electrochemical luminescence sensor based on double suppression for highly sensitive detection of glyphosate. Sensors Actuators B Chem 304:127364. https://doi.org/10.1016/j.snb.2019.127364

Lu Y, Yang Q, Wu J (2020) Recent advances in biosensor-integrated enrichment methods for preconcentrating and detecting the low-abundant analytes in agriculture and food samples. TrAC Trends Anal Chem 128:115914. https://doi.org/10.1016/j.trac.2020.115914

Lv J, Christie P, Zhang S (2019) Uptake, translocation, and transformation of metal-based nanoparticles in plants: recent advances and methodological challenges. Environ Sci Nano 6(1):41–59

Mancini MC, Kairdolf BA, Smith AM, Nie S (2008) Oxidative quenching and degradation of polymer-encapsulated quantum dots: new insights into the long-term fate and toxicity of nanocrystals in vivo. J Am Chem Soc 130(33):10836–10837. https://doi.org/10.1021/ja8040477

Marathe K, Naik J, Maheshwari V (2021) Biogenic synthesis of silver nanoparticles using Streptomyces spp. and their antifungal activity against *Fusarium verticillioides*. J Clust Sci 32(5):1299–1309

Marooufpour N, Alizadeh M, Hatami M, Asgari Lajayer B (2019) Biological synthesis of nanoparticles by different groups of bacteria in microbial nanobionics. Springer International Publishing, Cham, pp 63–85

Marrone PG (2019) Pesticidal natural products–status and future potential. Pest Manag Sci 75(9):2325–2340

Mokhtarzadeh A, Dolatabadi JE, Abnous K, de la Guardia M, Ramezani M (2015) Nanomaterial-based cocaine aptasensors. Biosens Bioelectron 68:95–106. https://doi.org/10.1016/j.bios.2014.12.052

Moustafa-Farag M, Almoneafy A, Mahmoud A, Elkelish A, Arnao MB, Li L, Ai S (2019) Melatonin and its protective role against biotic stress impacts on plants. Biomolecules 10(1):54

Mubeen I, Mfarrej MF, Razaq Z, Iqbal S, Naqvi SA, Hakim F, Mosa WF, Moustafa M, Fang Y, Li B (2023) Nanopesticides in comparison with agrochemicals: outlook and future prospects for sustainable agriculture. Plant Physiol Biochem 198:107670

Naderi MR, Shahraki AD (2013) Nanofertilizers and their roles in sustainable agriculture. Int J Agric Crop Sci 5(19):2229–2232. https://www.researchgate.net/publication/271885138_Nanofertilizers_and_their_roles_in_sustainable_agriculture

Nosrati R, Golichenari B, Nezami A, Taghdisi SM, Karimi B, Ramezani M, Abnous K, Shaegh SA (2017) Helicobacter pylori point-of-care diagnosis: nano-scale biosensors and microfluidic systems. TrAC Trends Anal Chem 97:428–444. https://doi.org/10.1016/j.trac.2017.10.013

Phillips RL, Miranda OR, You CC, Rotello VM, Bunz UH (2008) Rapid and efficient identification of bacteria using gold-nanoparticle–poly (para-phenyleneethynylene) constructs. Angew Chem Int Ed 47(14):2590–2594. https://doi.org/10.1002/anie.200703369

Prasad R (2014) Synthesis of silver nanoparticles in photosynthetic plants. J Nanopart 2014(1):963961

Rastogi A, Zivcak M, Sytar O, Kalaji HM, He X, Mbarki S, Brestic M (2017) Impact of metal and metal oxide nanoparticles on plants: a critical review. Front Chem 5:78. https://doi.org/10.3389/fchem.2017.00078

Rauwel P, Küünal S, Ferdov S, Rauwel E (2015) A review on the green synthesis of silver nanoparticles and their morphologies studied via TEM. Adv Mater Sci Eng 2015(1):682749

Raza A, Razzaq A, Mehmood SS, Zou X, Zhang X, Lv Y, Xu J (2019) Impact of climate change on crops adaptation and strategies to tackle its outcome: a review. Plants 8(2):34

Roy S, Das TK (2015) Plant-mediated green synthesis of silver nanoparticles-A. Int J Plant Biol Res 3:1044–1055

Ruttkay-Nedecky B, Krystofova O, Nejdl L, Adam V (2017) Nanoparticles based on essential metals and their phytotoxicity. J Nanobiotechnol 15(1):33

Samuel MS, Ravikumar M, John JA, Selvarajan E, Patel H, Chander PS, Soundarya J, Vuppala S, Balaji R, Chandrasekar N (2022) A review on green synthesis of nanoparticles and their diverse biomedical and environmental applications. Catalysts 12(5):459

Sarker U, Oba S (2018) Drought stress enhances nutritional and bioactive compounds, phenolic acids, and antioxidant capacity of *Amaranthus* leafy vegetable. BMC Plant Biol 18(1):258

Satapathi S, Kumar V, Chini MK, Bera R, Halder KK, Patra A (2018) Highly sensitive detection and removal of mercury ion using a multimodal nanosensor. Nano-Structures & Nano-Objects 16:120–126. https://doi.org/10.1016/j.nanoso.2018.05.006

Servin A, Elmer W, Mukherjee A, De la Torre-Roche R, Hamdi H, White JC, Bindraban P, Dimkpa C (2015) A review of the use of engineered nanomaterials to suppress plant disease and enhance crop yield. J Nanopart Res 17(2):92

Shah M, Fawcett D, Sharma S, Tripathy SK, Poinern GE (2015) Green synthesis of metallic nanoparticles via biological entities. Materials 8(11):7278–7308

Sharma R, Kumar V (2024) Nano-enabled agriculture for sustainable soil. Waste Manag Bull 2(1):152–161

Sharma B, Tiwari S, Kumawat KC, Cardinale M (2023) Nano-biofertilizers as bio-emerging strategies for sustainable agriculture development: potentiality and their limitations. Sci Total Environ 860:160476

Singh A, Singh NB, Hussain I, Singh H, Singh SC (2015) Plant-nanoparticle interaction: an approach to improve agricultural practices and plant productivity. Int J Pharm Sci Invent 4(8):25–40

Singh P, Singh H, Kim YJ, Mathiyalagan R, Wang C, Yang DC (2016a) Extracellular synthesis of silver and gold nanoparticles by *Sporosarcina koreensis* DC4 and their biological applications. Enzym Microb Technol 86:75–83

Singh S, Tripathi DK, Dubey NK, Chauhan DK (2016b) Effects of nano-materials on seed germination and seedling growth: striking a slight balance between the concepts and controversies. Mater Focus 5(3):195–201

Singh J, Kumar S, Alok A, Upadhyay SK, Rawat M, Tsang DC, Bolan N, Kim KH (2019) The potential of green-synthesized zinc oxide nanoparticles as a nutrient source for plant growth. J Clean Prod 214:1061–1070

Singh P, Arif Y, Miszczuk E, Bajguz A, Hayat S (2022) Specific roles of lipoxygenases in development and responses to stress in plants. Plants 11(7):979

Tang Y, Zhao W, Zhu G, Tan Z, Huang L, Zhang P, Gao L, Rui Y (2023) Nano-pesticides and fertilizers: solutions for global food security. Nanomaterials 14(1):90

Taran N, Storozhenko V, Svietlova N, Batsmanova L, Shvartau V, Kovalenko M (2017) Effect of zinc and copper nanoparticles on drought resistance of wheat seedlings. Nanoscale Res Lett 12(1):60

Tijjani A, Bashir KA, Mohammed I, Muhammad A, Gambo A, Musa H (2016) Biopesticides for pest control: a review. J Biopestic Agric 3(1):6–13

Tripathi DK, Singh S, Singh VP, Prasad SM, Dubey NK, Chauhan DK (2017) Silicon nanoparticles more effectively alleviated UV-B stress than silicon in wheat (*Triticum aestivum*) seedlings. Plant Physiol Biochem 110:70–81

Wang Z, Xie X, Zhao J, Liu X, Feng W, White JC, Xing B (2012) Xylem-and phloem-based transport of CuO nanoparticles in maize (*Zea mays* L.). Environ Sci Technol 46(8):4434–4441

Wang A, Jin Q, Xu X, Miao A, White JC, Gardea-Torresdey JL, Ji R, Zhao L (2020) High-throughput screening for engineered nanoparticles that enhance photosynthesis using mesophyll protoplasts. J Agric Food Chem 68(11):3382–3389

Wu S, Lan X, Zhao W, Li Y, Zhang L, Wang H, Han M, Tao S (2011) Controlled immobilization of acetylcholinesterase on improved hydrophobic gold nanoparticle/Prussian blue modified surface for ultra-trace organophosphate pesticide detection. Biosens Bioelectron 27(1):82–87. https://doi.org/10.1016/j.bios.2011.06.020

Wuryantini S (2020) The prospect of nanoinsecticide in controlling Citrus Aphids Aphis gossypii. In: International Conference and the 10th Congress of the Entomological Society of Indonesia (ICCESI 2019). Atlantis Press, pp 182–187

Yadav A, Kon K, Kratosova G, Duran N, Ingle AP, Rai M (2015) Fungi as an efficient mycosystem for the synthesis of metal nanoparticles: progress and key aspects of research. Biotechnol Lett 37(11):2099–2120

Yang CH, Ding YL, Qian J (2018) Design of magnetic-fluorescent based nanosensor for highly sensitive determination and removal of HG2+. Ceram Int 44(8):9746–9752. https://doi.org/10.1016/j.ceramint.2018.02.209

Zhang P, Ma Y, Zhang Z, He X, Zhang J, Guo Z, Tai P, Zhao Y, Chai Z (2022) Effect of cerium oxide nanoparticles on the antioxidant and defense systems in maize seedlings under drought stress. Sci Total Environ 807:150827. https://doi.org/10.1016/j.scitotenv.2021.150827
Zhu ZJ, Wang H, Yan B, Zheng H, Jiang Y, Miranda OR, Rotello VM, Xing B, Vachet RW (2012) Effect of surface charge on the uptake and distribution of gold nanoparticles in four plant species. Environ Sci Technol 46(22):12391–12398

MIX
Papier aus verantwortungsvollen Quellen
Paper from responsible sources
FSC® C105338

If you have any concerns about our products,
you can contact us on
ProductSafety@springernature.com

In case Publisher is established outside the EU,
the EU authorized representative is:
Springer Nature Customer Service Center GmbH
Europaplatz 3, 69115 Heidelberg, Germany

Printed by Libri Plureos GmbH
in Hamburg, Germany